BASIC ELECTRONICS

Georgies Alene Asres

TAP

Toronto Academic Press

BASIC ELECTRONICS

Georgies Alene Asres

Toronto Academic Press
4164 Lakeshore Road
Burlington ON L7L 1A4
Canada
www.tap-books.com

Email: orders@arclereducation.com

ISBN: 978-1-77956-303-3

ABOUT THE AUTHOR

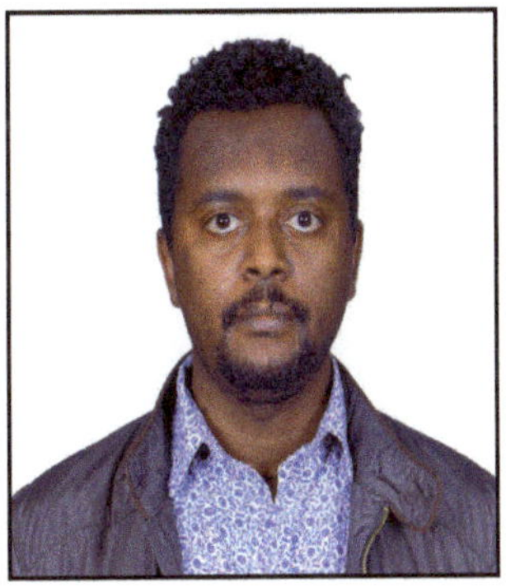

Georgies Alene Asres is a dedicated materials engineer with a diverse background in materials science and engineering, electrical engineering, physics, and research. He currently serves as an associate professor of Materials Engineering at Addis Ababa University in Ethiopia, where he is also the head of Center for Materials Engineering. Georgies earned his PhD in Electrical Engineering (Electronics Materials and Devices) from the University of Oulu, Finland. He holds dual master's degrees in Functionalized Advanced Materials and Engineering from Grenoble Institute of Technology (France) and Augsburg University (Germany). With over eight years of teaching and research experience, including his tenure at Addis Ababa University, Georgies has established himself as a respected academic and researcher. In his role as an associate professor, Georgies teaches undergraduate and graduate courses in materials science and engineering. His research focuses on developing new materials for energy applications, among other topics in materials engineering. He supervises MSc and PhD students, providing guidance and feedback on their research projects. Georgies is actively involved in collaborative research with faculty members and external partners, contributing to interdisciplinary projects and grant proposals. He enriches the academic community by publishing his research findings in peer-reviewed journals and conferences, reviewing papers for scholarly publications, and serving on academic committees.

Table of Contents

List of Figures *xi*

List of Tables *xix*

Preface *xxi*

1 SEMICONDUCTOR FUNDAMENTALS 1

Introduction **2**

1.1. Overview of Semiconductors **3**

1.1.1. Holes and Electrons in Semiconductors 3

1.1.2. Properties of Semiconductors 4

1.2. Semiconductor Materials **7**

1.2.1. Semiconductor Materials in Periodic Table 7

1.2.2. List of Semiconductor Materials 8

1.2.3. Advantages and Disadvantages 9

1.3. Energy Band Description of Semiconductors **11**

1.3.1. Effect of Temperature on Semiconductors 12

1.3.2. Hole Current 13

1.4. Intrinsic Semiconductor **15**

1.5. Extrinsic Semiconductor **16**

1.5.1. n-type Semiconductor 16

1.5.2. p-type Semiconductor 18

1.6. Electrical Conduction in Semiconductors **20**

Summary **33**

Review Questions **33**

References **34**

2 DIODE FUNDAMENTALS 35

Introduction **36**

2.1. Diodes **37**

2.1.1. Using Diode 38

2.2. Types of Diodes **41**

2.2.1. Classification of Diodes According to Their Applications and Characteristics 41

2.2.2. pn Diodes 42

2.2.3. Schottky Barrier Diodes 49

2.2.4. SiC Schottky Barrier Diodes 53

2.2.5. Zener Diodes 57

2.2.6. Zener Breakdown and Avalanche Breakdown 57

2.3. Basic Diode Circuits **61**

2.3.1. Half Wave Rectifier 61

2.3.2. Full Wave Rectifier 61

2.3.3. Voltage Multiplier 61

2.3.4. Clipping Circuit 61

2.3.5. Diode Functions in Circuit 62

Summary **68**

Review Questions **68**

References **68**

3 BJT FUNDAMENTALS 71

Introduction 72
3.1. Bipolar Junction Transistor 73
3.1.1. Why BJT Is Called Bipolar? 73
3.1.2. Construction of Bipolar Junction Transistor 74
3.1.3. Types of Bipolar Junction Transistor 74
3.1.4. Working Principle of BJT 75
3.1.5. Functions of BJT 76
3.1.6. Advantages of BJT 76
3.1.7. Disadvantages of BJT 77
3.1.8. Applications of BJT 77
3.2. Physical Principles and Operation Laws of BJT 78
3.2.1. Physical Structure: Terminal Voltages and Currents 78
3.2.2. Principle of Operation 79
3.3. Terminals of BJT 82
3.3.1. Emitter Terminal 82
3.3.2. Collector Terminal 82
3.3.3. Base Terminal 82
3.4. Operating Regions of BJT 83
3.4.1. Active Region 84
3.4.2. Saturation Region 86
3.4.3. Cutoff Region 88
3.5. Transistor v–i Dependencies 89
3.6. Early Effect 91
3.7. BJT Configuration and Characteristics 92
3.7.1. Common Base Configuration 92
3.7.2. Common Emitter Configuration 93
3.7.3. Common Collector Configuration 93
3.7.4. BJT Characteristics 94
3.8. Large-Signal Circuit Model of a BJT 95
3.8.1. Large-Signal DC Circuit Model of a BJT 97
3.8.2. Method of Assumed States 99
3.8.3. Transistor Circuit Analysis Using the Method of Assumed States 99
3.8.4. DC Transistor Bias Circuits 101
Summary 109
Review Questions 110
References 110

4 FIELD EFFECT TRANSISTORS 113

Introduction 114
4.1. Field Effect Transistors (FETs) Types and Working Principles 115
4.1.1. Field Effect Transistor (FET) History 116
4.1.2. Field Effect Transistor (FET) Types 117
4.2. Basics, Working Principle, and Applications of MOSFET 122
4.2.1. Two-Terminal MOS Structure 123
4.2.2. MOSFET Operation 125
4.2.3. n-Channel Enhancement-Mode MOSFET 127
4.2.4. Additional MOSFET Structures and Circuit Symbols 128
4.3. MOSFET DC Circuit Analysis 134
4.3.1. Common-Source Circuit 134
4.3.2. Load Line and Modes of Operation 137
4.3.3. Common MOSFET Configurations: DC Analysis 138
4.3.4. Constant-Current Source Biasing 142
4.3.5. Basic MOSFET Applications 143
4.4. Metal Oxide Semiconductor Field Effect Transistor (MOSFET) 146
4.4.1. Constructional Features of a Power MOSFET 147
4.4.2. Operating Principle of a MOSFET 149
Summary 154
Review Questions 154
References 155

5 SPECIAL SEMICONDUCTOR DEVICES 157

Introduction 158
5.1. Silicon Controlled Rectifier 159
5.1.1. Modes of Operation 160

5.1.2. Thyristor Turn-on Methods 161
5.1.3. Simple SCR Circuit 162
5.1.4. Applications 162
5.1.5. Comparison with SCS 162
5.2. Triode for Alternating Current (TRIAC) 163
5.2.1. Operation 164
5.2.2. Issues 167
5.2.3. Snubber Circuits 169
5.2.4. Application 170
5.3. DIAC (Diode for Alternating Current) 171
5.3.1. Circuit Symbol 171
5.3.2. Construction 172
5.3.3. Operation 173
5.3.4. Applications 173
5.4. Unijunction Transistor 175
5.4.1. Types 175
5.4.2. Applications 176
5.4.3. Construction 176
5.5. Insulated-Gate Bipolar Transistor (IGBT) 178
5.5.1. Basic Structure 178
5.5.2. IGBT Characteristics 179
5.5.3. Operation Modes 180
Summary 186
Review Questions 187
References 187

6 FEEDBACK AMPLIFIER 189

Introduction 190
6.1. Conceptual Development Through Block Diagrams 191
6.2. Negative Feedback 193
6.2.1. Advantage of Negative Feedback in Op-Amps 195
6.2.2. How Does the Circuit in the Op-Amp Works? 195
6.3. Calculations of Open-Loop Gain, Closed-Loop Gain and Feedback Factors 197
6.4. Topologies of the Feedback Amplifier 199
6.4.1. Voltage Series Feedback Amplifier 199
6.4.2. Voltage Shunt Feedback Amplifier 200
6.4.3. Current Series Feedback Amplifier 200
6.4.4. Current Shunt Feedback Amplifier 200
6.5. Effect of Feedback on Input and Output Impedance 201
6.6. Sensitivity 209
6.6.1. Sensitivity of the Negative Feedback Amplifier (S) 209
6.7. Effect of Positive Feedback 211
6.7.1. Positive Feedback in OP-AMPs 212
6.7.2. The Oscillation Mechanism of Feedback Networks 215
Summary 225
Review Questions 225
References 226

7 FUNDAMENTALS OF ELECTRONIC CIRCUIT FABRICATION 227

Introduction 228
7.1. Electronic Circuit Design 229
7.1.1. Circuit Board Simulator 229
7.1.2. Basics of an Electronics Circuit Design Process 229
7.1.3. Electronic Circuit Development Strategies 233
7.1.4. Factors to Consider When Creating a Circuitry Design 234
7.2. Ohm's Law, Kirchhoff's Laws, and Equivalent Circuits 236
7.2.1. Time and Frequency Domains 236
7.2.2. The Fourier Transform 238
7.2.3. Discrete and Digital Signals 239
7.2.4. Feedback and Frequency Response 241
7.2.5. Loop Gain and the Stability Problem 248
7.3. Noise in Electronic Circuits 252
7.3.1. Noise in Passive Components 252
7.3.2. Effect of Circuit Capacitance 254
7.3.3. Noise in Semiconductors and Amplifiers 254
7.3.4. Circuit Noise Calculations and

Noise Bandwidth 258
7.3.5. Noise Figure and Noise Temperature 258
Summary **263**
Review Questions **264**
References **264**

8 OPERATIONAL AMPLIFIERS 265

Introduction **266**
8.1. An Overview of Operation **267**
8.1.1. Open-Loop Amplifier 268
8.1.2. Closed-Loop Amplifier 268
8.1.3. Op-Amp Characteristics 269
8.2. Internal Circuitry of 741-Type Op Amp **274**
8.2.1. Architecture 274
8.2.2. Varactor Bridge Op Amps 275
8.2.3. Biasing Circuits 277
8.2.4. Small-Signal Differential Mode 278
8.2.5. Other Linear Characteristics 279
8.2.6. Non-Linear Characteristics 279
8.2.7. Applicability Considerations 280
8.2.8. Classification 280
8.3. Operational Amplifier Applications **282**
8.3.1. Practical Considerations 282
8.3.2. Amplifiers 283
8.3.3. Filters 287
8.3.4. Comparator 288
8.3.5. Integration and Differentiation 288
Summary **291**
Review Questions **291**
References **292**

9 OSCILLATORS 295

Introduction **296**
9.1. What Is Oscillator? **297**
9.1.1. Positive Feedback Oscillators 297
9.1.2. Oscillator Design Process 298
9.1.3. Oscillator Design Procedure 299
9.2. Armstrong Oscillator **307**
9.2.1. Armstrong Oscillator Circuit and Its Working 307
9.3. Tuned Collector Oscillator **312**
9.3.1. Tuned Collector Oscillator Circuit Working 312
9.4. The Hartley Oscillator **314**
9.4.1. Mutual Inductance in Hartley Oscillators 315
9.4.2. The Hartley Circuit 316
9.4.3. The Tuned LC (Tank) Circuit 317
9.4.4. Automatic 'Sliding' Class C Bias 317
9.4.5. How the Hartley Oscillator Works 318
9.4.6. Alternative Hartley Designs 319
9.5. Crystal Oscillator **320**
9.5.1. What is a Quartz Crystal? 320
9.5.2. Crystal Oscillator Circuit Diagram 321
9.5.3. Crystal Oscillator Working 321
9.5.4. Uses of Crystal Oscillator 322
9.5.5. Application of Crystal Oscillator 323
Summary **327**
Review Questions **328**
References **328**

10 DIGITAL ELECTRONIC PRINCIPLES 331

Introduction **332**
10.1. Analog vs. Digital **333**
10.2. Number Systems **334**
10.2.1. Decimal (Base 10) Number System 334
10.2.2. Binary (Base 2) Number System 334
10.2.3. Octal (Base 8) Number System 334
10.2.4. Hexadecimal (Base 16) Number System 334
10.2.5. Number System Conversions 335
10.2.6. 1's and 2's Complement 339
10.2.7. Fixed Point Number Representation 340
10.2.8. Floating Point Number Representation 341
10.2.9. Binary Arithmetic 342
10.2.10. Digital Codes 344
10.3. Digital Electronics **348**

10.3.1. Digital Electronic Circuits 348
10.3.2. Asynchronous Circuits 349
10.3.3. Properties 350
10.3.4. Difference between Digital and Analog Electronics 352
10.4. Boolean Algebra and Logic Gates 354
10.4.1. De-Morgan's Theorem 354
10.4.2. The AND Gate 356
10.4.3. The OR Gate 360
10.4.4. The NOT Gate (INVERTER) 361
10.4.5. The NAND, NOR, EX-OR, and EX-NOR Gates 362
10.5. Combinational and Sequential Circuits 367
10.5.1. Combinational Circuit 367
10.5.2. Sequential Circuit 368
10.5.3. Difference between Combinational and Sequential Circuit 368
10.5.4. Flip-Flops and Registers 369
Summary 373
Review Questions 373
References 373

INDEX 375

List of Figures

Figure 1.1. Mobility of Electrons and Holes
Figure 1.2. Semiconductor materials
Figure 1.3. Semiconductor materials in periodic table
Figure 1.4. Energy band diagrams of germanium
Figure 1.5. Energy band diagrams of germanium Silicon
Figure 1.6. Semiconductor crystal
Figure 1.7. Applied across the semiconductor crystal
Figure 1.8. Valence electron
Figure 1.9. Conduction band
Figure 1.10. Free electrons and holes
Figure 1.11. Valence electron of arsenic atom
Figure 1.12. The energy band description of n-type semi-conductor
Figure 1.13. N-type conductivity
Figure 1.14. Three single covalent bonds with three germanium atoms
Figure 1.15. The energy band description of the p-type semiconductor
Figure 1.16. p-type conductivity
Figure 1.17. (a) Metal-oxide-semiconductor field effect transistor (MOSFET) symbol and structure; (b) bipolar junction transistor (BJT) symbol and structure
Figure 1.18. The electrical conductivity characteristics of a semiconductor (silicon) vs. those of a metal (tungsten)
Figure 1.19. Silicon structure and electrical conduction
Figure 1.20. Elemental semiconductors and the most common n- and p-type dopants used to create extrinsic semiconductors
Figure 1.21. The effect of impurities on the band gap in extrinsic silicon
Figure 1.22. The dependency of the electrical resistivity at 298K on the doping concentration
Figure 2.1. Diode in forward bias
Figure 2.2. Diode in reverse bias
Figure 2.3. A diode for reverse current protection
Figure 2.4. Zener diode as voltage regulator
Figure 2.5. The rectification of AC power to DC power
Figure 2.6. Types of diodes
Figure 2.7. Classification of diodes according to their internal connections
Figure 2.8. Fermi level of intrinsic semiconductors, Fermi level of p-type semiconductors, Fermi level of n-type semiconductors from left to right respectively

Figure 2.9. P-type and n-type semiconductors before they are joined together
Figure 2.10. P-type and n-type semiconductors after they are joined together
Figure 2.11. Band diagram of a pn junction
Figure 2.12. Band diagram of a pn junction in an unbiased state
Figure 2.13. Band diagram of a pn junction in the forward-biased state
Figure 2.14. Band diagram of a pn junction in the reverse-biased state
Figure 2.15. Current-voltage curve of a pn diode
Figure 2.16. Example of a pn diode's reverse current-vs-ambient temperature curve
Figure 2.17. Structure of a PIN diode
Figure 2.18. Increasing the breakdown voltage of a PIN diode
Figure 2.19. Change in the dopant concentration of the n layer due to conductivity modulation
Figure 2.20. Example of a structure of a Schottky barrier diode
Figure 2.21. Band diagram of a metal {normally the VB and CB overlap in metals, the schematic doesn't depict that please modify}
Figure 2.22. Band diagram of an n-type semiconductor
Figure 2.23. Band diagram of a Schottky junction in an unbiased state
Figure 2.24. Band diagram of a Schottky junction in the forward-biased state
Figure 2.25. Band diagram of a Schottky junction in the reverse-biased state
Figure 2.26. JBS structure
Figure 2.27. Formation of depletion layers by the application of a reverse bias across the Schottky junction
Figure 2.28. Expansion of depletion layers by an increase in reverse bias across the Schottky junction
Figure 2.29. Comparison of the reverse current characteristics of SBDs with typical and JBS structures
Figure 2.30. Bandgap of Si
Figure 2.31. Bandgap of SiC
Figure 2.32. Comparison of electric fields in Si and SiC SBDs
Figure 2.33. Example of a waveform of a charging current flowing to a capacitor for full-wave rectification
Figure 2.34. Cross-sectional view of an SBD with an improved JBS structure
Figure 2.35. Forward current-vs-forward voltage curves of a typical SBD, a pn diode, and an SBD with an improved JBS structure
Figure 2.36. Zener breakdown
Figure 2.37. Avalanche breakdown
Figure 2.38. Example of Zener voltage temperature coefficient vs zener voltage
Figure 2.39. Example of I_Z-V_Z curves at different zener voltages
Figure 2.40. Forward and reverse biased diodes
Figure 3.1. BJT or bipolar junction transistor symbol
Figure 3.2. Construction of bipolar junction transistor
Figure 3.3. PNP & NPN bipolar junction transistor
Figure 3.4. Working principle of BJT
Figure 3.5. Simplified physical structure of the npn-junction transistor
Figure 3.6. Circuit symbol for the npn-junction transistor, terminal currents, and terminal voltages
Figure 3.7. Operation of the npn-junction transistor for three values of base-emitter voltage $\boxtimes_{BE}$. The collector-base voltage υ_{CB} is fixed at zero volts
Figure 3.8. Conceptual test circuits for a npn transistor in common-emitter configuration
Figure 3.9. Transistor test circuit of the common-emitter configuration in the active region
Figure 3.10. Current flow in the transistor when EBJ is at zero volts and CBJ is forward biased

Figure 3.11. Typical dependence of the collector current on $\boxtimes_{BE}$
Figure 3.12. Typical common-emitter characteristics of the npn BJT. The collector current is independent of the collector-to-emitter voltage in the active region in (a). (b) shows a more accurate representation which includes the early effect
Figure 3.13. CB configuration
Figure 3.14. CE configuration
Figure 3.15. CC configuration
Figure 3.16. First-order π-type BJT model in the active region of operation
Figure 3.17. A base-bias transistor circuit and its equivalent representation
Figure 3.18. Large-signal DC circuit model of the EBJ
Figure 3.19. Large-signal DC circuit model of a npn BJT transistor for three regions of operation
Figure 3.20. Large-signal DC circuit model of a pnp BJT transistor
Figure 3.21. Transistor circuit solved using the method of assumed states
Figure 3.22. Transistor circuit solved using the large-signal exponential model
Figure 3.23. A "puppet master" analogy of the transistor bias circuit
Figure 4.1. Junction FET circuit symbol
Figure 4.2. Junction FET working below saturation
Figure 4.3. Types of FET
Figure 4.4. The basic MOS capacitor structure
Figure 4.5. (a) A parallel plate capacitor, showing the electric field and conductor charges, (b) a corresponding MOS capacitor with a negative gate bias, showing the electric field and charge flow, and (c) the MOS capacitor with an accumulation layer of holes
Figure 4.6. The MOS capacitor with p-type substrate: (a) effect of positive gate bias, showing the electric field and charge flow, (b) the MOS capacitor with an induced space-charge region due to a moderate gate bias, and (c) the MOS capacitor with an induced space-charge region and electron inversion layer due to a larger gate bias
Figure 4.6. The MOS capacitor with p-type substrate: (a) effect of positive gate bias, showing the electric field and charge flow, (b) the MOS capacitor with an induced space-charge region due to a moderate gate bias, and (c) the MOS capacitor with an induced space-charge region and electron inversion layer due to a larger gate bias
Figure 4.7. The MOS capacitor with n-type substrate for: (a) a positive gate bias, (b) a moderate negative bias, and (c) a larger negative bias
Figure 4.8. P-channel MOSFET
Figure 4.9. N-channel MOSFET
Figure 4.10. (a) Schematic diagram of an n-channel enhancement mode MOSEFT and (b) an n-channel MOSFET, showing the field oxide and poly silicon gate
Figure 4.11. (a) Cross section of the n-channel MOSFET prior to the formation of an electron inversion layer, (b) equivalent back-to-back diodes between sources and drain when the transistor is in cutoff, and (c) cross section after the formation of an electron inversion layer
Figure 4.12. Cross section of n-channel depletion mode MOSFET for (a) $v_{GS}=0$, (b) $v_{GS}<0$, (c) $v_{GS}>0$
Figure 4.13. Family of i_D versus v_{DS} curves for an n-channel depletion mode MOSFET
Figure 4.14. The n-channel depletion mode MOSFET: (a) conventional circuit symbol and (b) simplified circuit symbol
Figure 4.15. Cross section of p-channel MOSFETs: (a) enhancement mode and (b) depletion mode
Figure 4.16. The p-channel enhancement mode MOSFET: (a) conventional circuit symbol and (v) simplified circuit symbol
Figure 4.17. The p-channel depletion mode MOSFET: (a) conventional circuit symbol and (b) simplified circuit symbol

Figure 4.18. An NMOS common source circuit
Figure 4.19. (a) An NMOS common source circuit and (b) the NMOS circuit
Figure 4.20. (a) A PMOS common source circuit, (b) result when saturation region bias assumption in incorrect, and (c) result when non saturation region bias assumption is correct
Figure 4.21. Transistor characteristics, v_{DS} (sat) curve, load line, and Q-point for the NMOS common source circuit in Figure 18 (b)
Figure 4.22. Enhancement mode NMOS device with the gate connected to the drain
Figure 4.23. Current voltage characteristics of an enhancement load device
Figure 4.24. Circuit with enhancement load device and NMOS driver
Figure 4.25. (a) Depletion mode NMOS device with the gate connected to the source and (b) current voltage characteristics
Figure 4.26. Circuit containing a depletion load device
Figure 4.27. Example of CMOS inverter
Figure 4.28. Implementation of a MOSFET constant current source
Figure 4.29. NMOS inverter circuit
Figure 4.30. A two input NMOS NOR logic gate
Figure 4.31. Different types of power MOSFET. (a) Circuit symbols and transfer characteristics (b) Photograph of n-channel enhancement type MOSFET
Figure 4.32. Schematic construction of a power MOSFET (a) Construction of a single cell. (b) Arrangement of cells in a device
Figure 4.33. Parasitic BJT in a MOSFET cell
Figure 4.34. Gate control of MOSFET conduction. (a) Depletion layer formation; (b) Free electron accumulation; (c) Formation of inversion layer
Figure 5.1. The schematic symbol for a silicon-controlled rectifier (SCR)
Figure 5.2. Forward blocking mode of SCR
Figure 5.3. Forward conduction mode of SCR
Figure 5.4. Reverse blocking mode of SCR
Figure 5.5. TRIAC schematic symbol
Figure 5.6. Triggering modes. Quadrants, 1 (top right), 2 (top left), 3 (bottom left), 4 (bottom right); voltages relative to MT1
Figure 5.7. TRIAC semiconductor construction
Figure 5.8. Operation in quadrant 1
Figure 5.9. Equivalent electric circuit for a TRIAC operating in quadrant 1
Figure 5.10. Operation in quadrant 2
Figure 5.11. Operation in quadrant 3
Figure 5.12. Operation in quadrant 4
Figure 5.13. Circuit symbol
Figure 5.14. The structure of a DIAC
Figure 5.15. The V- I characteristics of a DIAC
Figure 5.16. Heat Control
Figure 5.17. Light Dimmer
Figure 5.18. Symbol of UJT
Figure 5.19. Structure of a p-type UJT
Figure 5.20. Basic structure of insulated-gate bipolar transistor
Figure 6.1. A block diagram for a feedback amplifier
Figure 6.2. Output voltage of the op-amp closely follows that input voltage

Figure 6.3. The effects of negative feedback
Figure 6.4. Circuit in the op-amp
Figure 6.5. Bode plot of an op-amp
Figure 6.6. Example of a negative feedback circuit (inverting amplifier)
Figure 6.7. Feedback amplifier topologies
Figure 6.8. Two possible output topologies. (a) Feedback of load-voltage information. (b) Feedback of load-current information
Figure 6.9. Two possible input topologies. (a) Input signal applied in series with amplifier input. (b) Input signal applied in parallel with amplifier input
Figure 6.10. Amplifier with high input and output resistances
Figure 6.11. Block diagram for amplifier of Figure 6.10
Figure 6.12. Unity-gain amplifiers. (a) Follower connection. (b) Amplifier with input attenuator. (c) Amplifier with input transformer
Figure 6.13. Comparison of 2 types of feedback configurations containing negative feedback (a) and positive feedback (b)
Figure 6.14. An op-amp circuit with simple positive feedback
Figure 6.15. Configuration of a positive feedback op-amp circuit as a comparator
Figure 6.16. Waveforms of the input signal (Vs) and the output signal (Vout) versus time
Figure 6.17. Hysteresis and threshold voltages in positive-feedback comparator
Figure 6.18. A general feedback system without external excitation
Figure 6.19. Generation of the periodic waveform due to oscillation in the circuit
Figure 7.1. Unit circle with N = 8
Figure 7.2. General structure of the feedback amplifier
Figure 7.3. Concept of increasing the SNR of a circuit: (a) basic amplifier indicating the case of input referred noise; (b) with an additional noise-free amplifier at the front end for SNR improvement
Figure 7.4. The four basic feedback topologies: (a) voltage-sampling series mixing (series-shunt) topology; (b) current-sampling shunt mixing (shunt-series) topology; (c) current-sampling series-mixing (series-series) topology; (d) voltage-sampling shunt-mixing (shunt-shunt) topology
Figure 7.5. Derivation of the A circuit and β circuit for the series-shunt feedback amplifier: (a) the circuit in Figure 7.4a with the feedback represented by h parameters; (b) the circuit in Figure 7.5a, neglecting h_{21} and modifying the basic amplifier to form the A circuit
Figure 7.6. Derivation of the A circuit and the β circuit for a series-series feedback amplifier: (a) the circuit in Figure 7.4c with the feedback represented by z parameters; (b) the same circuit neglecting z_{21} and modifying the basic amplifier to form the A circuit
Figure 7.7. The series-shunt feedback amplifier
Figure 7.8. The series-series feedback amplifier: (a) ideal structure; (b) equivalent circuit with modified input and output impedances
Figure 7.9. Relationship between pole location and transient response: (a) poles on the left-hand side of the imaginary axis; (b) poles on the right-hand side of the imaginary axis; (c) poles on the imaginary axis
Figure 7.10. Thermal noise in a resistor: (a) noise as a voltage source; (b) noise as a current source; (c) noise in a practical resistor shunted by its parasitic capacitance
Figure 7.11. Small-signal equivalent of junction diode with noise
Figure 7.12. Burst noise: (a) typical waveform; (b) spectral density
Figure 7.13. Combined spectral density due to burst and flicker noise
Figure 7.14. Base current noise density of a BJT
Figure 8.1. The op amp is one type of differential amplifier
Figure 8.2. Operational amplifier open-loop frequency response curve

Figure 8.3. Generalized block diagram for a varactor bridge solid-state op amp
Figure 8.4. A component-level diagram of the common 741 op amp
Figure 9.1. Practical oscillator circuit
Figure 9.2. Feedback oscillator block diagram
Figure 9.3. The use of OSCTEST for linear analysis of oscillators
Figure 9.4. The use of Oscaprobe for nonlinear analysis of oscillators
Figure 9.5. Circuit for determining the I-V characteristic of a transistor
Figure 9.6. Amplifier for the Colpitts oscillator
Figure 9.7. Amplifier performance
Figure 9.8. Colpitts resonator network
Figure 9.9. Linear analysis using OCSTEST
Figure 9.10. Linear analysis using OCSTEST Resistor added to increase Q and reduce gain
Figure 9.11. Non-linear analysis using OSCAPROBE, circuit and output spectrum
Figure 9.12. Non-linear analysis using OSCAPROBE, Tuning range and output waveform
Figure 9.13. Phase noise of the oscillator using non-linear (harmonic balance) analysis
Figure 9.14. Armstrong oscillator circuit and its working
Figure 9.15. A-DC bias is determined primarily by Rb, B-DC collector emitter circuit current
Figure 9.16. The input and output signal circuit
Figure 9.17. Tuned collector oscillator circuit
Figure 9.18. The Hartley Oscillator
Figure 9.19. Centre tapped inductors on a common core
Figure 9.20. Inductors with adding fields
Figure 9.21. Inductors with opposing fields
Figure 9.22. Adjustable Ferrite Core
Figure 9.23. The tank circuit
Figure 9.24. Sliding Class C Bias
Figure 9.25. Common Emitter Hartley Oscillator
Figure 9.26. Hartley Oscillator with Tuned Feedback
Figure 9.27. Quartz Crystal Circuit
Figure 9.28. Crystal Oscillator
Figure 9.29. Crystal Oscillator Circuit Diagram
Figure 9.30. Crystal oscillator working
Figure 9.31. Colpitts crystal oscillator
Figure 9.32. Armstrong crystal oscillator
Figure 10.1. The logic symbol for an AND gate
Figure 10.2. AND gate logic symbol
Figure 10.3. Digital electronic signals
Figure 10.4. Integrated circuit (IC)
Figure 10.5. The output (X) (motor ON or motor OFF) should respond according to the position of each door (OPENED or CLOSED)
Figure 10.6. The AND gate truth table
Figure 10.7. The electrical circuit and light (X) will turn ON
Figure 10.8. A simple alarm system based on the OR gate
Figure 10.9. OR gate logic truth table

Figure 10.10. The NOT gate
Figure 10.11. The NAND gate
Figure 10.12. A truth table of NAND gate
Figure 10.13. A NOR gate are 0 (LOW), the output of a NOR gate will be 1 (HIGH)
Figure 10.14. The EX-OR Gate reverse is also true
Figure 10.15. EX-NOR gates
Figure 10.16. EX-NOR gate circuit

List of Tables

Table 1.1. Relative electrical resistivities and conductivities of some metals
Table 2.1. Physical properties of typical semiconductor materials
Table 3.1. Operating regions of an npn-junction transistor
Table 3.2. The different transistor configuration characteristics
Table 4.1. NMOS NOR logic gate response
Table 5.1. IGBT comparison table
Table 7.1. Comparison of Approximate Computational Loads for Direct DFT and Decimations in Time FFT
Table 10.1. Precision representation
Table 10.2. The binary and BCD codes for the decimal numbers 0 to 15
Table 10.3. Difference between Analog and Digital Electronics

PREFACE

Electronics is a significant branch of science within the fields of physics and technology. This area focuses on various components such as resistors, conductors, semiconductors, circuits, and microchips, all of which involve the motion of electric charge. Electronics play a crucial role in engineering, where these components are essential for operating machines and managing a wide array of tasks. Electronics is the study and management of electron flow. Electric current, defined as the movement of charges, is fundamental to the field of electronics. In our daily lives, virtually every appliance and device we use relies on electric current and integrated circuits. These integrated circuits and microchips function based on the flow of charges within them, enabling the operation of everything from household gadgets to sophisticated industrial machinery.

The history of electronics dates back several centuries. The term "electronic" was first used by Luigi Galvani in 1792 when he demonstrated the presence of electricity in animal bodies through experiments on frogs. This early discovery laid the groundwork for future developments in the field. In 1799, Charles-Augustin de Coulomb introduced Coulomb's Law of Electrostatics, further advancing the understanding of electric charge. That same year, Italian scientist Alessandro Volta invented the battery, providing a reliable source of electric power.

In 1827, Georg Simon Ohm proposed Ohm's Law, establishing a fundamental principle in the study of electric circuits. This was followed by Thomas Alva Edison's invention of the electric bulb, which revolutionized the way people used electric light. These milestones marked significant advancements in the field of electronics, paving the way for the technological innovations we rely on today.

Contents

This book is a comprehensive guide to understanding the essential concepts and components that form the foundation of electronics. Spanning ten chapters, it systematically covers the critical aspects of the field, providing readers with a solid grasp of both theoretical principles and practical applications.

Chapter 1 begins with an in-depth exploration of semiconductors, the backbone of modern electronics. This chapter introduces the basic properties of semiconductors, differentiating them from conductors and insulators. It explains concepts such as energy bands, charge carriers (electrons and holes), and intrinsic and extrinsic semiconductors, setting the stage for understanding more complex devices. Building on the knowledge of semiconductors,

Chapter 2 covers diodes, one of the simplest yet most essential semiconductor devices. The chapter covers the principles of diode operation, including the p-n junction, forward and reverse bias, and the various types of diodes like Zener diodes and light-emitting diodes (LEDs). Practical applications of diodes in rectification and signal modulation are also discussed.

Chapter 3 focuses on Bipolar Junction Transistors (BJTs), explaining their structure, operation, and characteristics. This chapter details the different configurations (common base, common emitter, and common collector) and their respective applications in amplification and switching.

Chapter 4 shifts the focus to Field Effect Transistors (FETs), including both Junction FETs (JFETs) and Metal-Oxide-Semiconductor FETs (MOSFETs). It covers their construction, working principles, and the distinctions between them. The chapter also highlights their significance in integrated circuits and digital electronics.

In Chapter 5, readers are introduced to a variety of special semiconductor devices such as thyristors, TRIACs, and Varactors. Each device is discussed in terms of its unique properties, functions, and applications in electronics, including power control and frequency tuning.

Chapter 6 explores the concept of feedback in amplifiers, a crucial technique for enhancing performance and stability. It explains positive and negative feedback, their effects on gain and bandwidth, and practical feedback amplifier circuits.

Chapter 7 provides insight into the fabrication of electronic circuits. It covers the processes involved in creating integrated circuits (ICs), including photolithography, doping, and etching, as well as the materials and technologies used in modern electronics manufacturing.

Operational amplifiers (op-amps) are the focus of Chapter 8. This chapter details their characteristics, configurations (inverting and non-inverting), and applications in analog signal processing, such as filtering, amplification, and mathematical operations.

Chapter 9 discusses oscillators, circuits that generate periodic waveforms. It explains the principles behind various types of oscillators, including RC, LC, and crystal oscillators, and their applications in clocks, radios, and signal generators.

Chapter 10 transitions from analog to digital electronics, covering the fundamental principles of digital logic, binary systems, and Boolean algebra. It introduces basic digital components like logic gates, flip-flops, counters, and their role in digital systems and microprocessors.

Audience

The readers of this book is individuals with little to no prior knowledge of electronics who are interested in learning about the fundamental principles and components of electronic circuits. This book serves as a beginner's guide, providing clear explanations and examples to help readers grasp key concepts such as voltage, current, resistance, and circuit analysis. Whether you are a hobbyist looking to understand how electronic devices work or a student seeking to build a foundation in electronics, this book aims to make the complex world of electronics accessible and engaging.

Learning Features

- Start-of-chapter learning objectives
- Key terms in each chapter
- Summary Points at the end of each chapter
- Review Questions at the end of each chapter
- Case Studies and Activities in each chapter for solving real problems
- References at the end of each chapter

CHAPTER 1

Semiconductor Fundamentals

LEARNING OBJECTIVES

After studying this chapter, you will be able to:

- Understand the overview of semiconductors;
- Explain the semiconductor materials;
- Define energy band description of semiconductors;
- Discuss intrinsic semiconductor and extrinsic semiconductor;
- Explain the electrical conduction in semiconductors.

KEY TERMS FROM THIS CHAPTER

Band theory
Conduction band
Electron-hole pair
Extrinsic
Intrinsic
Photovoltaic cell
Silicon
Charge carriers
Doping
Energy gap
Germanium
N-type
P-type
Valence band

INTRODUCTION

Semiconductors are materials that have electrical conductivity between that of conductors and insulators, and their conductivity can be controlled making them essential in modern electronic devices. The most common semiconductor material is silicon, which is abundant and has ideal properties for the fabrication of integrated circuits. Semiconductors are characterized by their ability to control the flow of electrical current, which is achieved through the manipulation of their electrical properties. This control is fundamental to the operation of various electronic components such as transistors, diodes, and solar cells.

The intrinsic electrical properties of semiconductors can be modified by adding impurities in a process known as doping. Doping involves introducing small amounts of other elements into the semiconductor crystal to create either an excess of electrons (n-type) or an excess of holes (p-type). This alteration enhances the material's ability to conduct electricity in a controlled manner. The interaction between n-type and p-type semiconductors forms the basis of many semiconductor devices, such as p-n junctions, which are crucial for the operation of diodes and transistors.

Semiconductors have revolutionized the field of electronics, enabling the development of compact, efficient, and powerful devices. The invention of the transistor in the mid-20th century marked a significant milestone, paving the way for the miniaturization of electronic circuits and the advancement of modern computing. Today, semiconductors are integral to a wide range of applications, from consumer electronics like smartphones and laptops to advanced technologies in telecommunications, medical devices, and renewable energy systems. The ongoing research and development in semiconductor technology continue to drive innovation, leading to faster, smaller, and more energy-efficient electronic devices.

1.1. OVERVIEW OF SEMICONDUCTORS

Certain electrical characteristics of semiconductor materials are necessary for the functioning of some electronic devices. In this case, resistivity decreases with temperature, but metal exhibits the opposite behavior in this context. Semiconductors, integrated circuits, and diodes are components that aid in the conduction of electricity under specific conditions, but not always. In addition to conducting electricity, it can also respond to light and heat (Worzyk & Thomas, 2009).

1.1.1. Holes and Electrons in Semiconductors

The semiconductor's charge carriers, or holes and electronics, are essentially what allow current or electricity to flow through them. Electrons are negatively charged particles that revolve around an atom's nucleus. They are considered to be the main electric charge carriers in semiconductors. Electrons are confined to atoms within the semiconductor's valence band, where they have minimal effect on current flow. A hole is the space left behind in a semiconductor when an electron leaves due to energy absorption. A hole in a semiconductor is an area of positive charge caused by an opening in the covalent bond between atoms caused by the absence of an electron.

- *Mobility of Electrons and Holes:*

Because of their underlying behavioral differences within the structure of semiconductors like silicon, electrons have greater mobility than holes in these materials.

The semiconductor's valence band contains holes, which are created when electrons move from lower energy levels to higher ones. Excited electrons live and move within the semiconductor's conduction band. Because of their greater freedom within the conduction band, electrons are comparatively less hindered in their movement when an electric field is applied than holes. Additionally, because electrons have a negative charge, they move through the lattice with less resistance from the positively charged atomic nuclei than do holes, which have a positive charge and are therefore more strongly repelled by them.

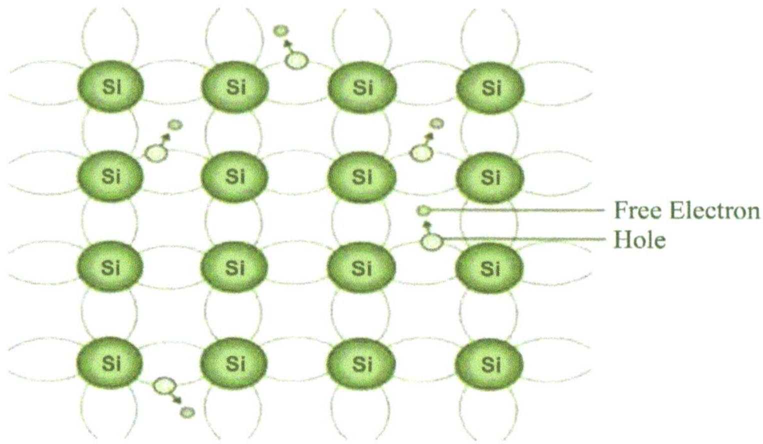

Figure 1.1. *Mobility of electrons and holes.*

Source: https://www.geeksforgeeks.org/semiconductors/.

A free electron in the specified Silicon Bond Model leaves behind a hole with the opposite charge when it moves from its lattice position. In the lattice, these holes carry positive charges (Figure 1.1)

1.1.2. Properties of Semiconductors

We delve into the various properties of semiconductors, exploring their electrical, thermal, optical, and mechanical characteristics.

Electrical Properties

1. **Conductivity and Resistivity:**
   - Semiconductors have moderate electrical conductivity, which can be significantly altered by doping, temperature and the application of electric fields. Their resistivity is higher than that of conductors but lower than insulators.
   - Intrinsic semiconductors (pure forms, like silicon or germanium) have their electrical properties determined by the electrons and holes generated within the material.
   - Extrinsic semiconductors result from doping, where impurities are added to increase free carriers (electrons or holes), thereby enhancing conductivity.
2. **Band Structure:**
   - The electronic band structure of semiconductors features a valence band full of electrons and an empty conduction band separated by a band gap.
   - The size of the band gap is crucial: small enough to allow electron excitation at room temperature but large enough to prevent significant conduction without external energy.

3. **Carrier Concentration:**
   - The density of free charge carriers (electrons and holes) is a critical parameter, influenced by temperature and doping.
   - At absolute zero, an intrinsic semiconductor acts as a perfect insulator. As temperature increases, thermal energy generates electron-hole pairs, enhancing conductivity.
4. **Mobility:**
   - Carrier mobility reflects how quickly electrons and holes can move through the semiconductor when subjected to an electric field. Higher mobility implies faster electronic devices.
5. **Doping and Types of Semiconductors:**
   - N-type: Doped with elements having more valence electrons than the semiconductor (e.g., phosphorus in silicon), creating excess electrons.
   - P-type: Doped with elements having fewer valence electrons (e.g., boron in silicon), creating holes.
   - Doping controls the concentration and type of carriers, tailoring materials for specific electronic functions.

Thermal Properties

1. **Thermal Conductivity:**
   - Semiconductors typically exhibit lower thermal conductivity compared to metals. For example, silicon has a thermal conductivity of about 149 W/m·K.
   - Thermal management is critical in semiconductor devices to prevent overheating and ensure performance reliability.
2. **Thermal Generation of Carriers:**
   - As temperature increases, thermal energy excites electrons from the valence to the conduction band, increasing intrinsic carrier concentration.
3. **Thermal Expansion:**
   - Semiconductors expand when heated. The coefficient of thermal expansion must be considered in device design to avoid mechanical stress and failure.

Optical Properties

1. **Absorption and Emission:**
   - Semiconductors can absorb photons with energy equal to or greater than their band gap, leading to electron excitation.
   - Upon recombination of electrons and holes, semiconductors might emit photons, a principle underlying light-emitting diodes (LEDs) and laser diodes.
2. **Photoconductivity:**
   - Exposure to light increases the number of free carriers in a semiconductor, enhancing its conductivity. This property is utilized in photodetectors and solar cells.
3. **Band Gap and Photons:**
   - The band gap determines the wavelength of light a semiconductor can absorb or emit. For example, gallium

arsenide (GaAs) has a direct band gap suitable for efficient light emission.

Mechanical Properties

1. **Crystal Structure and Defects:**
   - Semiconductors typically form crystalline structures. The arrangement and bonding of atoms in the lattice influence their electronic and mechanical properties.
   - Defects in the crystal lattice, such as vacancies and dislocations, affect electrical properties and can degrade device performance.
2. **Mechanical Strength:**
   - Semiconductors must possess adequate mechanical strength to withstand processing and operational stresses. Silicon, for instance, has good mechanical hardness and brittleness.
3. **Stress and Strain Effects:**
   - Mechanical stress can alter the electronic properties of semiconductors (piezoelectric effect in materials like GaN).
   - Strain engineering is used to modify the band structure and enhance carrier mobility in semiconductor devices.

1.2. SEMICONDUCTOR MATERIALS

Semiconductor materials are those that possess conductivity between conductors and insulators. These are compounds like gallium arsenide (GaAs) or cadmium selenide (CdSe), or pure elements like silicon (Si) or germanium (Ga) (Figure 1.2). Tiny amounts of impurities can be added to pure semiconductors in a process called doping to drastically alter the material's conductivity (Shockley, William (1950)).

Figure 1.2. *Semiconductor materials.*

Source: https://www.elprocus.com/semiconductor-material/.

The conductivity of a semiconductor material primarily falls between that of conductors and insulators. Semiconductor materials are used in the design of solid-state devices such as diodes and transistors because of their useful properties.

1.2.1. Semiconductor Materials in Periodic Table

Semiconductor materials fall into several groups in the periodic table of elements. Thus, the atomic characteristics of these materials are linked to their properties, which also differ amongst groups. The advantages of these differences are used by researchers and designers to improve designs and choose the best material for a given application (Figure 1.3).

Atoms that are available from various groups in the periodic table, such as combinations of group II and VI, group IV, or group III and V, make up semiconductor materials. Although the materials used to design the various semiconductors in the periodic table come from various groups, each semiconductor has a unique set of properties. Group IV elements in the periodic table represent silicon (Si) material, which is the most widely used semiconductor material since it serves as the foundation for integrated circuits, or ICs. Si (Silicon) is also the primary material used in solar cells.

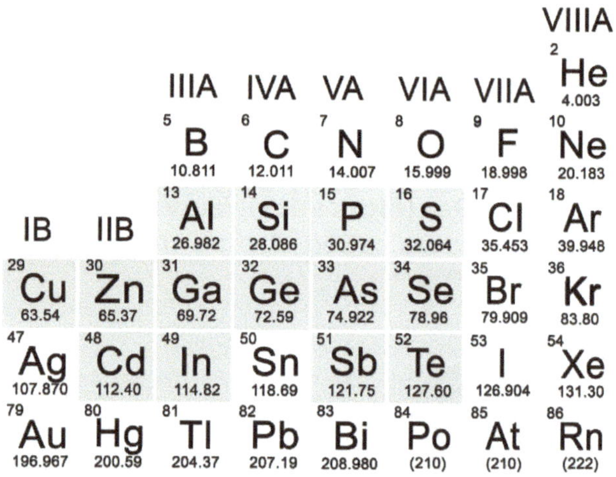

Figure 1.3. *Semiconductor materials in periodic table.*

Source: https://www.elprocus.com/semiconductor-material/.

Generally, a semiconductor is a single element like silicon, germanium, a compound like GaAs, CdTe/ InP, or an alloy like AlxGa(1-x) As/ SixGe(1-x), where 'x in these materials are the part of the specific element that ranges from zero to one.

1.2.2. List of Semiconductor Materials

The list of semiconductor materials is discussed below:

- *Germanium (Ge):* In the periodic table, group IV contains semiconductors like germanium. This substance was utilized in the earliest diodes and transistors, among other devices. Because diodes have a higher reverse conductivity and temperature coefficient than transistors, thermal runaway may occur in early transistors. Because it offers greater charge carrier mobility than silicon, some radio frequency (RF)-based devices employ it.
- *Silicon (S):* The most widely used semiconductor material is silicon, which is a group IV element in the periodic table of chemical elements. These materials have the best mechanical and electrical qualities and are very easy to fabricate. When these materials are utilized in integrated circuits (ICs), high-quality silicon dioxide is formed to serve as insulating layers between the chip's various active components, which is one of the reason that makes silicon a very important semiconductor.
- *Gallium Arsenide (GaAs):* The semiconductor gallium arsenide, which belongs to the III-V group of the periodic table, is the most widely used material after silicon. Due in large part to this element's high electron mobility, high-performance RF devices use it. It is also utilized as substrate-like GaInNAs and InGaAs in other III-V semiconductors. Compared to silicon, this material has less hole mobility. It also raises the cost of GaAs devices and is fairly complex to fabricate.
- *Silicon Carbide (SiC):* The periodic table classifies silicon carbide material as an IV group element. Compared to Si-based devices, these elements have higher operating temperatures and significantly lower losses, which is why they

are utilized in power devices. This material has a breakdown capacity that is more than ten times greater than silicon's. LEDs with blue and yellow colors use silicon carbide material forms.

- *Gallium Nitride (GaN):* In the periodic table, gallium nitride, or GaN material, belongs to the III-V group. It is most frequently utilized in microwave ICs and transistors when the highest temperatures and powers are needed. This semiconductor material responds to electrostatic discharge and is difficult to dope to produce p-type regions in, but it is not very sensitive to ionizing radiation. Blue LEDs have been made with this material.
- *Gallium Phosphide (GaP):* In the periodic table, gallium phosphorus, or GaP, is a semiconductor material belonging to the III-V group. This substance is utilized in early low brightness to medium based LEDs, which produce various colors depending on the dopant addition. In pure GaP, the light is green; in nitrogen-doped, it is yellow-green; in ZnO-doped, it is red.
- *Cadmium Sulphide (CdS):* In the periodic table, cadmium sulphide, or CdS, is a semiconductor material belonging to the II–VI group. Photoresistors and solar cells both employ this substance.
- *Lead Sulphide (PbS):* An IV-VI group element in the periodic table, lead sulphide, or PbS, is a semiconductor material used in the early radio detectors known as "Cat's Whiskers," in which a tip contact was created by employing thin wire on the galena to provide signal rectification.

1.2.3. Advantages and Disadvantages

There are different semiconductor materials are available in the market like silicon, germanium, gallium arsenide, silicon carbide, gallium nitride, etc. Each material has some advantages and disadvantages. So here, some of the material's advantages and disadvantages are discussed below.

The advantages of silicon material are:

- Silicon material cost is low
- Silicon has high efficiency
- In silicon, large wafers can be handled very securely without any harm because of their hardness
- Silicon is stable thermally up to 1100°C.

The disadvantages of a silicon material are:

- It requires crystalline
- Easily broken
- Substrates are limited
- It is costly to produce Si crystals
- Toxic substances are generated.

The advantages of a germanium material are:

- Pure germanium is gray-white, hard, and brittle metalloid
- Its crystalline structure is like a diamond
- Germanium is constant in water and air
- It is not affected by acids and alkalis apart from nitric acid.
- By using Germanium material, first, the transistor action was observed.

So, it is one of the most essential semiconductor materials.

- High atomic number, so it enhances the gamma-ray interaction probability.
- Large crystals
- The material used in this is grey metallic

The disadvantages of a germanium material are:

- Its energy resolution is responsive to radiation damage
- Less efficiency for higher energy-based gamma rays.

1.3. ENERGY BAND DESCRIPTION OF SEMICONDUCTORS

The resistivity is of the order of 10^{-4}-to-0.5-ohm meter. However, a semiconductor can be defined much more comprehensively on the basis of energy bands as under:

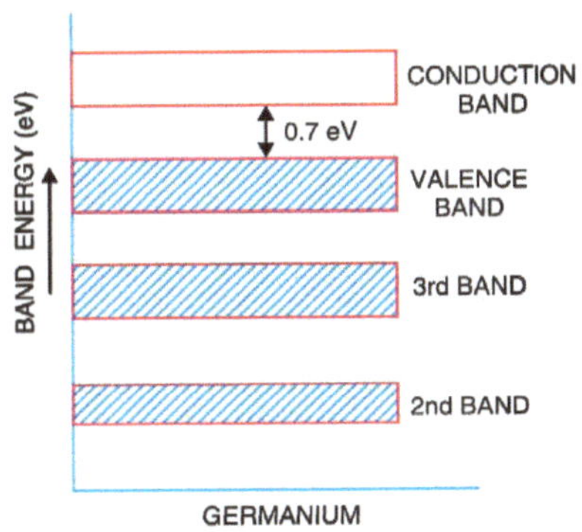

Figure 1.4. *Energy band diagrams of germanium.*

Source: https://mpsasc.edu.in/storage/app/application/semiconductor%20physics%20unit%202.pdf.

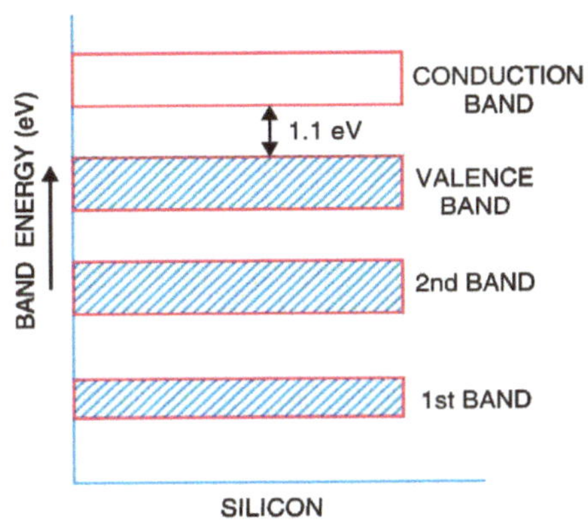

Figure 1.5. *Energy band diagrams of silicon.*

Source: https://mpsasc.edu.in/storage/app/application/semiconductor%20physics%20unit%202.pdf.

A material with a nearly empty conduction band and an almost filled valence band with energy gap (0.1eV-4 eV) separating them is called a semiconductor. Figure 1.4 and 1.5 show the energy band diagrams of germanium and silicon respectively. It may be seen that forbidden energy gap is very small; being 1.1 eV for silicon and 0.7 eV for germanium.

Their valence electrons therefore require comparatively little energy to move from the ground state to the conduction band. Some of the valence electrons may gain enough energy, even at room temperature, to move into the conduction band and become free electrons. Nonetheless, there are relatively few free electrons accessible at this temperature. Consequently, a piece of silicon or germanium is not a good insulator or conductor at room temperature. This is why these materials are referred to as semiconductors (Nashelsky & Boylestad, 2006).

1.3.1. Effect of Temperature on Semiconductors

Temperature variations have a noticeable impact on a semiconductor's electrical conductivity. It is imperative to bear this in mind.

- **At Absolute Zero:** The semiconductor atoms firmly hold onto all of the electrons at absolute zero temperature. While the valence electrons are participating in covalent bonding, the electrons in the inner orbit are bound. There are no free electrons and very strong covalent bonds at this temperature. As a result, the semiconductor crystal exhibits ideal insulator properties (See Figure. 1.6 (i)).

The valence band is filled and the energy gap between it and the conduction band is substantial when it comes to energy band description. Consequently, no valence electron can enter the conduction band and liberate itself. A semiconductor acts like an insulator because there are no free electrons in the material (Figure 1.6 (ii)).

- **Above Absolute Zero:** The thermal energy present in the semiconductor causes some of its covalent bonds to break when the temperature is raised. The electrons involved in the formation of these bonds are released when the bonds break. As a result, the semiconductor contains a small number of free electrons. If a potential difference is applied across the semiconductor crystal, these free electrons may form a tiny electric current (See Figure. 1.7 (i)).

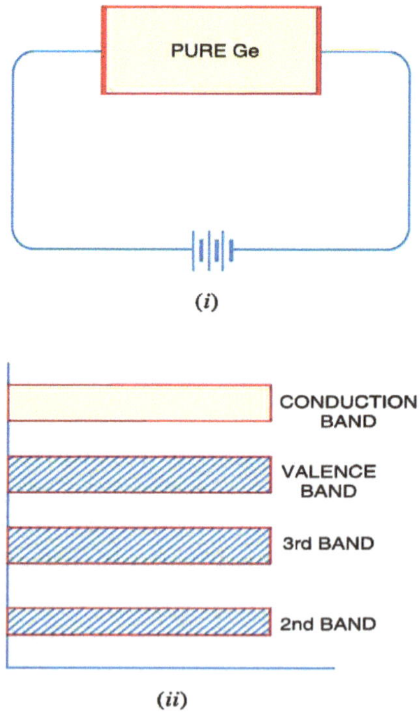

Figure 1.6. *Semiconductor crystal.*

Source: https://mpsasc.edu.in/storage/app/application/semiconductor%20physics%20unit%202.pdf.

This demonstrates how a semiconductor's resistance lowers as temperature rises. Its temperature coefficient of resistance is negative. It should be noted that a semiconductor's current is too tiny to be useful at room temperature.

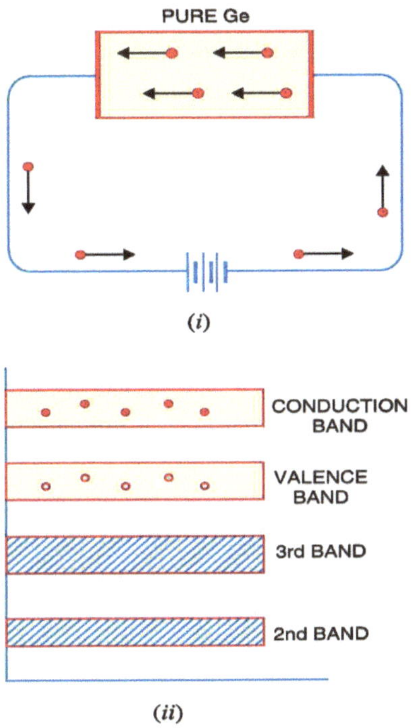

Figure 1.7. *Applied across the semiconductor crystal.*

Source: https://mpsasc.edu.in/storage/app/application/semiconductor%20physics%20unit%202.pdf.

The energy band diagram is displayed in Figure 1.7(ii). A portion of the valence electrons gain enough energy to move into the conduction band and become free electrons when the temperature rises.

These free electrons will form electric current when subjected to an electric field. It should be noted that a hole is made in the valence band each time a valence electron moves into the conduction band. As the upcoming article will demonstrate, holes also contribute to current. In actuality, the most important idea in semiconductors is hole current (Hulls & McMillan, 1972).

1.3.2. Hole Current

At room temperature, some of the covalent bonds in pure semiconductor break, setting up free electrons. Under the influence of electric field, these free electrons constitute electric current. At the same time, another current – the hole current – also flows in the semiconductor.

When a covalent bond is broken due to thermal energy, the removal of one electron leaves a vacancy, i.e., a missing electron in the covalent bond. This missing electron is called a hole which acts as a positive charge. For one electron set free, one hole is created. Therefore, thermal energy creates hole-electron pairs; there being as many holes as the free electrons. The current conduction by holes can be explained as follows:

The hole shows a missing electron. Suppose the valence electron at L (See Figure. 1.8) has become free electron due to thermal energy. This creates a hole in the covalent bond at L. The hole is a strong center of attraction for the electron. A valence electron (say at M) from nearby covalent bond comes to fill in the hole at L. This results in the creation of hole at M. Another valence electron (say at N) in turn may leave its bond to fill the hole at M, thus creating a hole at N. Thus, the hole having a positive charge has moved from L to N, i.e., towards the negative terminal of supply. This constitutes hole current.

It should be mentioned that the flow of valence electrons from one covalent bond to another is what causes the entire current. When electrons (valence electrons, of course) are once again conducting the current, the reader might wonder why it is referred to as a hole current. The existence of holes in the covalent bonds is the fundamental cause of current flow, is the response. Thus, thinking of the current as the movement of holes is more appropriate.

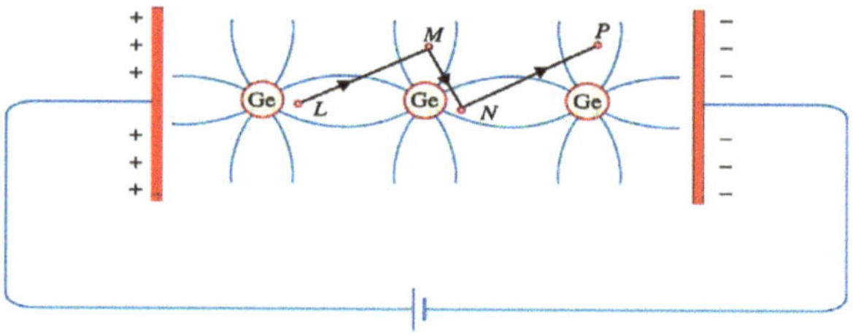

Figure 1.8. *Valence electron.*

Source: https://mpsasc.edu.in/storage/app/application/semiconductor%20physics%20unit%202.pdf.

Energy band description. The hole current can be beautifully explained in terms of energy bands. Suppose due to thermal energy, an electron leaves the valence band to enter into the conduction band as shown in Figure 1.9. This leaves a vacancy at L. Now the valence electron at M comes to fill the hole at L. The result is that hole disappears at L and appears at M. Next, the valence electron at N moves into the hole at M. Consequently, hole is created at N. It is clear that valence electrons move along the path PNML whereas holes move in the opposite direction, i.e., along the path LMNP.

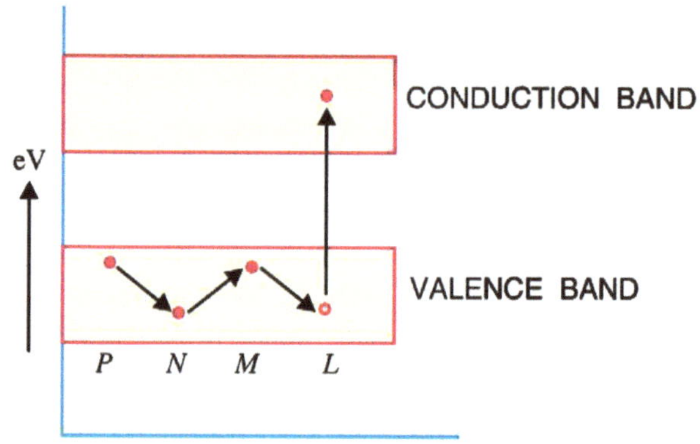

Figure 1.9. *Conduction band.*

Source: https://mpsasc.edu.in/storage/app/application/semiconductor%20physics%20unit%202.pdf.

1.4. INTRINSIC SEMICONDUCTOR

An intrinsic semiconductor is a semiconductor that is incredibly pure. Even at room temperature, hole-electron pairs are generated in an intrinsic semiconductor. As illustrated in Figure 1.10, current conduction via free electrons and holes occurs when an electric field is applied across an intrinsic semiconductor. Some covalent bonds are broken by thermal energy, which results in the production of free electrons. The covalent bonds develop holes at the same time. Both free electrons and holes can conduct through semiconductors when subjected to an electric field. As a result, the sum of the currents caused by free electrons and holes inside the semiconductor equals the total current inside.

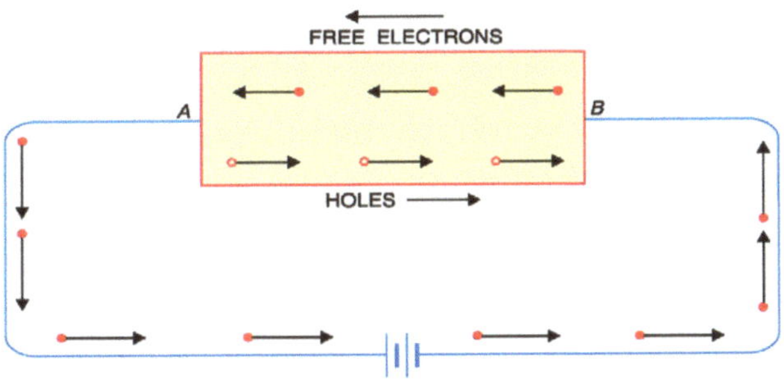

Figure 1.10. *Free electrons and holes.*

Source: https://mpsasc.edu.in/storage/app/application/semiconductor%20physics%20unit%202.pdf.

It should be mentioned that the external wires' current is entirely electronic, i.e., through electrons. Regarding the holes, as shown in Figure 1.10, positively charged holes travel in the direction of the supply's negative terminal. Electrons enter the semiconductor crystal close to the negative terminal B as the holes approach it, combining with the holes to cancel them out. The loosely held electrons close to positive terminal A are drawn into the positive terminal at the same time as they are drawn away from their atoms. New holes are made as a result, and they move back toward the negative terminal in the direction of the positive terminal (Moskowitz & Sanford, 2016).

1.5. EXTRINSIC SEMICONDUCTOR

The intrinsic semiconductor has little current conduction capability at room temperature. To be useful in electronic devices, the pure semiconductor must be altered so as to significantly increase its conducting properties. This is achieved by adding a small amount of suitable impurity to a semiconductor. It is then called impurity or extrinsic semiconductor. The process of adding impurities to a semiconductor is known as doping. The amount and type of such impurities have to be closely controlled during the preparation of extrinsic semiconductor. Generally, for 108 atoms of semiconductor, one impurity atom is added.

The purpose of adding impurity is to increase either the number of free electrons or holes in the semiconductor crystal. As we shall see, if a pentavalent impurity (having 5 valence electrons) is added to the silicon semiconductor, a large number of free electrons are produced in the semiconductor. On the other hand, addition of trivalent impurity (having 3 valence electrons) creates a large number of holes in the silicon semiconductor crystal. Depending upon the type of impurity added, extrinsic semiconductors are classified into:

(i) n-type semiconductor

(ii) p-type semiconductor

1.5.1. n-type Semiconductor

When a small amount of pentavalent impurity is added to a pure semiconductor with 4 valence electron, it is known as n-type semiconductor.

The addition of pentavalent impurity provides a large number of free electrons in the semiconductor crystal. Typical examples of pentavalent impurities are arsenic and antimony. Such impurities which produce n-type semiconductor are known as donor impurities because they donate or provide free electrons to the semiconductor crystal. To explain the formation of n-type semiconductor, consider a pure germanium crystal. We know that germanium atom has four valence electrons. When a small amount of pentavalent impurity like arsenic is added to germanium crystal, a large number of free electrons become available in the crystal. The reason is simple.

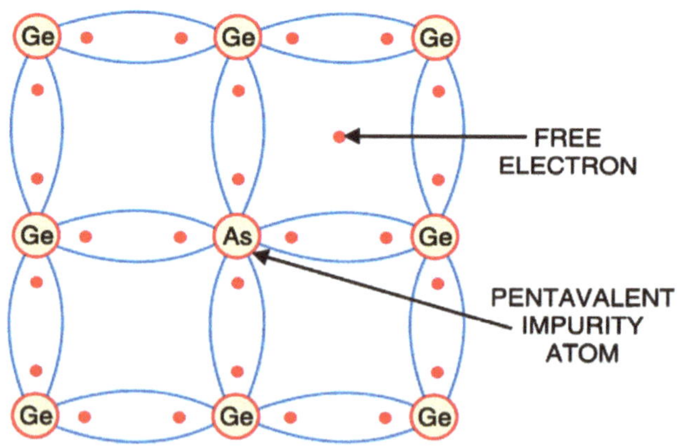

Figure 1.11. *Valence electron of arsenic atom.*

Source: https://mpsasc.edu.in/storage/app/application/semiconductor%20physics%20unit%202.pdf.

Arsenic is pentavalent, i.e., its atom has five valence electrons. An arsenic atom fits in the germanium crystal in such a way that its four valence electrons form covalent bonds with four germanium atoms. The fifth valence electron of arsenic atom finds no place in covalent bonds and is thus free as shown in Figure 1.11. Therefore, for each arsenic atom added, one free electron will be available in the germanium crystal. Though each arsenic atom provides one free electron, yet an extremely small amount of arsenic impurity provides enough atoms to supply millions of free electrons.

Figure 1.12 shows the energy band description of n-type semi-conductor. The addition of pentavalent impurity has produced a number of conduction band electrons, i.e., free electrons. The four valence electrons of pentavalent atom form covalent bonds with four neighboring germanium atoms. The fifth left over valence electron of the pentavalent atom cannot be accommodated in the valence band and travels to the conduction band. The following points may be noted carefully:

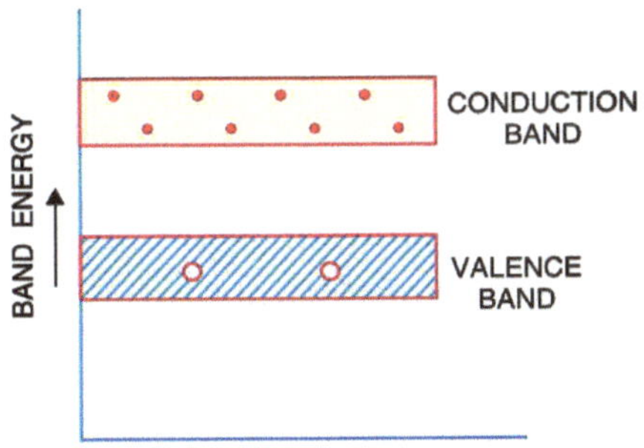

Figure 1.12. *The energy band description of n-type semi-conductor.*

Source: https://mpsasc.edu.in/storage/app/application/semiconductor%20physics%20unit%202.pdf.

i. Many new free electrons are produced by the addition of pentavalent impurity.

ii. Thermal energy of room temperature still generates a few hole-electron pairs. However, the number of free electrons provided by the pentavalent impurity far exceeds the number of holes. It is due to this predominance of electrons over holes that it is called n-type semiconductor (n stands for negative).

– *N-type conductivity*: The current conduction in an n-type semiconductor is predominantly by free electrons, i.e., negative charges and is called n-type or electron type conductivity. To understand n-type conductivity, refer to Figure. 1.13. When potential difference is applied across the n-type semiconductor, the free electrons (donated by impurity) in the crystal will be directed towards the positive terminal, constituting electric current. As the current flow through the crystal is by free electrons which are carriers of negative charge, therefore, this type of conductivity is called negative or n-type conductivity. It may be noted that conduction is just as in ordinary metals like copper.

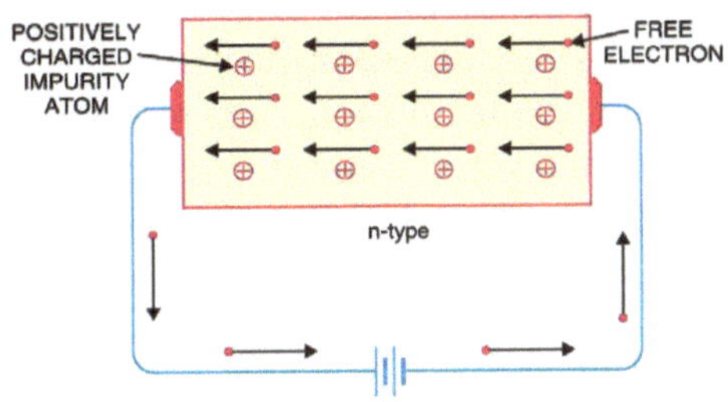

Figure 1.13. *N-type conductivity.*

Source: https://mpsasc.edu.in/storage/app/application/semiconductor%20physics%20unit%202.pdf.

1.5.2. p-type Semiconductor

When a small amount of trivalent impurity is added to a pure semiconductor, it is called p-type semiconductor. The addition of trivalent impurity provides a large number of holes in the semiconductor. Typical examples of trivalent impurities are gallium and indium. Such impurities which produce p-type semiconductor are known as acceptor impurities because the holes created can accept the electrons.

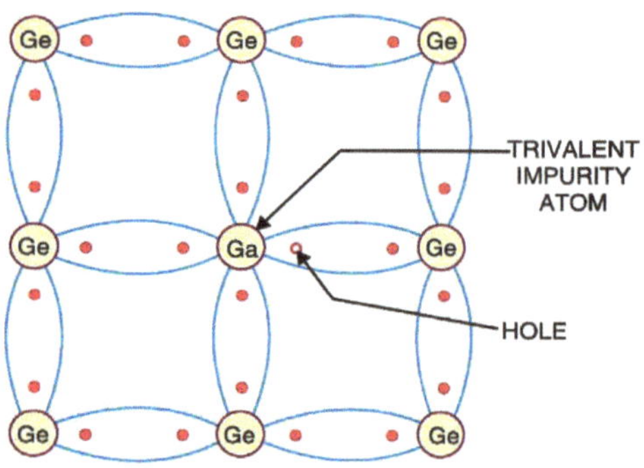

Figure 1.14. *Three single covalent bonds with three germanium atoms.*

Source: https://mpsasc.edu.in/storage/app/application/semiconductor%20physics%20unit%202.pdf.

To explain the formation of p-type semiconductor, consider a pure germanium crystal. When a small amount of trivalent impurity like gallium is added to germanium crystal, there exists a large number of holes in the crystal. The reason is simple. Gallium is trivalent, i.e., its atom has three valence electrons. Each atom of gallium fits into the germanium crystal but now only three covalent bonds can be formed. It is because three valence electrons of gallium atom can form only three single covalent bonds with three germanium atoms as shown in Figure 1.14.

In the fourth covalent bond, only germanium atom contributes one valence electron while gallium has no valence electron to contribute as all its three valence electrons are already engaged in the covalent bonds with neighboring germanium atoms. In other words, fourth bond is incomplete; being short of one electron. This missing electron is called a hole. Therefore, for each gallium atom added, one hole is created. A small amount of gallium provides millions of holes.

Figure 1.15 shows the energy band description of the p-type semiconductor. The addition of trivalent impurity has produced a large number of holes. However, there are a few conduction band electrons due to thermal energy associated with room temperature. But the holes far outnumber the conduction band electrons. It is due to the predominance of holes over free electrons that it is called p-type semiconductor (p stands for positive).

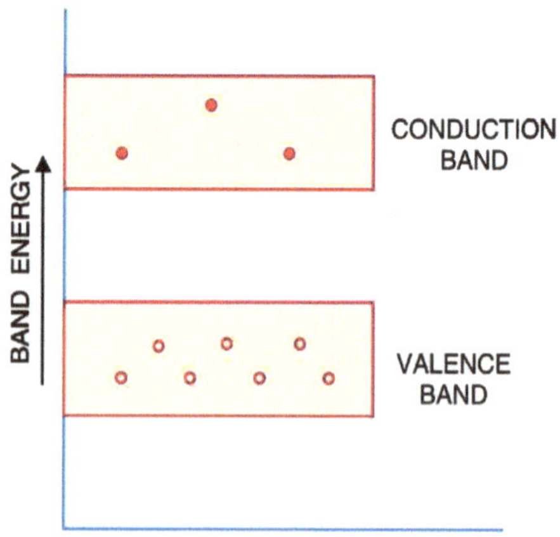

Figure 1.15. *The energy band description of the p-type semiconductor.*

Source: https://mpsasc.edu.in/storage/app/application/semiconductor%20physics%20unit%202.pdf.

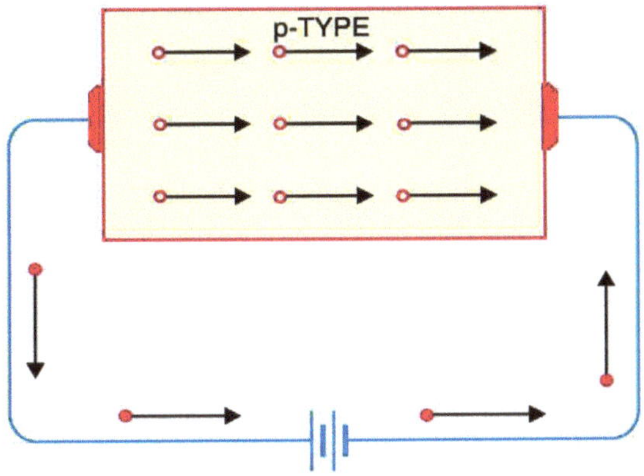

Figure 1.16. *p-type conductivity.*

Source: https://mpsasc.edu.in/storage/app/application/semiconductor%20physics%20unit%202.pdf.

- *P-type conductivity*: The current conduction in p-type semiconductor is predominantly by holes, i.e., positive charges and is called p-type or hole-type conductivity. To understand p-type conductivity, refer to Figure 1.16. When potential difference is applied to the p-type semiconductor, the holes (donated by the impurity) are shifted from one covalent bond to another. As the holes are positively charged, therefore, they are directed towards the negative terminal, constituting what is known as whole current. It may be noted that in p-type conductivity, the valence electrons move from one covalent bond to another unlike the n-type where current conduction is by free electrons.

1.6. ELECTRICAL CONDUCTION IN SEMICONDUCTORS

Despite having substantially higher conductivities than insulators, pure semiconductor crystals are not very good electrical conductors (Table 1.1). Because dopant atoms can be carefully added to the crystal lattice to vary the electrical conductivity of semiconducting materials—particularly silicon—this electrical property makes semiconducting materials so valuable in electronics and other device applications.

This characteristic makes it possible to control the electrical current in a wide range of electronic devices, such as diodes, capacitors, and transistors, using silicon, both doped and undoped. Particularly, transistors are crucial components of contemporary technology. A Metal-Oxide-Semiconductor Field Effect Transistor (MOSFET) device, Figure 1.17(a), behaves as an electrical on/off switch, and a Bipolar Junction Transistor (BJT), Figure 1.17(b), primarily acts as an amplifier (controlling signals) in its linear range. Together, they can be compared to a valve that regulates the flow of electricity. Semiconductor materials are the sole means of fabricating such transistors.

Table 1.1. *Relative Electrical Resistivities and Conductivities of Some Metals*

Material	Resistivity (Ω-cm)	Conductivity (Ω^{-1}-m^{-1})
Insulators		
Hard Rubber	1-100 x 10^{13}	1×10^{-15} to 1×10^{-13}
Glass	1-10000 x 10^{9}	1×10^{-13} to 1×10^{-9}
Quartz (fused)	7.5 x 10^{17}	1.33×10^{-18}
Semiconductors		
Carbon (Graphite)	3-60 x 10^{-5}	1.67×10^{3} to 3.33×10^{4}
Germanium	1-500 x 10^{-3}	2.0 to 1.00×10^{3}
Silicon	0.10- 60	1.67×10^{-2} to 10
Metals		
Silver	1.63 x 10^{-8}	6.17×10^{7}
Copper (annealed)	1.72 x 10^{-8}	5.95×10^{7}
Aluminum	2.65 x 10^{-8}	3.77×10^{7}

Source: https://www.mks.com/n/electrical-conduction-semiconductors.

We will use silicon to demonstrate the electrical characteristics of semiconductors since it is currently the main semiconducting material used in the large-scale production of electronic devices.

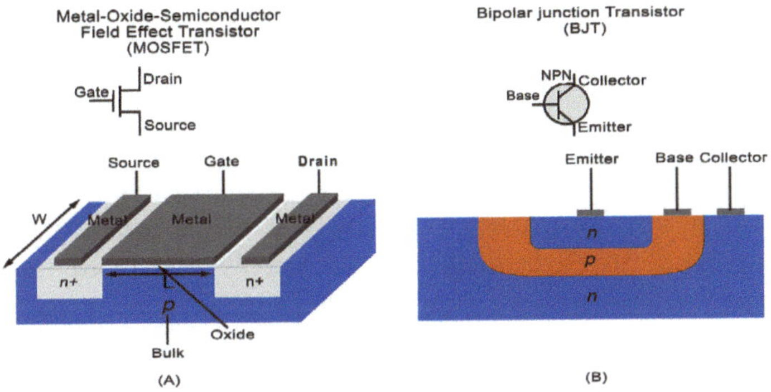

Figure 1.17. *(a) Metal-Oxide-Semiconductor Field Effect Transistor (MOSFET) symbol and structure; (b) Bipolar Junction Transistor (BJT) symbol and structure.*

Source: https://www.mks.com/n/electrical-conduction-semiconductors.

Remember that the characteristics listed for silicon also apply to the other semiconducting elements in Table 1.1 and the different compound semiconductors (such as GaAs, InP, etc.) that are employed in fewer and more specific contexts (though with minor variations in the specifics).

Intrinsic semiconductors are generally used to refer to pure silicon and other elemental semiconductors. Intrinsic refers to the fact that electrical conductivity is a feature of semiconducting materials that exists independently of additives. Extrinsic semiconductors, which rely on the presence of dopant impurities to increase the electrical conductivity of the material, are also used in semiconductor device fabrication in addition to intrinsic semiconductors (see Figure 1.18). The room-temperature resistivities and conductivities for a variety of metals, intrinsic semiconductors, and insulators are displayed in Table 1.1. The resistivity of elemental silicon is 0.10–60 Ω-cm, which is much higher than metallic resistivities but orders of magnitude lower than that of a typical insulator.

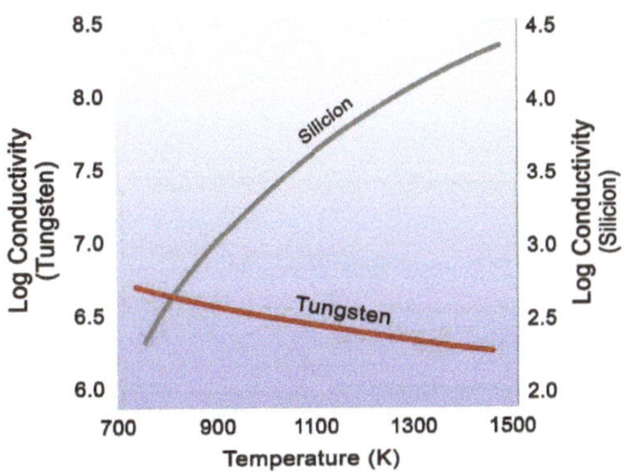

Figure 1.18. *The electrical conductivity characteristics of a semiconductor (silicon) vs. those of a metal (tungsten).*

Source: https://www.mks.com/n/electrical-conduction-semiconductors.

The way the conductivities of metals and semiconductors change with temperature is another way that the two types of materials differ electrically. A semiconductor's electrical conductivity is zero at absolute zero temperature (0 K) (i.e., the conductivity is at its lowest) while a metal shows its maximum electrical conductivity at absolute zero; additionally, in a semiconductor, conductivity rises with temperature while in a metal, it decreases with temperature. A graph comparing the temperature variation of a metal (tungsten) and a semiconductor (silicon)'s electrical conductivity is presented in Figure 1.18. The various mechanisms that control electrical conductivity in a metal versus a semiconductor can be used to explain the differences in electrical conductivity behaviors shown in Figure 1.18.

Metals are electrically conductive due to their abundance of free electrons. Because of the overlap between the valence and conduction bands in metals, electrons are free to move into a wide range of available and vacant energy levels. These electrons are free to move from a higher potential to a lower potential when an electrical potential is applied across a piece of metal. The amount of electron scattering caused by collisions between the flowing electrons and fixed atoms in the metal lattice is the only factor limiting the conductivity. The metal lattice's atoms are at rest and electron scattering is at its minimum at 0 K. Thermal energy causes the atoms in the lattice to vibrate as the metal's temperature rises. Each atom's capacity to scatter electrons is increased by these vibrations because they broaden the effective diameter that each atom presents to the electron flow. As a result, the metal atoms vibrate at amplitudes that increase with temperature above 0 K, increasing the degree of electron scattering and causing the conductivity to decrease as shown in Figure 1.18.

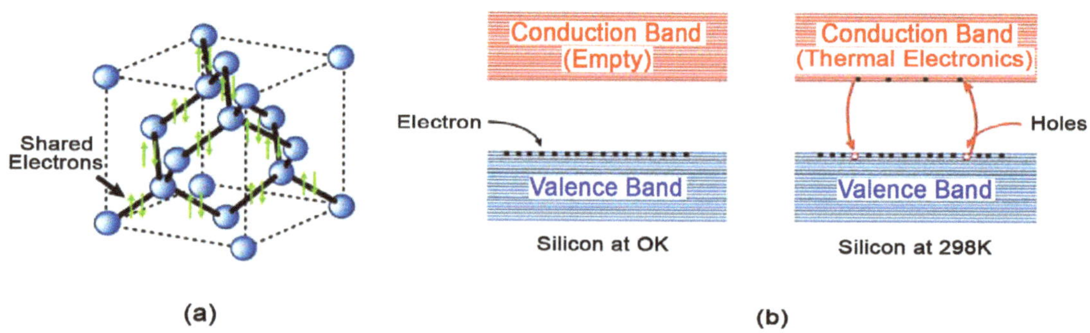

Figure 1.19. *Silicon structure and electrical conduction.*

Source: https://www.mks.com/n/electrical-conduction-semiconductors.

Understanding how the mechanism of electrical current flow in semiconductors varies from that of current flow in metals can be aided by looking at Figure 1.19. Silicon atoms in a crystal are depicted in Figure 1.19(a), along with a rough schematic of how each silicon atom's valence electrons are shared to create four bonds with eight electrons in each atom's valence shell. Eight electrons are necessary for the stability of covalently bound atoms in their valence shell, as you may remember from basic chemistry. Silicon is an insulator at absolute zero because, as Figure 1.19(b) illustrates, every valence electron in an atom of silicon is bound by a covalent bond; in contrast, none of the valence electrons in a metal are free for electrical conduction.

An unfinished schematic of silicon's band structure at 0 K vs is displayed in Figure 1.19(b). the same configuration at 25°C, or 298 K. There is no thermal energy available to raise a valence electron into the conduction band at 0 K, meaning that the valence band is full and the conduction band is empty. Thermal energy is absorbed by silicon as temperature rises. Even though this energy causes the silicon atoms to vibrate more thermally, which increases atom-electron scattering, the increase in electrical conductivity brought about by the thermal promotion of electrons from the valence to the conduction band more than offsets any loss in electrical conductivity caused by this phenomenon. Metals do not exhibit this effect because the valence band and conduction band overlap. An electron-hole pair, which makes up mobile carriers in both the valence band (holes) and the conduction band (electrons), is created when an electron from the valence band is promoted to the conduction band. When a potential is applied, electrical current will now flow through the material. The number of electrons in the conduction band and the number of holes in the valence band in an intrinsic semiconductor are balanced, as can be seen in Figure 1.19(b). Since the movement of holes is typically used to represent electrical current in discussions of electronic devices, the idea of holes is crucial.

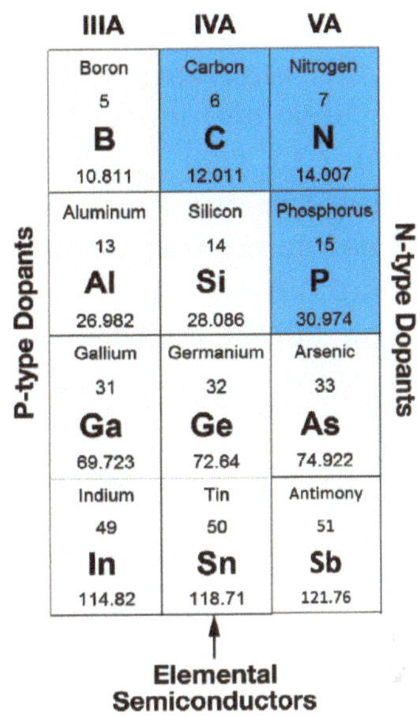

Figure 1.20. *Elemental semiconductors and the most common n- and p-type dopants used to create extrinsic semiconductors.*

Source: https://www.mks.com/n/electrical-conduction-semiconductors.

It is helpful to think of holes as positive charges since they are the absence of negative charges, or electrons. Conventionally, regions of positive charge are represented by a vector that extends outward to represent electric fields and, consequently, electric potential. As a result, electrons flow in the opposite direction to the direction of the applied electric field when considering current flow (i.e., holes flow in the direction of the electric field, whereas positive charges move in the opposite direction.

When compared with metals, semiconductor electrical conductivities are not very high, as can be seen from the values in Table 1.1. However, semiconductor conductivities can be significantly increased (and, in fact, fine-tuned to target values) by adding impurity atoms known as dopants.

"Doped" semiconductor materials are commonly referred to as extrinsic semiconductors. Figure 1.20 shows the area of the Periodic Table that contains the elemental semiconductors and the dopant elements that can be used to create an extrinsic semiconductor. The dopants are those elements on either side of Column IVA in the Periodic Table. A dopant atom is similar in size to the intrinsic semiconductor atom in the same row but has either one fewer or one greater electron in its valence shell. The dopant atoms can easily replace the semiconductor atom in its crystal lattice. Figure 1.21 shows the insertion of either a Group VA (n-type dopant, phosphorus) or a Group IIIA (p-type dopant, boron) atom into the silicon crystal lattice as well as the effect that the substitution of the silicon atom by a dopant atom has on the band structure of the silicon.

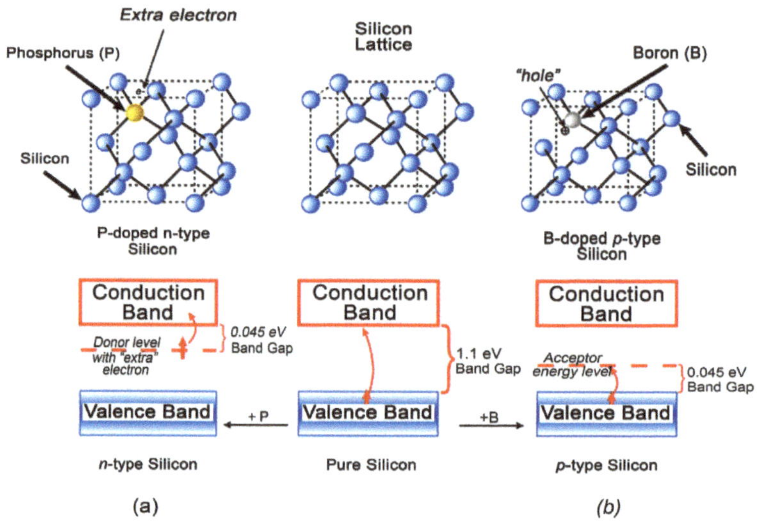

Figure 1.21. *The effect of impurities on the band gap in extrinsic silicon.*

Source: https://www.mks.com/n/electrical-conduction-semiconductors.

When phosphorus, an n-type (donor) dopant is substituted for silicon, the extra electron in phosphorus's valence shell is at a higher energy level (the donor level in Figure 1.21(a)) than the electrons in the filled silicon valence band. This dopant electron can easily jump the small energy gap (0.045 eV for P, much less than the 1.1 eV band gap for pure silicon at room temperature; typical donor band gaps range from.039 to.054 eV) to the empty conduction band and become a free carrier. The concentration of free carriers (and therefore the electrical conductivity) in an n-type extrinsic semiconductor is roughly proportional to the dopant concentration in the material. Since the free electron generated from dopant atoms in an n-type semiconductor does not produce a corresponding hole in the filled valence band, the dominant charge carriers in an n-type semiconductor are electrons. In a similar manner, boron can be easily substituted for silicon in the bulk crystal lattice, as shown in Figure 1.21(b). Since boron has one less electron than silicon, this creates a hole (an acceptor level) at an energy just greater than the top of the filled silicon valence band. Under the influence of an electric field, it takes little energy (0.045 eV for B; typical acceptor band gaps range from 0.045 to

0.160 eV) for an electron in the valence band to be promoted to this acceptor level, leaving behind a hole in the valence band. Other bound electrons in the valence band can then jump into this hole and the subsequent hole movement carries the electrical current. Thus, for p-type extrinsic semiconductors, holes are the majority charge carriers for current flow. Figure 1.22 shows the relationship between dopant concentration and resistivity (the inverse of conductivity) for n-and p-type dopants in crystalline silicon.

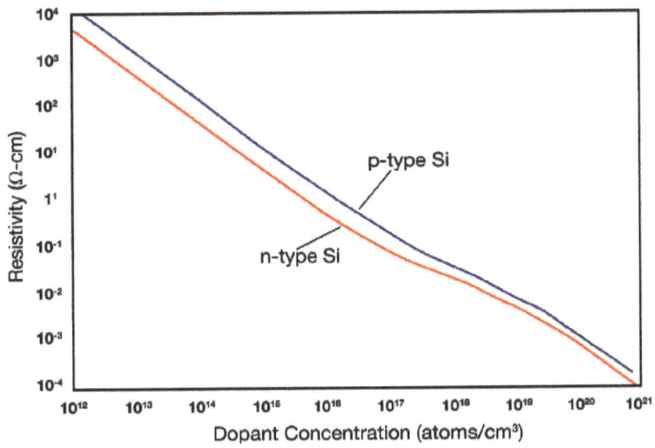

Figure 1.22. *The dependency of the electrical resistivity at 298K on the doping concentration.*

Source: https://www.mks.com/n/electrical-conduction-semiconductors.

ROLE MODEL

WILLIAM B. SHOCKLEY: AMERICAN PHYSICIST AND INVENTOR

William Shockley was an American physicist and inventor. His study of semiconductors earned him a Nobel Prize in Physics along with John Bardeen and Walter Brattain. A graduate of California Institute of Technology and the Massachusetts Institute of Technology, he worked at Bell Labs for a considerable amount of time to work on what would become his making. He headed numerous research teams and worked on many notable scientific studies. During the latter part of his life, Shockley dabbled in eugenics. His studies and theories of black inferiority made him an academic pariah. His personal life was marred by estrangement. His children from his first marriage, whom he barely spoke to during his lifetime, did not even attend his funeral when he passed away. His long-term companion was his second wife, Emmy. His extreme mood swings and stubborn nature took away the essence of what made him a great scientist and left him a man whose legacy lived on in his work and not as a human being.

Childhood & Early Life

- William Bradford Shockley Jr. was born on February 13, 1910, in London, England. His father, William Hillman Shockley was a mining engineer and his mother, Mary Bradford was the first female US Deputy mining surveyor. His parents were American and he grew up in Palo Alto, California from age 3.
- Shockley studied at the California Institute of Technology and received his Bachelors of Science in 1932.
- He studied for his Ph.D. at the Massachusetts Institute of Technology under Professor J.C. Slater. His thesis was on the energy band structure of Sodium Chloride. He received his doctorate in 1936. The same year, he joined Bell Labs in New Jersey and began research on semiconductors. The research group was headed by Clinton Davisson. He wrote many fundamental papers and had them published in "Physical Review".

Career

- At Bell Labs, Shockley was involved in radar research. This was when World War II broke out. During the war, in May 1942, he served as the director of research for the U.S. Navy's Antisubmarine Warfare Operations Research Group.
- In 1944, he organized a training program for B-29 bomber pilots and took tours around the world to analyze results. The training involved the usage of new

radar bomb sights. For this, he was awarded the "Medal for Merit" on October 17, 1946.

- He was asked to prepare a report on the casualties from Japan's invasion in July 1945 by the War Department. His report laid the foundation for the Hiroshima and Nagasaki bombings and Japan's eventual surrender.
- In 1945, after the war ended, he was invited to form a solid state group involving Stanley Morgan, John Bardeen, Walter Brattain, Gerald Pearson, Robert Gibney, and Hilbert Moore. They were handed the responsibility of finding a solid-state alternative to glass vacuum tubes. After many failures and attempts, the group was able to submit a paper on their findings in 1946.
- Shockley, Bardeen, and Brattain invented the point-contact transistor in 1947 which replaced traditional bulky transistors. This work earned them the Nobel Prize in Physics.
- He published a 558-page treatise, 'Electrons and Holes in Semiconductors', a collection of his research and work in 1950. This would go on to become a reference for other scientists working on the development of semiconductors and their variants.
- He invented the junction transistor in 1951 and announced his invention at a press conference on July 4 of that year. The same year, he was elected a member of "National Academy of Sciences" (NAS), a post that was too high for someone as young as 41. He received the "Comstock Prize" in 1953 from the NAS as well as many other honors.
- He moved to Mountain View in California in 1956 and set up the "Shockley Semiconductor Laboratory". After his Nobel victory, he became paranoid and autocratic forcing 8 colleagues at his lab to resign. These 8 went on to form the "Fairchild Semiconductor" without him.
- He was appointed to the President's Scientific Advisory Committee in 1962. In 1963, he received the Holley Medal of the American Society of Mechanical Engineers.
- During the 1960s he actively started questioning the intellectual differences between races. Many of his shocking proposals included paid sterilization of individuals with IQ below 100 and stated that a high rate of reproduction among blacks was retrogressive.
- He was appointed the first Alexander M. Poniatoff professor of engineering science in 1963 at Stanford University, a promotion from being a lecturer when he joined in 1958. He remained in this position until his retirement in 1975.

Major Works

- Shockley's "point-contact transistor" was a major influence in helping usher an age of micro-miniature electronics. He managed a research team that consisted of himself, John Bardeen, and Walter H. Brattain and used semiconductors to amplify electronic signals. The transistor was further improved which replaced the bulky and less-efficient vacuum tubes.

Awards & Achievements

- His crowning glory was the "Nobel Prize for Physics" in 1956. It was awarded to him for the invention of the "point-contact transistor" in 1947. He was the co-recipient of the award along with his colleagues John Bardeen and Walter Brattain.
- He had to his name over 90 US patents, the first of which was the "Electron Discharge Device" on electron multipliers.
- His contribution to science earned him many honors and medals. The Institute of Radio Engineers (now, Institute of Electrical and Electronics Engineers, IEEE) awarded him the Maurice Liebman Memorial Prize in 1980.

Personal Life & Legacy

- At the age of 23, he married Jean Alberta Bailey from Iowa in August 1933. In March 1934 they had their first child, Alison. Later the couple divorced.
- He married Emmy Lanning, a psychiatric nurse, later. She outlived him by 18 years and passed away on April 28, 2007.
- He was an accomplished climber and made many hikes and climbs in the "Shawangunks" in the Hudson River Valley. A route there is named "Shockley's Ceiling". He was an amateur magician, speaker, as well as lecturer.
- He passed away on August 12, 1989, due to prostate cancer. He was estranged from his friends and family and his children came to know about his death through media. His resting place is "Alta Mesa Memorial Park" in Palo Alto, California.

Trivia

- Though his most significant work included semiconductors, he considered genetics his primary field.
- He was a sperm donor at the infamous "Repository for Germinal Choice," popularly known as "The Nobel Prize sperm bank". He was the only one to publicly acknowledge his contribution to the bank.
- He was home-schooled until the age of 8. This was because his parents thought they could provide an education better than any school. Another reason could have been that they often moved from one place to another.
- As a child, he was quite ill-tempered and spoiled; traits he inherited from his parents. But he also possessed a funny bone and was a practical joker in college.

CASE STUDY

THE EVOLUTION OF SEMICONDUCTORS IN THE COMPUTING INDUSTRY

- **Introduction**

  Semiconductors have been the cornerstone of the modern computing industry, enabling the development of increasingly powerful and efficient electronic devices. This case study explores the evolution of semiconductor technology, focusing on its impact on the computing industry, key innovations, and future prospects.

- **Background**

  Semiconductors, materials with electrical conductivity between conductors and insulators, are the foundation of modern electronics. Silicon, the most commonly used semiconductor material, has enabled the creation of integrated circuits (ICs), which are essential for computers, smartphones, and countless other devices. The journey from the invention of the transistor to the current state of semiconductor technology has been marked by continuous innovation and miniaturization.

- **The Problem:**

  In the early days of computing, electronic devices were large, expensive, and inefficient. The vacuum tubes used in early computers were prone to failure and consumed significant power. There was a pressing need for a more reliable, efficient, and scalable solution to meet the growing demand for computational power.

- **Solution: Development of Semiconductor Technology**

1. **Invention of the Transistor:**
   - In 1947, John Bardeen, Walter Brattain, and William Shockley at Bell Labs invented the transistor, a semiconductor device that could amplify and switch electronic signals.
   - Transistors were smaller, more reliable, and more energy-efficient than vacuum tubes.
2. **Integration and Miniaturization:**
   - The development of integrated circuits (ICs) in the late 1950s by Jack Kilby and Robert Noyce allowed multiple transistors to be fabricated on a single silicon chip.

- o This breakthrough led to the miniaturization of electronic circuits and paved the way for modern microprocessors.

3. **Moore's Law:**
   - o In 1965, Gordon Moore predicted that the number of transistors on a chip would double approximately every two years, leading to exponential increases in computing power.
   - o This prediction, known as Moore's Law, became a guiding principle for the semiconductor industry, driving continuous innovation and scaling.
4. **Advances in Manufacturing:**
   - o The transition from micron-scale to nanometer-scale manufacturing processes has enabled the production of chips with billions of transistors.
   - o Techniques such as photolithography, chemical vapor deposition, and atomic layer deposition have been critical in achieving these advancements.

- **Implementation: Impact on the Computing Industry**

1. **Personal Computers:**
   - o The development of microprocessors, starting with the Intel 4004 in 1971, revolutionized the computing industry.
   - o Affordable and powerful personal computers became accessible to consumers and businesses, transforming how people work, learn, and communicate.
2. **Mobile Devices:**
   - o The integration of semiconductors in mobile devices has enabled the proliferation of smartphones and tablets.
   - o These devices combine communication, computing, and entertainment functions in a compact form factor, powered by advanced semiconductor technology.
3. **Data Centers and Cloud Computing:**
   - o Semiconductors have driven the growth of data centers and cloud computing, providing the computational power and storage capacity needed for big data and artificial intelligence applications.
   - o High-performance processors and specialized chips, such as GPUs and TPUs, are at the core of these infrastructures.
4. **Internet of Things (IoT):**
   - o Semiconductors are essential for the IoT, enabling the development of smart devices and sensors that connect and interact with each other.
   - o The proliferation of IoT devices is transforming industries such as healthcare, manufacturing, and transportation.

- **Results**

The evolution of semiconductor technology has had profound implications for the computing industry and beyond. Key outcomes include:

1. **Increased Computational Power:**
   - The exponential growth in transistor density has led to unprecedented increases in computational power, enabling complex simulations, data analysis, and machine learning.
2. **Energy Efficiency:**
   - Advances in semiconductor materials and manufacturing have resulted in more energy-efficient devices, reducing power consumption and extending battery life in portable devices.
3. **Cost Reduction:**
   - The scalability of semiconductor manufacturing has driven down the cost of electronic components, making technology more accessible to a broader audience.
4. **Innovation and** New Applications:
   - Continuous improvements in semiconductor technology have spurred innovation, leading to new applications such as autonomous vehicles, augmented reality, and personalized medicine.

CLASS ACTIVITY

Investigating Doping in Semiconductors

1. **Materials Needed:**
   - o A diagram or model of a silicon crystal lattice
   - o Small colored beads to represent different atoms (e.g., silicon, phosphorus, boron)
   - o Worksheet with questions
2. **Instructions:**
   - o **Step 1:** Review the basic structure of a pure silicon crystal lattice.
- Discuss how silicon atoms form a covalent bond with four neighboring silicon atoms.
   - o **Step 2:** Introduce the concept of doping.
- Explain how adding a small amount of another element can change the electrical properties of silicon.
   - o **Step 3:** Using the model, add colored beads to represent dopant atoms in the silicon lattice.
- Use one color for phosphorus (n-type dopant) and another color for boron (p-type dopant).
   - o **Step 4:** Discuss the effects of each type of doping on the silicon lattice.
     - For n-type doping, how does adding phosphorus change the number of free electrons?
     - For p-type doping, how does adding boron create "holes" in the lattice?
3. **Questions:**
   - o **Q1:** Explain in your own words how n-type and p-type doping alter the electrical properties of silicon.
   - o **Q2:** Predict what would happen to the electrical conductivity of silicon if it were doped with an equal amount of phosphorus and boron. Justify your answer.
   - o **Q3:** How do the free electrons in n-type semiconductors and the holes in p-type semiconductors contribute to the flow of current?
   - o **Q4:** Describe a real-world application where semiconductors are crucial. How does the doping process play a role in this application?
4. **Extension Activi**ty:
   - o Research and present a brief report on a modern semiconductor device (e.g., a transistor, solar cell, or LED). Explain the role of doping in its operation and why specific materials are chosen.

SUMMARY

- Semiconductor materials have some electrical properties that contribute to the operation of some electronic devices. In this, the resistivity falls as the temperature increases, whereas metal behaves differently in this term which is oppositely.
- Holes and electronics are basically the charge carriers of the Semiconductor which results in the flow of current or electricity through it. Electrons, which carry a negative charge, orbit the nucleus of an atom.
- Semiconductors have moderate electrical conductivity, which can be significantly altered by doping, temperature and the application of electric fields. Their resistivity is higher than that of conductors but lower than insulators.
- The materials which have conductivity between conductors and insulators are known as semiconductor materials. These are pure elements like silicon (Si) or germanium (Ga), or compounds like gallium arsenide (GaAs) or cadmium selenide (CdSe).
- Silicon material is a group IV element in the periodic table of chemical elements and it is the most frequently used semiconductor material. These materials are very simple to fabricate and offer the best mechanical & electrical properties.
- A semiconductor in an extremely pure form is known as an intrinsic semiconductor. In an intrinsic semiconductor, even at room temperature, hole-electron pairs are created.
- The intrinsic semiconductor has little current conduction capability at room temperature. To be useful in electronic devices, the pure semiconductor must be altered so as to significantly increase its conducting properties. This is achieved by adding a small amount of suitable impurity to a semiconductor. It is then called impurity or extrinsic semiconductor.
- Pure semiconductor crystals are not particularly good electrical conductors, although their conductivities are much greater than those of insulators. The electrical property that makes semiconducting materials, and especially silicon, so valuable in electronics and other device applications arises from the fact that their electrical conductivity can be continuously varied through the controlled incorporation of dopant atoms into the crystal lattice.

REVIEW QUESTIONS

1. What is a semiconductor and how does it differ from conductors and insulators in terms of electrical conductivity?
2. Identify two common semiconductor materials and describe one key property that makes each suitable for use in electronic devices.
3. Explain the concept of energy bands in semiconductors, including the significance of the valence band and conduction band.

4. What is the energy band gap, and why is it important in determining the electrical properties of a semiconductor?
5. Define an extrinsic semiconductor and explain how doping affects its electrical properties.
6. Differentiate between n-type and p-type semiconductors, including the type of dopants used and their impact on charge carriers.
7. Describe the process of electrical conduction in a semiconductor, including the roles of electrons and holes.

REFERENCES

1. Gösele, U. M., & Tan, T. Y. (1991). Point defects and diffusion in semiconductors. *MRS Bulletin, 16*(11), 42–46. http://dx.doi.org/10.1557/s0883769400055512.
2. Hulls, K., & McMillan, P. W. (1972). Amorphous semiconductors: A review of current theories. *Journal of Physics D: Applied Physics, 5*(5), 865–882. https://doi.org/10.1088/0022-3727/5/5/205.
3. Kittel, C. (1995). *Introduction to Solid State Physics* (7th ed.). Wiley. ISBN 0-471-11181-3.
4. Morris, P. R. (1990). *A History of the World Semiconductor Industry*. IET. ISBN 0-86341-227-0, pp. 11–25.
5. Moskowitz, S. L. (2016). *Advanced Materials Innovation: Managing Global Technology in the 21st century*. John Wiley & Sons. p. 168. ISBN 9780470508923.
6. Nashelsky, L., & Boylestad, R. L. (2006). *Electronic Devices and Circuit Theory* (9th ed.). Prentice-Hall of India Private Limited. pp. 7–10. ISBN 978-81-203-2967-6.
7. Shockley, W. (1950). *Electrons and Holes in Semiconductors: With Applications to Transistor Electronics*. R. E. Krieger Pub. Co. ISBN 978-0-88275-382-9.
8. Worzyk, T. (2009). *Submarine Power Cables: Design, Installation, Repair, Environmental Aspects*. Springer. ISBN 978-3-642-01270-9.
9. Yacobi, B. G. (2003). *Semiconductor Materials: An Introduction to Basic Principles*. Springer. ISBN 0-306-47361-5, pp. 1–3.
10. Yang, J.-P., Chen, H.-T., & Tang, G.-B. (2022). Modeling of thickness-dependent energy level alignment at organic and inorganic semiconductor interfaces. *Journal of Applied Physics, 131*(24), 245501. http://dx.doi.org/10.1063/5.0096697.

CHAPTER

2 Diode Fundamentals

LEARNING OBJECTIVES

After studying this chapter, you will be able to:

- Define diodes;
- List the types of diodes;
- Discuss about the basic diode circuits.

KEY TERMS FROM THIS CHAPTER

Blocking diode
Diode
pn junction
Resistance
Signal limiters
Zener diode
Conduction band
Dopant concentration
Positive voltage
Schottky diodes
Signal mixers
Current flow
pn diode
Rectifiers
Schottky junction
Voltage regulator

INTRODUCTION

A diode is a fundamental electronic component that allows current to flow in one direction while blocking it in the opposite direction, making it an essential building block in various electronic circuits. It is a two-terminal device composed of a p-n junction, which is created by joining p-type and n-type semiconductor materials. This junction forms a depletion region where charge carriers are depleted, establishing an electric field that controls the movement of electrons and holes.

The most basic type of diode is the semiconductor diode, which operates under two conditions: forward bias and reverse bias. In forward bias, the positive voltage is applied to the p-type material and the negative voltage to the n-type material, reducing the width of the depletion region and allowing current to flow. In reverse bias, the positive voltage is applied to the n-type material and the negative to the p-type, widening the depletion region and preventing current flow, except for a very small leakage current.

Diodes have various applications due to their ability to control the direction of current flow. One of the primary uses is rectification, where diodes convert alternating current (AC) to direct current (DC), crucial in power supply units. Another application is in signal demodulation, where diodes extract audio signals from radio waves. Additionally, diodes are employed in protection circuits to prevent damage from voltage spikes by redirecting excess voltage away from sensitive components.

Specialized diodes include light-emitting diodes (LEDs), which emit light when forward biased, and Zener diodes, which allow current to flow in reverse once a certain breakdown voltage is reached, making them useful for voltage regulation. Schottky diodes, known for their fast-switching speed and low forward voltage drop, are utilized in high-frequency applications. Understanding diode fundamentals is essential for anyone involved in electronics, as these devices play critical roles in modern technology, from simple rectifiers to complex communication systems. Their unique properties and diverse applications underscore the importance of diodes in electronic circuit design and functionality.

2.1. DIODES

A diode is a specialized electronic component that functions as a one-way switch, allowing electric current to flow in only one direction and blocking it in the opposite direction. When reverse biased, the diode acts as an insulator, while it is forward biased when allowing current to pass through. Diodes have two terminals – the anode and the cathode. They are used in various applications such as switches, signal modulators, signal mixers, rectifiers, signal limiters, voltage regulators, oscillators, and signal demodulators.

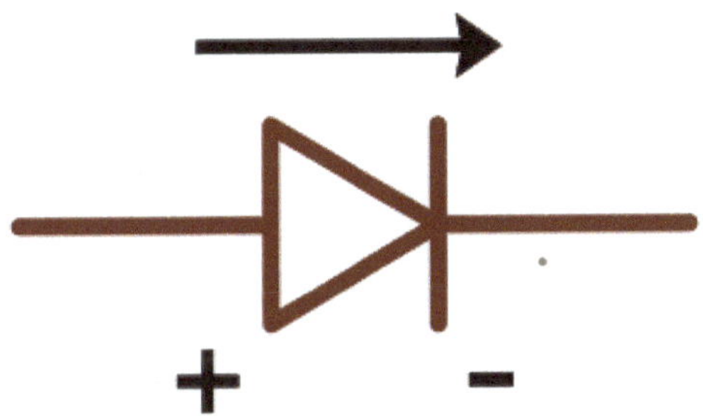

Figure 2.1. *Diode in forward bias.*

Source: https://www.circuitbasics.com/wp-content/uploads/2020/08/What-is-a-Diode-Diode-in-Forward-Bias-300x177.png.

In relation to the cathode, the voltage applied to the anode is positive. Additionally, because the diode's voltage is higher than the threshold voltage, current can flow through it as a short circuit.

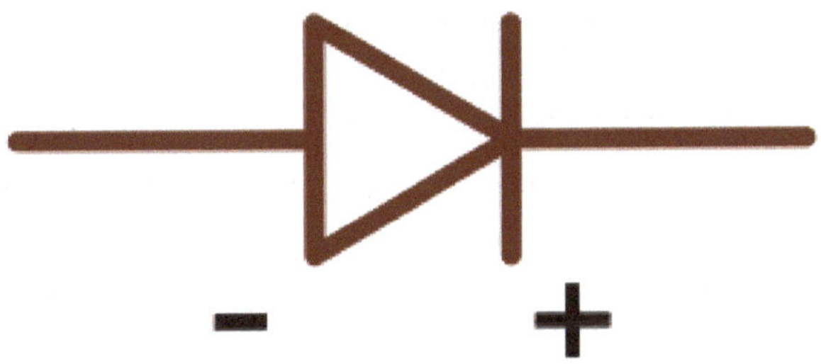

Figure 2.2. *Diode in reverse bias.*

Source: https://www.circuitbasics.com/wp-content/uploads/2020/08/What-is-a-Diode-Diode-in-Reverse-Bias-300x136.png.

When the cathode is given a positive charge in relation to the anode, the diode will be in a state of reverse bias. This causes it to behave like an open circuit, preventing any current from flowing.

2.1.1. Using Diode

- **Reverse Current Protection:** In certain circuits, a blocking diode serves as protection in the event of an unintentional reverse connection error, such as connecting the DC supply incorrectly or reversing its polarity. Other circuit components may sustain damage from an incorrectly directed current flow.

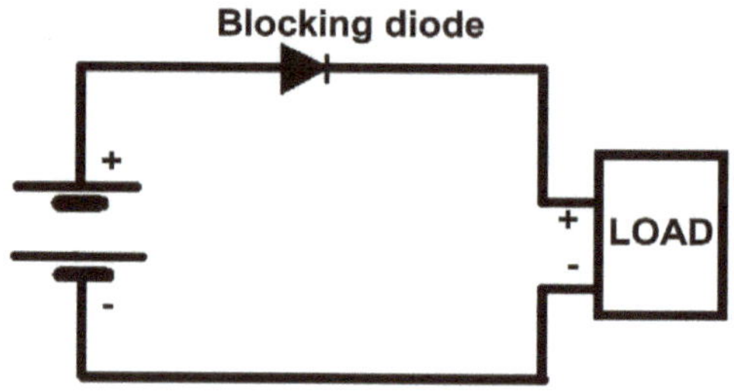

Figure 2.3. *A diode for reverse current protection.*

Source: https://www.circuitbasics.com/wp-content/uploads/2020/06/image-40.png.

The diagram demonstrates that the blocking diode is positioned in series with both the load and the positive end of the power supply. If there is a reverse connection, the diode will prevent current from flowing as it will be in reverse bias. This protects the load from any potential reverse current. On the other hand, if the polarity is correct, the diode will be in forward bias allowing the load current to pass through it.

- **Simple Voltage Regulators:** To step down the input voltage to the necessary level and maintain it there despite supply variations, a voltage regulator is utilized. It can be applied to control the output voltage as well. Because a Zener diode can operate in reversed bias conditions, it is typically used as a voltage regulator. When in forward bias, it functions similarly to a typical signal diode. On the other hand, when the reverse voltage is applied to it, the voltage stays constant over a large current range.

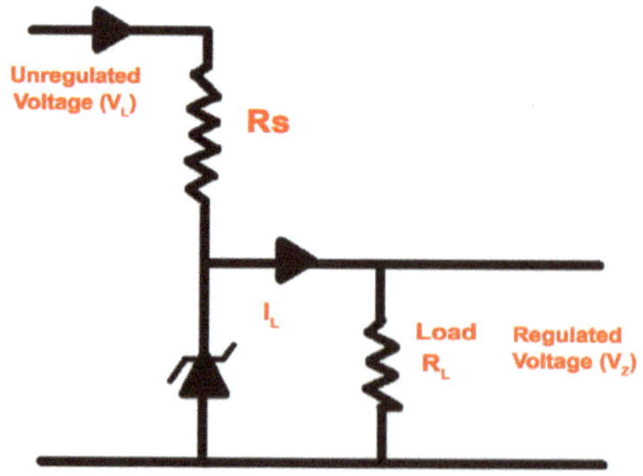

Figure 2.4. *Zener diode as voltage regulator.*

Source: https://www.circuitbasics.com/wp-content/uploads/2020/06/image-41.png.

The series resistor attached to the circuit in the above figure limits the current flowing through the diode. The diode functions as a reverse bias and can operate in breakdown conditions because it is connected to the positive terminal of the supply. Because high power rating diodes can withstand reverse bias above their breakdown voltage, they are typically used. If the maximum load current and the minimum input voltage are applied, the Zener diode current will always be at its lowest. We can use a Zener diode with a voltage roughly equal to the load voltage given the input voltage and the necessary output voltage.

- **Voltage Stabilizers:** When the load resistor is connected in parallel to the Zener diode, the current flowing through the diode is lowered in favor of the load current. Because it is essential to stabilization, the amount of current flowing through it is significant. The fact that Zener diodes exhibit a sharp increase above the breakdown voltage in their current-voltage curve indicates that they are the most effective at stabilizing small direct voltages. The diode's resistance falls as the current rises. As a result, the Zener diode's voltage is nearly constant. To ensure that the maximum permissible dissipation power is not exceeded, a resistor is typically connected.
- **Converting AC To DC:** Diodes are typically used to build center-tapped, full bridge, half-wave, and full wave rectifier circuits, among other types of rectifier circuits. Rectification of AC power to DC power is one of its primary uses.

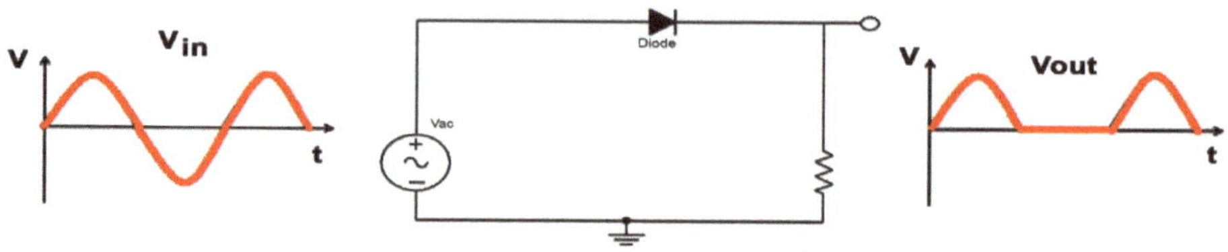

Figure 2.5. *The rectification of AC power to DC power.*

Source: https://www.circuitbasics.com/wp-content/uploads/2020/06/image-42.png.

The anode becomes positive in relation to the cathode during the input supply's positive half-cycle. Since the diode is in forward bias, current flows to the load as a result. Nevertheless, the anode becomes negative in relation to the cathode during the input sinusoidal wave's negative half-cycle. As a result, no current will flow to the load since the diode will be in reverse bias. When the load side voltage and current are both same polarity, the output voltage will pulse DC. Because the load is resistive during the positive half cycle, the supply voltage and the voltage across the load resistor will be equal. The input sinusoidal voltage will be at the load, and the load current flow will be proportionate to the applied voltage.

2.2. TYPES OF DIODES

2.2.1. Classification of Diodes According to Their Applications and Characteristics

There are various types of diodes used for different purposes. Diodes can be categorized into two main groups: pn diodes which are created by the junction of p-type and n-type semiconductors, and metal-semiconductor diodes (also known as Schottky barrier diodes or SBDs) which are formed by a metal junction with either an n-type or p-type semiconductor. Figure 2.6 illustrates the classification of diodes based on their characteristics and applications.

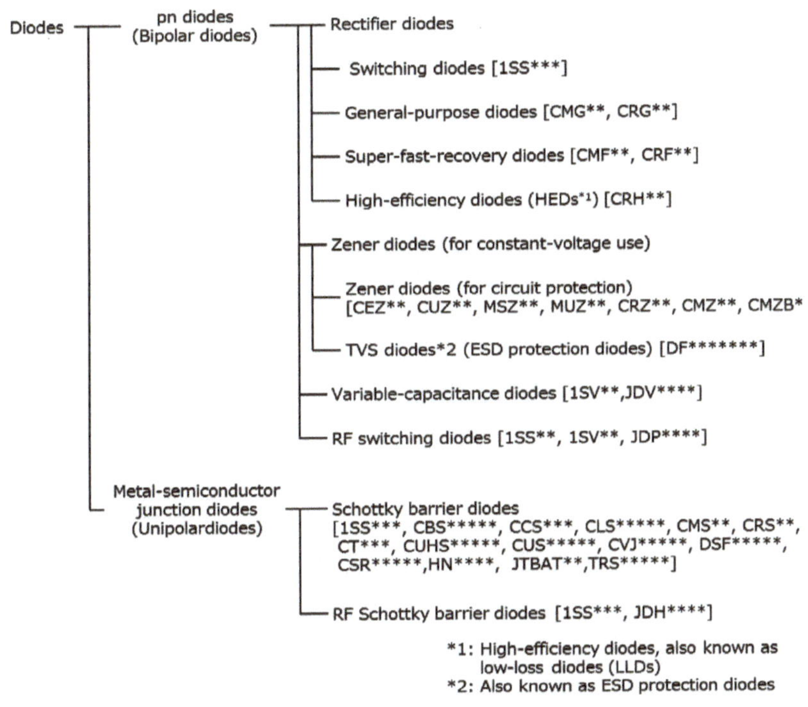

Figure 2.6. *Types of diodes.*

Source: https://toshiba.semicon-storage.com/info/application_note_en_20210831_AKX00997.pdf?did=141365.

Figure 2.7 shows the classification of diodes according to their internal connections.

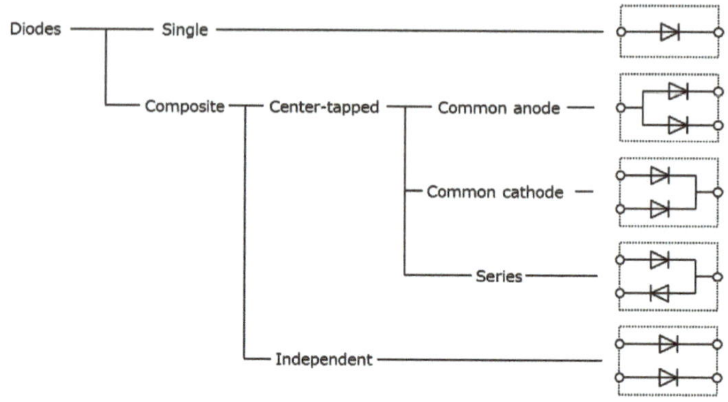

Figure 2.7. *Classification of diodes according to their internal connections.*

Source: https://toshiba.semicon-storage.com/info/application_note_en_20210831_AKX00997.pdf?did=141365.

2.2.2. pn Diodes

- *Ipn Junction:* An interface that preserves the orderly crystal lattice arrangements of two different types of semiconductor materials—p-type and n-type—is known as a pn junction. The Fermi levels of these semiconductors vary. The p-type region's Fermi level is in proximity to the valence band's highest energy level, while the n-type region's Fermi level is in proximity to the conduction band's lowest energy level.

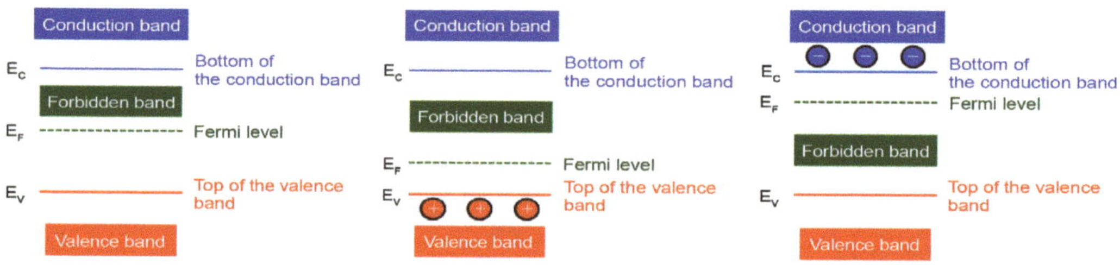

Figure 2.8. *Fermi level of intrinsic semiconductors, Fermi level of p-type semiconductors, Fermi level of n-type semiconductors from left to right respectively.*

Source: https://toshiba.semicon-storage.com/info/application_note_en_20210831_AKX00997.pdf?did=141365.

- **pn Junction in an Unbiased State:** Diffuse electrons from the n-type semiconductor near the pn junction recombine with holes in the p-type semiconductor. On the other hand, holes from the p-type semiconductor recombine with electrons in the n-type semiconductor through diffusion. Figure 2.10 illustrates how the recombination of carriers (free electrons and holes) results in the formation of a region devoid of carriers. We refer to this area as the depletion layer.

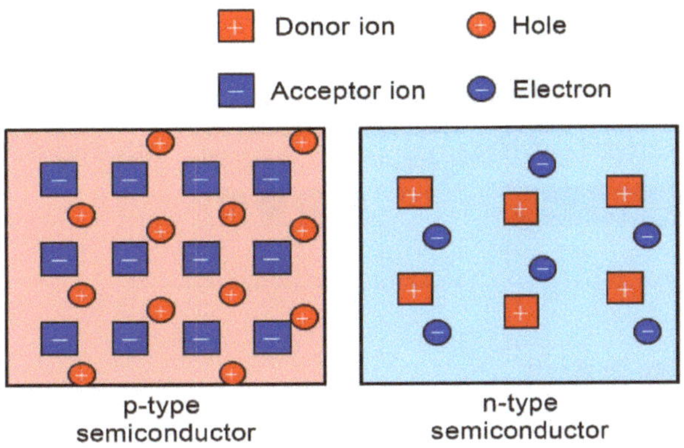

Figure 2.9. *P-type and n-type semiconductors before they are joined together.*

Source: https://toshiba.semicon-storage.com/info/application_note_en_20210831_AKX00997.pdf?did=141365.

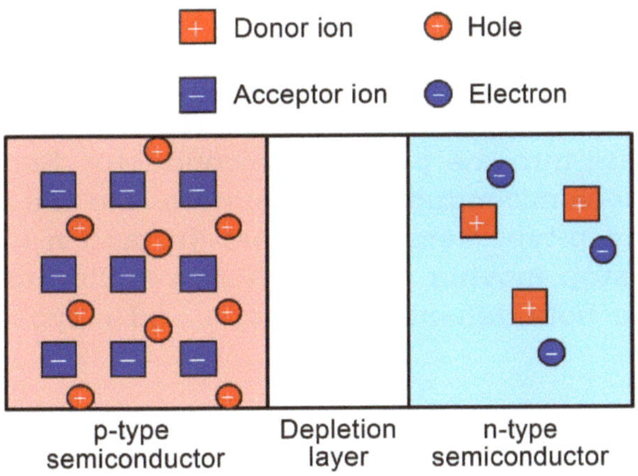

Figure 2.10. *P-type and n-type semiconductors after they are joined together.*

Source: https://toshiba.semicon-storage.com/info/application_note_en_20210831_AKX00997.pdf?did=141365.

New Fermi levels are formed when p-type and n-type semiconductors are bonded together due to carrier diffusion, which causes the p-type semiconductor's Fermi level to move upward and the n-type semiconductor's to move downward, as seen in Figure 2.11. The p-type and n-type regions' work functions are equal to the Fermi level decrease in the conduction band. A diffusion potential (qV_D) develops between them as a result.

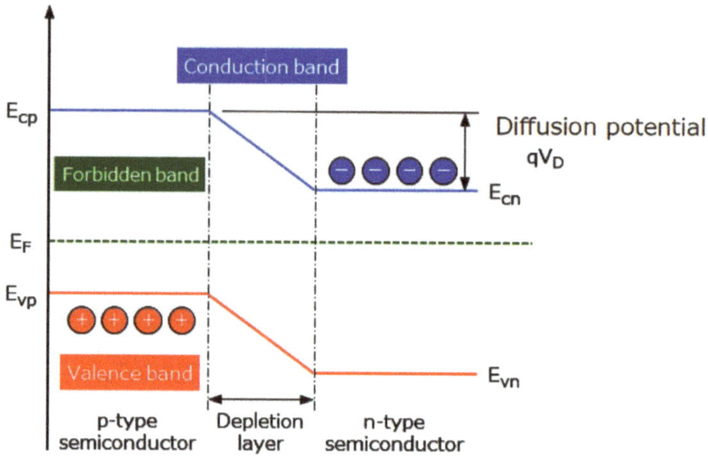

Figure 2.11. *Band diagram of a pn junction.*

Source: https://toshiba.semicon-storage.com/info/application_note_en_20210831_AKX00997.pdf?did=141365.

When the pn junction is unbiased, Figure 2.12 displays the electron and hole densities in the p-type and n-type semiconductors. Higher energy electrons in the n-type semiconductor can migrate into the p-type semiconductor. Similarly, holes with energy less than E_{vn} in the p-type semiconductor can migrate into the n-type semiconductor. At this point, the n-type and p-type semiconductors' (n_n and n_p) electron densities nearly equalize. The electrons stop moving and reach an equilibrium state when $n_n \approx n_p$. The relationship between the hole densities in n-type and p-type semiconductors (p_n and p_p) is the same.

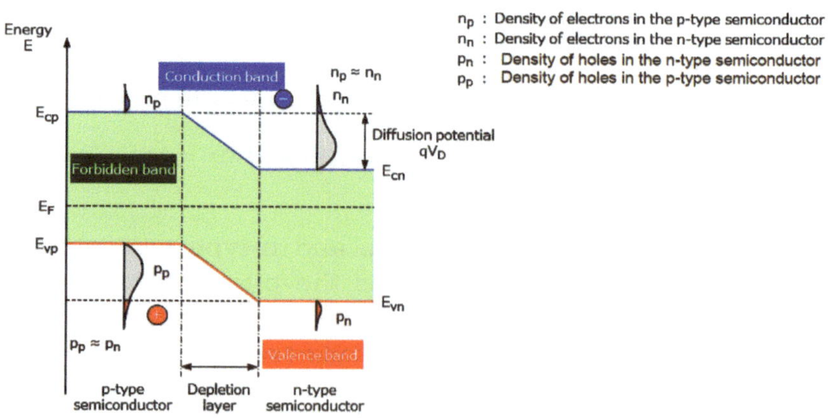

Figure 2.12. *Band diagram of a pn junction in an unbiased state.*

Source: https://toshiba.semicon-storage.com/info/application_note_en_20210831_AKX00997.pdf?did=141365.

- **pn Junction in the Forward-Biased State:** The pn junction's diffusion potential drops from qV_D in the unbiased state to $q(V_D\text{-}V_B)$ when a forward bias of qV_B is applied across it (where the p-type semiconductor is more positive than the n-type semiconductor). A significant number of electrons migrate from the n-type semiconductor to the p-type semiconductor as a result of the electron density in the n-type semiconductor having energy higher than E_{cp} becoming higher than the electron density in the p-type semiconductor. Numerous holes in the p-type semiconductor diffuse into the n-type semiconductor at the same time. Electric current flows as a result.

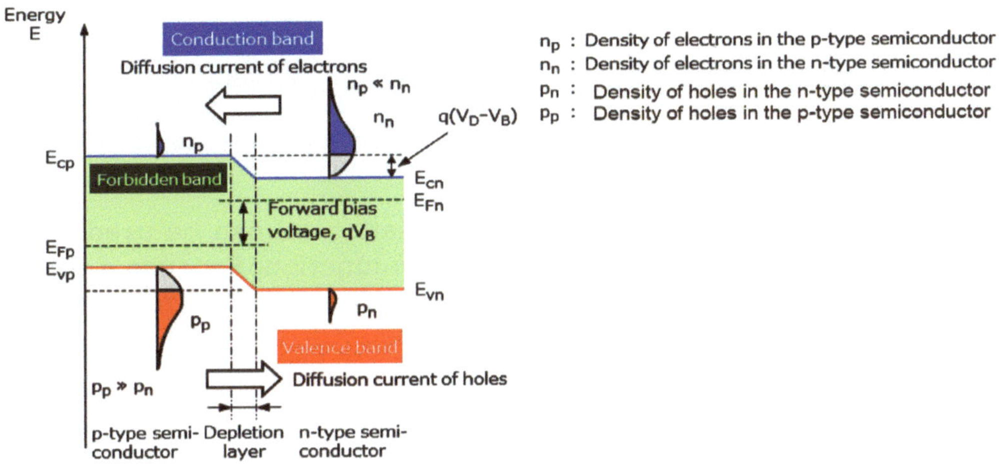

Figure 2.13. *Band diagram of a pn junction in the forward-biased state.*

Source: https://toshiba.semicon-storage.com/info/application_note_en_20210831_AKX00997.pdf?did=141365.

- **pn Junction in the Reverse-Biased State:** The diffusion potential increases from qV_D in the unbiased state to $q(V_D+V_R)$ when a reverse bias of qV_R is applied across the pn junction (where the p-type semiconductor is more negative than the n-type semiconductor, the p-type semiconductor is applied negatively, and the n-type semiconductor is applied positively). The electron density in the n-type semiconductor that has energy greater than E_{cp} consequently approaches zero. Similarly, the current flow stops when holes in the p-type semiconductor that can diffuse into the n-type semiconductor almost completely vanish.

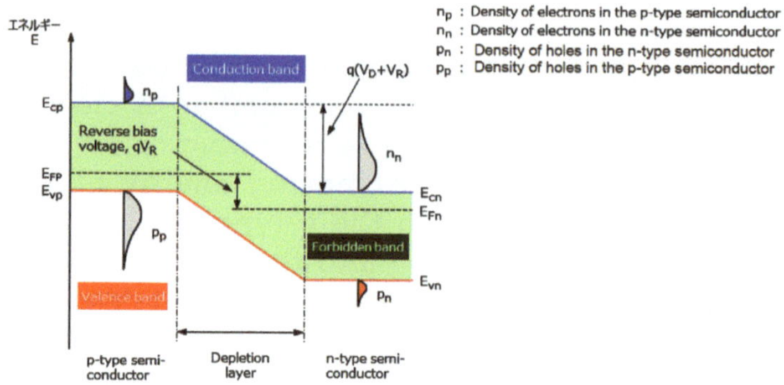

Figure 2.14. *Band diagram of a pn junction in the reverse-biased state.*

Source: https://toshiba.semicon-storage.com/info/application_note_en_20210831_AKX00997.pdf?did=141365.

- **Electric Characteristics of pn Diodes:** Equation (1) can be used to express the forward current (I_F) in a diode made of a pn junction. Because a diode's internal resistance results in a voltage drop in the high-current region, this equation only applies to the low-current region.

$$I_F = I_S \cdot \left(e^{\frac{q \cdot V_F}{K \cdot T}} - 1\right) \tag{1}$$

I_F	:	Forward current	(A)
I_S	:	Saturation current	(A)
q	:	Electron charge (1.602×10^{-19})	(C)
V_F	:	Forward voltage	(V)
K	:	Boltzmann constant (1.381×10^{-23})	($J\text{-}K^{-1}$)
T	:	Junction temperature (in Kelvin)	(K)

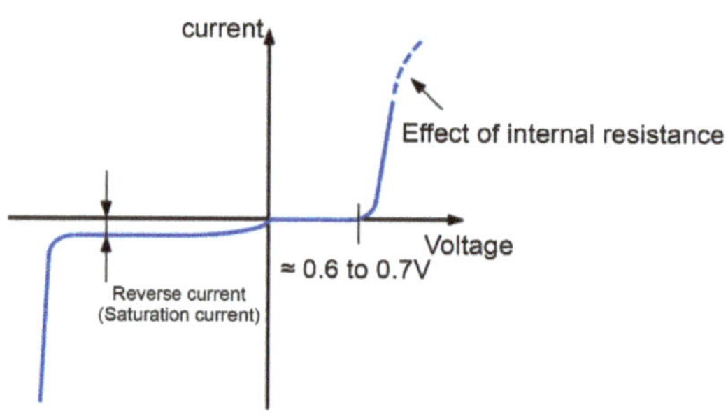

Figure 2.15. *Current-voltage curve of a pn diode.*

Source:https://toshiba.semicon-storage.com/info/application_note_en_20210831_AKX00997.pdf?did=141365.

A pn diode's current-voltage curve is displayed in Figure 2.15. The forward current of a Si diode increases sharply when a forward bias of 0.6 to 0.7 V is applied across it.

The Si diode's forward voltage (V_F) has a temperature coefficient of about -2 mV/°C. However, because an internal resistance-related voltage drop has a positive temperature coefficient, the temperature coefficient drops in the high-current region.

When a diode is reverse-biased, the current that passes through it is referred to as saturation current (IS) or reverse current (IR)*3. Equation (2): *3 can be used to approximate I_R, also known as leakage current.

$$I_R = I_{R0} \cdot e^{k(T-T_0)} \quad (2)$$

I_R	:	Reverse current at temperature T	(A)
I_{R0}	:	Reverse current at the reference temperature (T_0)	(A)
k	:	Constant determined by the semiconductor material used (approximately 0.1/°C in the case of Si)	
T	:	Temperature	(°C)
T_0	:	Reference temperature	(°C)

Figure 2.16 shows an example of the I_R curve over temperature. In this example, I_R exhibits a double-digit increase over a temperature increase (ΔT_a) of 75°C.

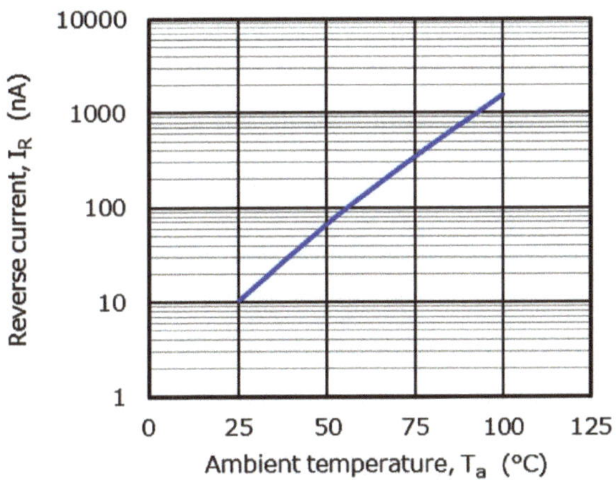

Figure 2.16. *Example of a pn diode's reverse current-vs-ambient temperature curve.*

Source: https://toshiba.semicon-storage.com/info/application_note_en_20210831_AKX00997.pdf?did=141365.

- **PIN Diode:** As seen in Figure 2.17, typical Si diodes employing a pn junction have a thick n-layer*4 with a low dopant concentration between p + and n+ layers. Because of its dopant concentration, which is similar to that of an intrinsic semiconductor, the n-layer is referred to as an intrinsic layer. This diode is known as a PIN diode because of its p-i-n structure. The special qualities of the PIN diode are listed below. *4 The n-layer is sometimes referred to as a drift layer or a high-resistance layer.

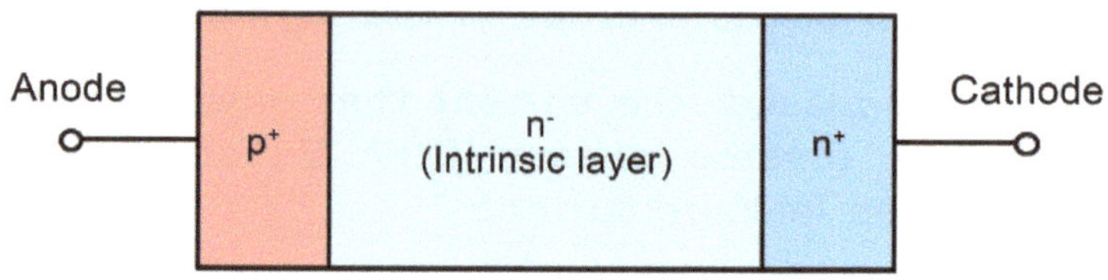

Figure 2.17. *Structure of a PIN diode.*

Source: https://toshiba.semicon-storage.com/info/application_note_en_20210831_AKX00997.pdf?did=141365.

1) *Increase in Breakdown Voltage:* In the intrinsic layer, a depletion layer grows when the pn junction is reverse-biased. At the pn junction, the electric field intensifies the greatest. Either an avalanche or a Zener breakdown takes place when it surpasses the maximum electric field strength. The triangular area in Figure 2.18 indicates the breakdown voltage of the pn junction. Thus, by increasing the n layer's thickness and reducing its dopant concentration to make the electric field strength's slope shallower, the breakdown voltage can be raised.

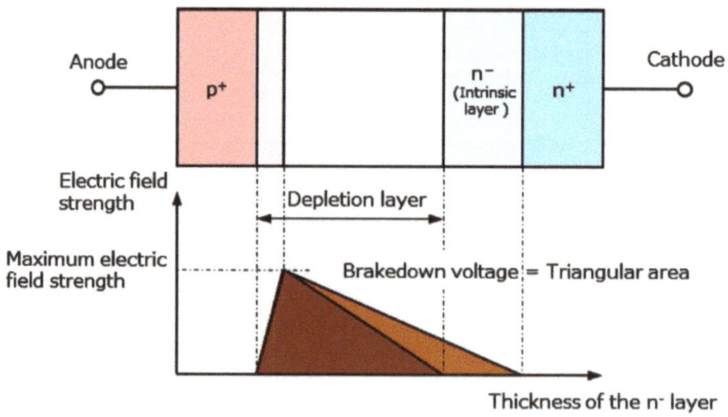

Figure 2.18. *Increasing the breakdown voltage of a PIN diode.*

Source: https://toshiba.semicon-storage.com/info/application_note_en_20210831_AKX00997.pdf?did=141365.

2) *Low Conduction Resistance:* Holes are injected from the p+ layer into the n- layer when the pn diode, which is composed of a p+ and an n- layer, conducts. On the other hand, the carrier density in the n- layer rises above the dopant concentration as a result of electrons being injected into it from the p+ layer. Consequently, when the pn diode conducts, the resistance of the n- layer drops. Conductivity

modulation is the phenomenon whereby an increase in the number of carriers in the n-layer results in a decrease in conduction resistance.

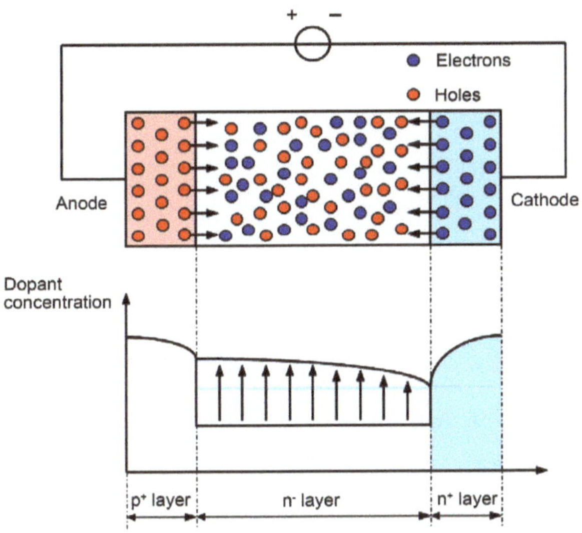

Figure 2.19. *Change in the dopant concentration of the nlayer due to conductivity modulation.*

Source: https://toshiba.semicon-storage.com/info/application_note_en_20210831_AKX00997.pdf?did=141365.

2.2.3. Schottky Barrier Diodes

- **Schottky Junction:** The Schottky barrier diode's structure is depicted in Figure 2.20. One kind of diode that functions as a rectifier is a Schottky barrier diode, which is created by the metal-to-semiconductor junction. Walter Hans Schottky is the inspiration behind the name Schottky barrier diode (SBD). An n-type semiconductor and a deposited metal combine to form a Schottky junction in an SBD.

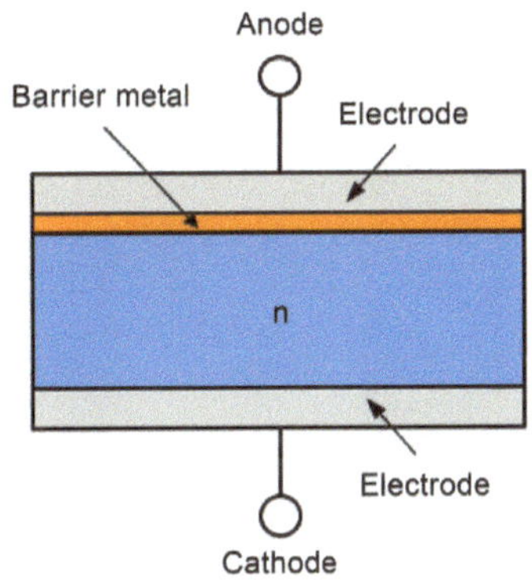

Figure 2.20. *Example of a structure of a Schottky barrier diode.*

Source: https://toshiba.semicon-storage.com/info/application_note_en_20210831_AKX00997.pdf?did=141365.

A metal lacks a forbidden band and has a Fermi level between the valence and conduction bands (i.e., as demonstrated in Figure 2.21, a bandgap). As a result, a metal has many electrons in the conduction band and a large number of holes in the valence band.

An n-type semiconductor's band diagram is displayed in Figure 2.22. Between the valance and conduction bands of an n-type semiconductor is a forbidden band where no electron state is permitted. The forbidden band contains the Fermi level.

When an n-type semiconductor and a metal with a lower Fermi level than the n-type semiconductor are combined, a Schottky junction is created. As a rectifier, the Schottky junction functions. On the other hand, when an n-type semiconductor is combined with a metal that has a higher Fermi level than the n-type semiconductor, an ohmic junction is created. Ohmic junctions function similarly to resistors.

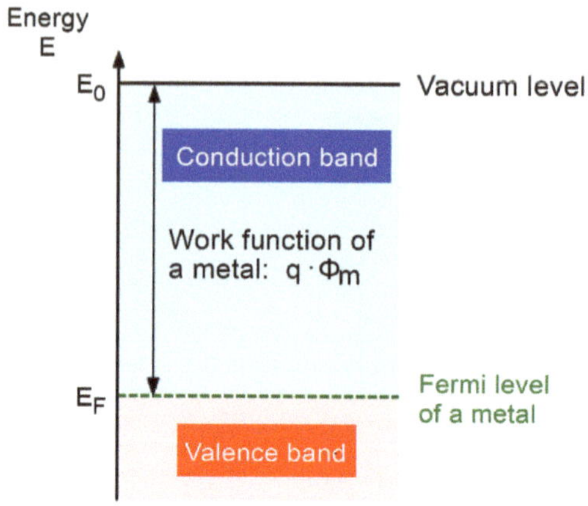

Figure 2.21. *Band diagram of a metal {normally the VB and CB overlap in metals, the schematic doesn't depict that please modify}.*

Source: https://toshiba.semicon-storage.com/info/application_note_en_20210831_AKX00997.pdf?did=141365.

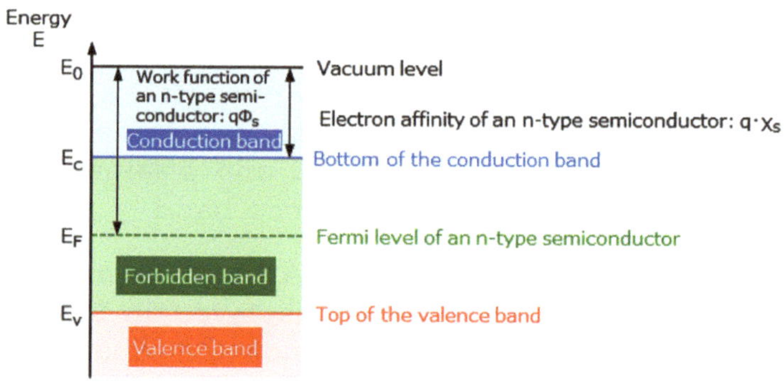

Figure 2.22. *Band diagram of an n-type semiconductor.*

Source: https://toshiba.semicon-storage.com/info/application_note_en_20210831_AKX00997.pdf?did=141365.

A metal's Fermi level remains constant when it is joined to an n-type semiconductor because some electrons from the latter diffuse into the metal. As seen in Figure 2.23, the n-type semiconductor's lowest energy level in the conduction band (EC) and highest energy level in the valence band (EV) change simultaneously. Positive ions form a depletion layer on the n-type semiconductor side at this point due to the generation of a diffusion energy of $q\Phi_{bi}=q(\Phi_m-\Phi_s)$ between the metal and the semiconductor. Because the metal has a large number of electrons, a depletion layer hardly expands into it even when electrons flow into it from the n-type semiconductor. No current flows when the Schottky junction is unbiased because electrons are unable to go above the depletion layer's diffusion energy level, $q(\Phi_m-\Phi_s)$.

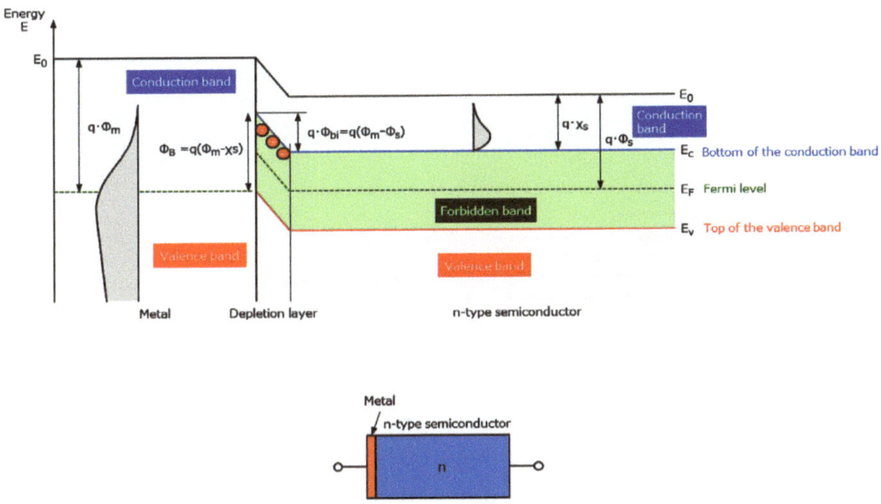

Figure 2.23. *Band diagram of a schottky junction in an unbiased state.*

Source: https://toshiba.semicon-storage.com/info/application_note_en_20210831_AKX00997.pdf?did=141365.

Applying a forward bias of V_B across the Schottky junction raises the energy level of the n-type semiconductor, causing an increase in the number of electrons exceeding the diffusion energy, $q(\Phi_m - \Phi_s - V_B)$. These excess electrons move into the metal, facilitating the flow of current.

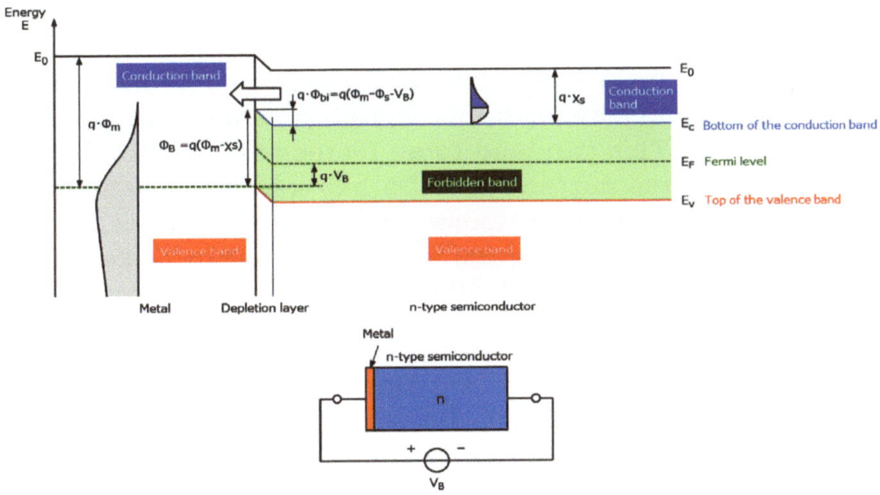

Figure 2.24. *Band diagram of a schottky junction in the forward-biased state.*

Source: https://toshiba.semicon-storage.com/info/application_note_en_20210831_AKX00997.pdf?did=141365.

When a reverse bias of V_R is applied across the Schottky junction, the energy level of the n-type semiconductor decreases, resulting in only a limited number of electrons being able to surpass the diffusion energy, $q(\Phi_m - \Phi_s + V_R)$, leading to the cessation of current flow.

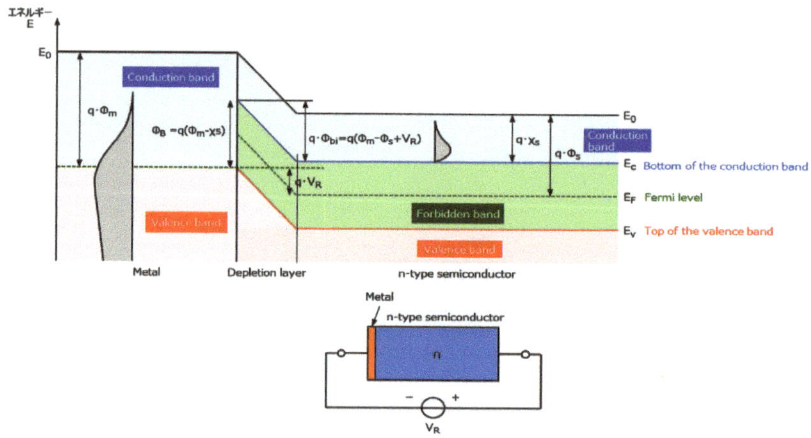

Figure 2.25. *Band diagram of a schottky junction in the reverse-biased state.*

Source: https://toshiba.semicon-storage.com/info/application_note_en_20210831_AKX00997.pdf?did=141365.

Molybdenum (Mo), titanium (Ti), and vanadium (V) are commonly utilized as barrier metals to create a Schottky junction. One notable feature of the Schottky Barrier Diode (SBD) is that it exhibits a lower voltage threshold for current to increase compared to a standard silicon diode.

Vanadium and titanium have lower diffusion potentials compared to molybdenum, resulting in SBDs using them having lower metal-to-semiconductor barrier potentials and larger reverse currents. This can lead to an increase in junction temperature due to power dissipation. The reverse current in SBDs has a positive temperature coefficient, meaning it increases as the junction temperature rises. This can eventually lead to thermal runaway and damage to the device. It is important to be cautious of the reverse bias voltage, ambient temperature, and other usage conditions when using SBDs.

- **JBS Structure:** Small forward voltage (VF) is accompanied by a high reverse current in an SBD. Reverse current rises with temperature because it has a positive temperature coefficient. Depending on the reverse voltage applied, ambient temperature, and other usage conditions, a rise in temperature eventually results in an increase in reverse current and thermal runaway. Degradation or device destruction could result from thermal runaway. Diodes with a Junction Barrier Schottky (JBS) structure are among the products offered by Toshiba that lower reverse current.

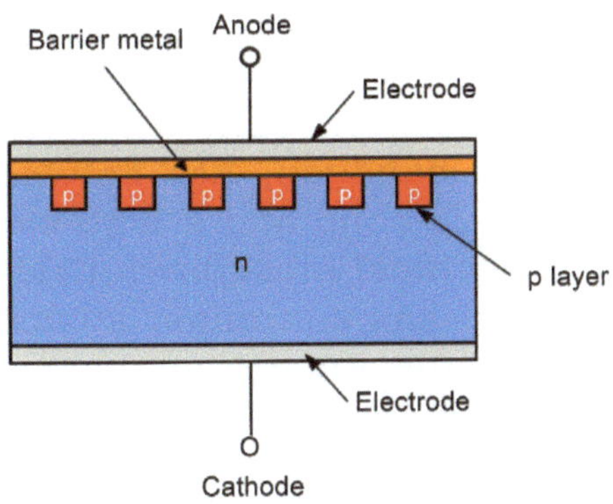

Figure 2.26. *JBS structure.*

Source: https://toshiba.semicon-storage.com/info/application_note_en_20210831_AKX00997.pdf?did=141365.

The cross-sectional view of an SBD with the JBS structure, which has p-type layers hidden in the n-layer, is displayed in Figure 2.26. Depletion layers spread over the n and diffused p layers in a reverse-biased SBD. Adjacent depletion layers combine as the reverse bias voltage rises, preventing an increase in reverse current.

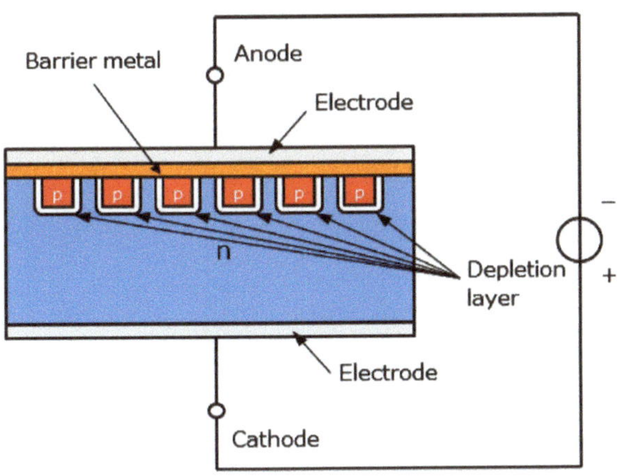

Figure 2.27. *Formation of depletion layers by the application of a reverse bias across the Schottky junction.*

Source: https://toshiba.semicon-storage.com/info/application_note_en_20210831_AKX00997.pdf?did=141365.

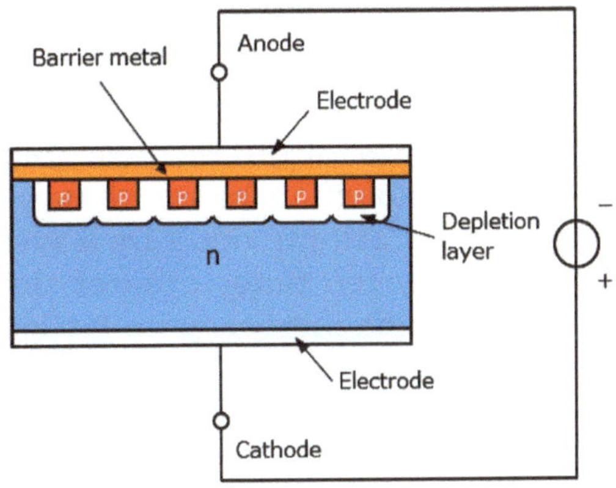

Figure 2.28. *Expansion of depletion layers by an increase in reverse bias across the Schottky junction.*

Source: https://toshiba.semicon-storage.com/info/application_note_en_20210831_AKX00997.pdf?did=141365.

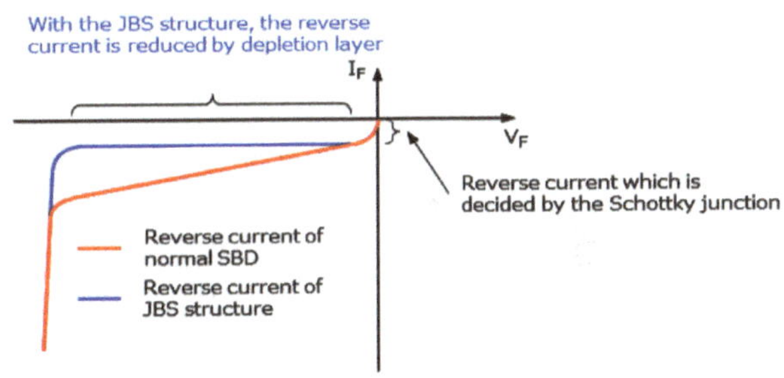

Figure 2.29. *Comparison of the reverse current characteristics of SBDs with typical and JBS structures.*

Source: https://toshiba.semicon-storage.com/info/application_note_en_20210831_AKX00997.pdf?did=141365.

2.2.4. SiC Schottky Barrier Diodes

2.2.4.1. What is SiC?

Silicon carbide (SiC) is made up of silicon (Si) and carbon (C) atoms, each surrounded by four different atoms in a regular tetrahedron formation. SiC is known as the compound semiconductor with the densest tetrahedral arrangement. It has multiple crystalline structures, known as polytypes, which have varying physical properties due to differences in the periodic arrangement of tetrahedrons.

SiC exhibits a larger energy gap (also known as a bandgap, Eg) between the valence and conduction bands than silicon. A strong atomic bond and a high electrical breakdown field are provided by a wide bandgap. The electrical breakdown field of SiC is approximately ten times greater than that of silicon. SiC has a stronger atomic bond than silicon, which increases its lattice vibration and facilitates energy conductivity. SiC has good thermal conduction, making it a semiconductor material. Three polytypes of SiC are known: 4H-SiC, 6H-SiC, and 3C-SiC, which have hexagonal crystal structures. The physical characteristics of silicon and other semiconductor materials are contrasted in Table 2.1. Since 4H-SiC offers a better balance than other polytypes of SiC in terms of electron mobility, electrical breakdown field, saturation electron drift velocity, and other physical properties, it is frequently used as a semiconductor material.

Table 2.1. *Physical Properties of Typical Semiconductor Materials*

Property	Symbol	Unit	Si	4H-SiC	6H-SiC	3C-SiC	GaN	GaAs	ダイヤモンド
Bandgap	E_g	eV	1.12	3.26	3.02	2.23	3.39	1.43	5.47
Electron mobility	μ_e	cm^2/ Vs	1400	1000/1200	450/ 100	1000	900	8500	2200
Hole mobility	μ_h		600	120	100	50	150	400	1600
Electric breakdown field	E_c	V/ cm	3.0×10^5	2.8×10^6	3.0×10^6	1.5×10^6	3.3×10^6	4.0×10^5	1.0×10^7
Thermal conductivity	λ	W/ cm·K	1.5	4.9	4.9	4.9	2.0	0.5	20
Saturation electron drift velocity	V_{sat}	cm/ s	1.0×10^7	2.2×10^7	1.9×10^7	2.7×10^7	2.7×10^7	2.0×10^7	2.7×10^7
Relative dielectric constant	ε_r	-	11.8	9.7/ 10.2	9.7/ 10.2	9.7	9.0	12.8	5.5

Source: https://toshiba.semicon-storage.com/info/application_note_en_20210831_AKX00997.pdf?did=141365.

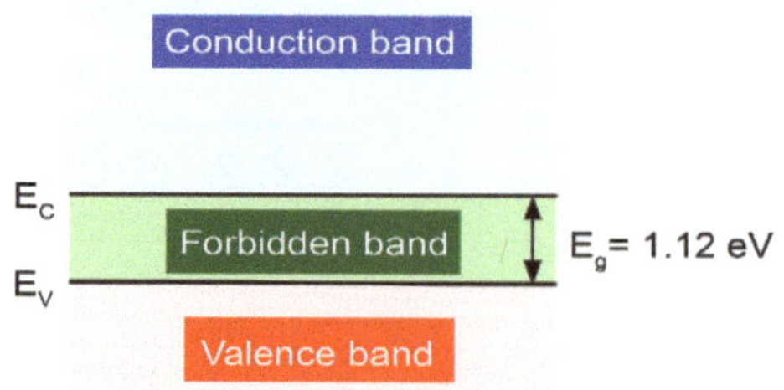

Figure 2.30. *Bandgap of Si.*

Source: https://toshiba.semicon-storage.com/info/application_note_en_20210831_AKX00997.pdf?did=141365.

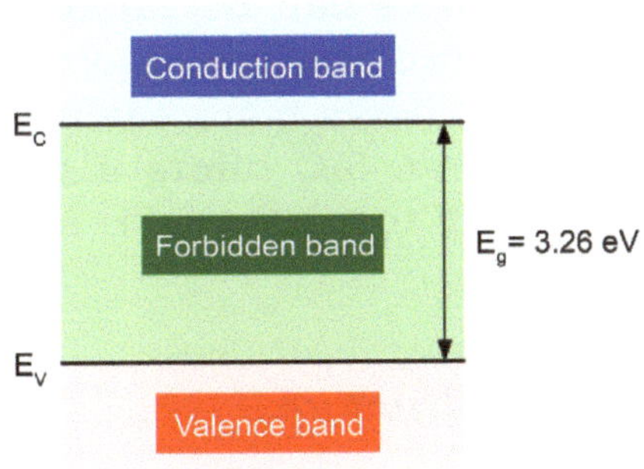

Figure 2.31. *Bandgap of SiC.*

Source: https://toshiba.semicon-storage.com/info/application_note_en_20210831_AKX00997.pdf?did=141365.

Si power devices with high voltage have a low electric breakdown field. As seen in Figure 2.32, a broad n-layer with low dopant concentration is required to produce a high breakdown voltage. However, a thinly doped n-layer has a high resistivity, which raises the conduction resistance of the diode.

Because SiC has an electric breakdown field that is almost ten times greater than silicon's, it is possible to achieve a breakdown voltage comparable to Si diodes by increasing the dopant concentration and decreasing the thickness of the n-layer.

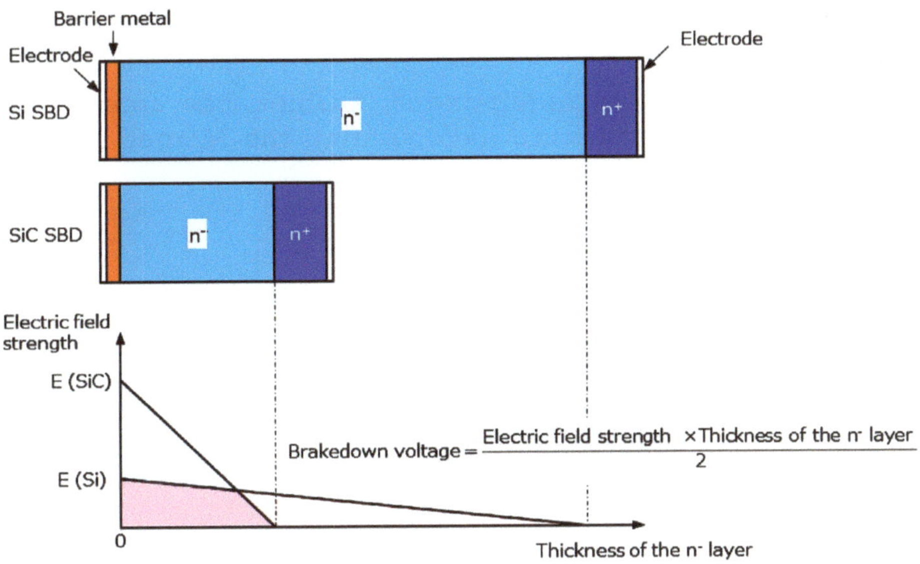

Figure 2.32. *Comparison of electric fields in Si and SiC SBDs.*

Source: https://toshiba.semicon-storage.com/info/application_note_en_20210831_AKX00997.pdf?did=141365.

The velocity of an electron or a hole can be rapidly increased by an electric field as indicated by their respective electron and hole mobilities (μ): velocity = mobility (μ) × electric field (E). An electric current flows more easily and encounters less resistance when the carrier mobility (μ) is higher. SiC offers a greater saturation electron drift velocity (i.e., SiC offers a higher limit to μ) than Si, allowing the fabrication of high-performance devices.

2.2.4.2. SBD with an Improved JBS Structure

When charging a capacitance load, SiC and other rectifier diodes have large current flows through them. As a result, these diodes must have high forward pulse current capability (IFP or IFSM).

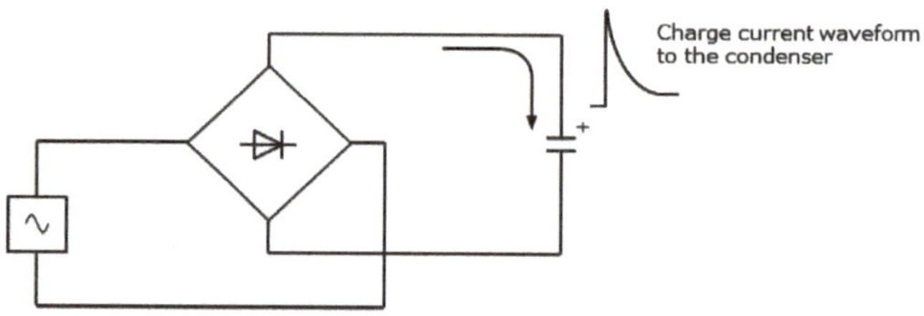

Figure 2.33. *Example of a waveform of a charging current flowing to a capacitor for full-wave rectification.*

Source: https://toshiba.semicon-storage.com/info/application_note_en_20210831_AKX00997.pdf?did=141365.

In order to comply with this requirement, Toshiba has created new SiC SBDs that feature an enhanced JBS structure incorporating the Merged PIN Schottky (MPS) structure concept.

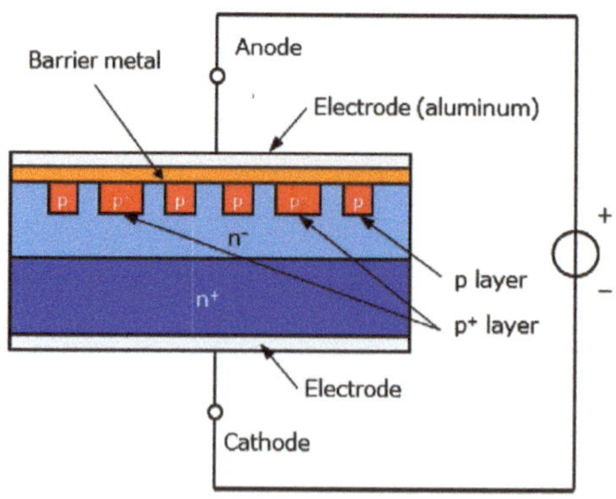

Figure 2.34. *Cross-sectional view of an SBD with an improved JBS structure.*

Source: https://toshiba.semicon-storage.com/info/application_note_en_20210831_AKX00997.pdf?did=141365.

Figure 2.34 illustrates the MPS structure with p+ layers embedded within the n-layer of SBD. When a high current is passed through the SBD, the pn diode created by the p+ and n- layers conducts alongside the Schottky junction. During this process, the p+ layers supply holes to the n- layer, while the n+ layer supplies electrons. This leads to conductivity modulation, boosting the carrier concentration in the n-layer. As a result, the resistance of the n-layer decreases, enabling the passage of high current and an increase in peak forward surge current capacity.

In Figure 2.35, the forward current-vs-forward voltage curve of an SBD with an enhanced JBS structure is depicted. This SBD offers high pulse current capacity and maintains a low forward voltage, particularly in high-current scenarios.

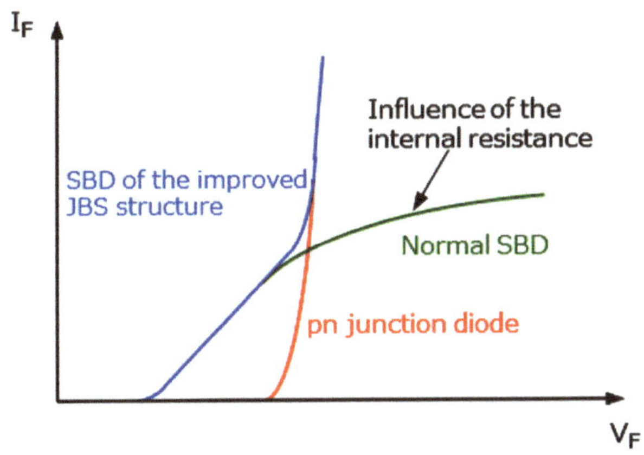

Figure 2.35. *Forward current-vs-forward voltage curves of a typical SBD, a pn diode, and an SBD with an improved JBS structure.*

Source: https://toshiba.semicon-storage.com/info/application_note_en_20210831_AKX00997.pdf?did=141365.

2.2.5. Zener Diodes

Zener diodes are employed in the breakdown region where the reverse characteristics of the pn junction are stable, which sets them apart from conventional rectifier and switching diodes. Unless otherwise specified, standard diodes are not allowed to be used in the reverse breakdown region. In the reverse breakdown region, the use of typical diodes results in a local breakdown phenomenon that may lead to current concentration and eventual destruction of the pn junction. However, the design of Zener diodes prevents localized concentration of current at the pn junction.

The dopant profile near the pn junction, along with its shape and surface conditions dictates the breakdown characteristics of a Zener diode. Thus, by adjusting these parameters, Zener diodes with various breakdown voltages can be created.

2.2.6. Zener Breakdown and Avalanche Breakdown

Avalanche breakdown and Zener breakdown are the two phenomena that can be used to explain how a Zener diode works.

- **Zener Breakdown:** When the reverse bias voltage across the pn junction increases, the crystal lattice bonding is disrupted as valence electrons are exposed to a strong electric field. This leads to the creation of free electrons that move to the conduction band through tunneling. As electrons and holes move towards the diode's electrodes, and current suddenly flows. If a diode is heavily doped the depletion layer does not easily expand even with reverse biasing, which leads to Zener breakdown. This is common in low-voltage Zener diodes where breakdown occurs below 5 to 6 V.

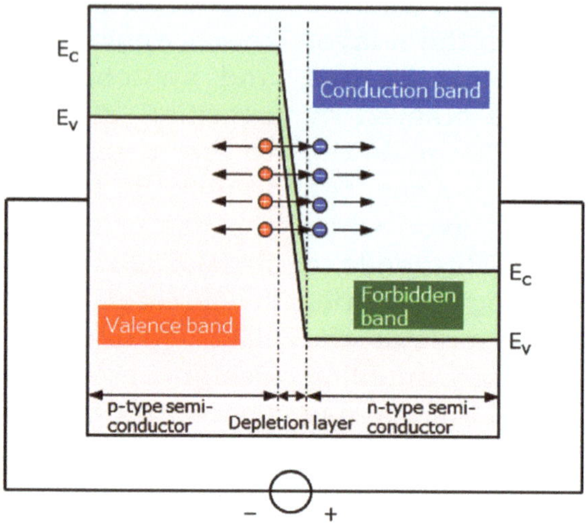

Figure 2.36. *Zener breakdown.*

Source: https://toshiba.semicon-storage.com/info/application_note_en_20210831_AKX00997.pdf?did=141365.

- **Avalanche Breakdown:** When the reverse bias voltage increases across the pn junction, it triggers avalanche breakdown, also known as electron avalanche breakdown. This process involves the acceleration of thermally generated electrons and holes by the strong electric field in the depletion layer. These accelerated particles collide with atoms in the crystal structure, freeing valence electrons and creating electron-hole pairs. The high electric field then accelerates these pairs, leading to further collisions and the creation of more electron-hole pairs. This cascading effect eventually attracts electrons and holes to the electrodes of a Zener diode, causing a sudden flow of current. Avalanche breakdown typically occurs at voltages exceeding 5 to 6 V.

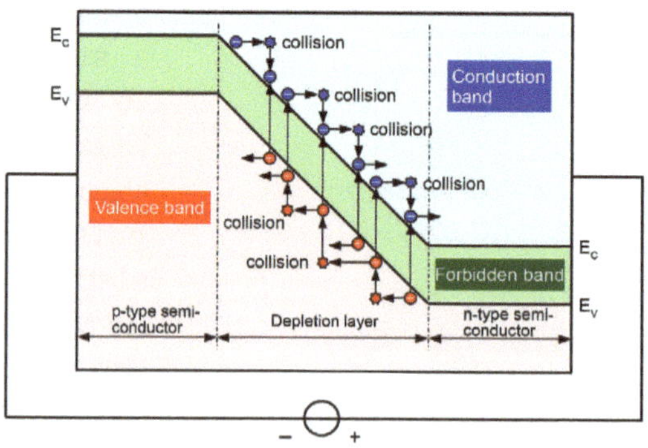

Figure 2.37. *Avalanche breakdown.*

Source: https://toshiba.semicon-storage.com/info/application_note_en_20210831_AKX00997.pdf?did=141365.

Zener breakdown and avalanche breakdown demonstrate distinct temperature characteristics due to their varying mechanisms. The Zener breakdown voltage decreases with increasing temperature, whereas the avalanche breakdown voltage increases with temperature.

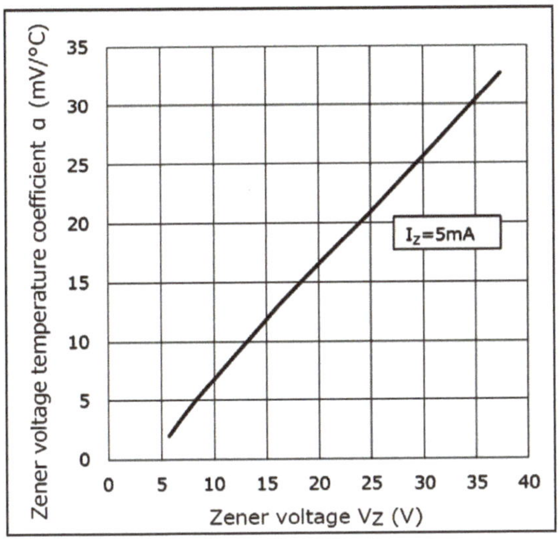

Figure 2.38. *Example of Zener voltage temperature coefficient vs zener voltage.*

Source: https://toshiba.semicon-storage.com/info/application_note_en_20210831_AKX00997.pdf?did=141365.

- **Forward characteristics of Zener diodes:** As is the case with typical pn diodes, the forward current of a Zener diode in the low-current region is expressed by Equation (3):

$$I_F = I_S \cdot \left(e^{\frac{q \cdot V_F}{K \cdot T}} - 1\right) \quad (3)$$

I_F	:	Forward current	(A)
I_S	:	Saturation current	(A)
q	:	Electron charge (1.602×10^{-19})	(C)
V_F	:	Forward voltage	(V)
K	:	Boltzmann constant (1.381×10^{-23})	($J \cdot K^{-1}$)
T	:	Junction temperature (in Kelvin)	(K)

- **Reverse Characteristics of Zener Diodes:** The voltage required for a Zener diode to conduct in the reverse direction is known as the Zener voltage (V_z), and is determined by the dopant concentration of the n-type semiconductor and the dopant concentration profile of the pn junction. Figure 2.39 illustrates different I_z-V_z curves at various Zener voltages. It is important to set the Zener current (I_z), appropriately based on the Zener voltage (V_z) to ensure proper operation.

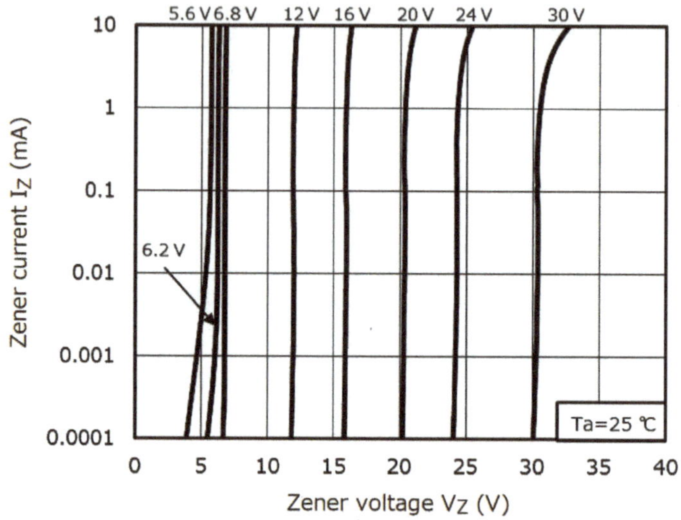

Figure 2.39. *Example of I_Z-V_Z curves at different zener voltages.*

Source: https://toshiba.semicon-storage.com/info/application_note_en_20210831_AKX00997.pdf?did=141365.

2.3. BASIC DIODE CIRCUITS

A diode is a crucial part of electronic circuits and finds use in many different devices, such as radio receivers and power supplies. The fundamental components of many of the electronic gadgets we use on a daily basis are diode circuits. To understand how electronic devices operate, one must understand basic diode circuits. A diode and a load resistor make up a basic diode circuit, which used to rectify an AC signal into a DC signal. Being a non-linear device, the diode permits current to flow in one direction while obstructing it from flowing in the other. Although fundamental diode circuits are straightforward, they serve as the basis for many intricate circuits.

2.3.1. Half Wave Rectifier

The simplest diode circuit is the half-wave rectifier, which transforms AC signals into DC signals. The diode permits current to flow in one direction while blocking current flow in the other when an AC signal is applied to it. Most applications cannot use the pulsating DC signal that the half-wave rectifier produces as its output.

2.3.2. Full Wave Rectifier

An enhanced half-wave rectifier, the full-wave rectifier is used to change an AC signal into a DC signal. The AC signal is rectified by the full-wave rectifier using a center-tapped transformer and two diodes. Most applications can benefit from the smoother DC signal that the full-wave rectifier produces as its output.

2.3.3. Voltage Multiplier

A diode circuit called a voltage multiplier is used to increase the input voltage. A diode and a capacitor make up the voltage multiplier, which is utilized in many different applications, including voltage doublers and triplers. Voltage multipliers are used in power supplies and electronic equipment.

2.3.4. Clipping Circuit

The input signal is clipped using a diode circuit, which is known as the clipping circuit. A diode and a load resistor make up the clipping circuit, which is utilized in many different circuits, including audio amplifiers and video circuits. The purpose of the clipping circuit is to prevent output signal distortion.

Basic diode circuits are utilized in many different applications and are crucial components of electronic devices. The basic diode circuits found in many electronic devices are the voltage multiplier, clipping circuit, half-wave rectifier, and full-wave rectifier. When building and debugging electronic circuits, it is essential to comprehend how simple diode circuits work.

2.3.5. Diode Functions in Circuit

A diode is a two-terminal passive electronic component that permits current to flow in only one direction when it is in forward bias mode. To achieve forward bias mode, the positive terminal of the voltage source is connected to the anode of the diode, while the negative terminal of the voltage source is connected to the diode's cathode.

The diode in this configuration allows current to flow through it and into the remaining circuit. The reverse bias mode is the opposite of this operation. In this state, the diode prevents any current flow. A diode functions essentially as a digital switch that can be turned on in forward bias and off in reverse bias, if you look closely (Figure 2.40).

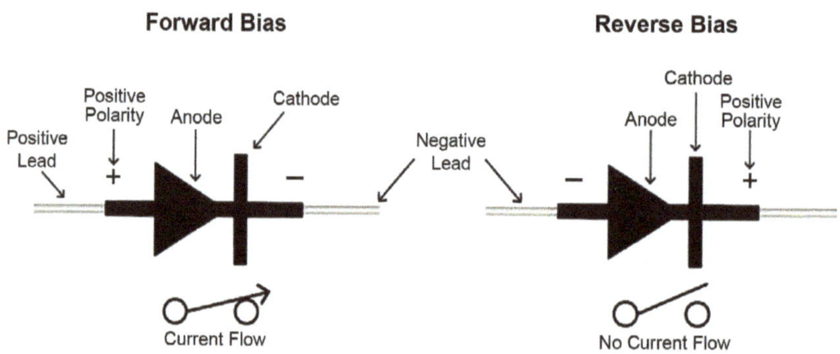

Figure 2.40. *Forward and reverse biased diodes.*

Source: https://www.yamanelectronics.com/wp-content/uploads/2018/05/Forward-and-Reverse-Biased-Diodes.jpg.

Let's see those functions and applications of a diode in any electronics or electrical circuit.

1. **As a Digital Switch:** Circuits employ diodes for switching purposes. They are known as the digital switches. The primary function of its usage as a switch is its automatic ON/OFF capability, which is achieved by applying the appropriate voltages.

Its switching speed is also quicker than that of a conventional switch. It is also inexpensive and extremely small. Its small size allows to save a significant amount of circuit space, which enables the creation of portable and intelligent devices.

2. **Handling High Power Signals:** One large area of study within electronics is power electronics. We deal with high voltage, current, and power signals in power electronics. Low-loss switches are necessary for the efficient handling of electricity. And so, the use of diodes is necessary. Depending on the application, a switch diode can handle power signals ranging from mW to KW. A signal can be the product of voltage and current.
3. **Safety:** In reverse bias mode, the diode does not completely stop the current flow, i.e., upon receiving a negative signal. Thus, we employ it to shield our circuit from undesirable, negative signals.
4. **Half Wave Rectification:** The majority of the electronics that we use or work with today rely on DC voltages, as you may know. However, AC is the primary voltage used in industry, schools, and our homes.

Because AC has significantly less heat loss than DC, it is utilized over longer distances. However, because we are unable to store AC, we must convert it to DC in order to store it in batteries.

Diodes are used to carry out the aforementioned necessary action: they convert AC to DC. And it is because of this crucial diode function that the world of modern electronics is possible.

AC now consists of two parts: the portion that is both positive and negative. Because these sections alternate (change over time), these signals are known as alternating current (AC).

Direct current, or DC, is unidirectional, i.e., There aren't any sections that alternate. As previously mentioned, a diode functions by blocking current in reverse bias, which means that when a signal with a negative component is applied to it, the diode will stop it.

Therefore, to convert AC to DC, we need to block the negative part, and diodes are used for this purpose. To put it briefly, the diode's primary job in an electronic circuit is to convert AC energy into DC.

5. **Full Wave Rectification:** Full wave rectification is like half wave rectification, with a few minor differences. In short, full wave rectification involves converting AC into DC.

Full wave rectifiers are more efficient than half wave rectifiers, although they require more diodes. A half wave rectifier only needs one diode, while a full wave rectifier requires four. These four diodes can be arranged manually or purchased as a single package known as a rectifier bridge.

You know, the fundamental component of any power supply is this full wave rectifier. You can now imagine that every device we use requires a power source. Such gadgets do not exist. They all require a power source. In summary, nearly all our electronic devices require diodes in order to function properly.

6. **Voltage Multipliers:** Higher voltage levels are occasionally required for a variety of reasons, depending on the project or application we are

working on. One method is to use battery sources that are designed with such high voltage levels. However, creating such a battery is very expensive, heavy, and not a suitable solution for use with contemporary smart electronics.

Imagine a large, unappealing gadget these days; nobody will value it. And that is, a far better solution is required. Therefore, when higher voltage levels are needed, we can multiply the supplied voltages by a diode to make them into double or quadruple levels while maintaining the same and unaltered circuit size.

7. **Signal Clampers:** One of the main uses of the diode in the circuit is its ability to clamp a signal to the levels we need without changing the signal's original information. Adding DC value to a signal involves clamping it. Depending on what you are designing, you can now add both positive and negative DC.

Consider a Cartesian coordinate system with a sine signal on the x-axis. At this point, all you need to do is shift the signal on Y-axis by, say, 3. In a particular electronics circuit, diodes are used to carry out this kind of operation. Diodes enable us to increase the sine signal by 3.

This addition is necessary for several reasons. Using the extra DC to power up the device that will use this signal is one of the primary reasons. so that we can save a ton of money and circuit space by not needing to install a separate power supply there specifically to power that device. It is evident that a basic diode serves a multitude of essential purposes in nearly all electronic devices, circuits, and even systems at the industrial level.

8. **Signal Clippers:** To clip a signal is to remove a portion of it. Depending on what you are designing, the percentage could be positive or negative. Assume that your system operates within a specific voltage amplitude range. Let's assume that 0. 5 V is the highest voltage it can withstand.

You might end up breaking that system if your signal happens to cross that threshold. This is where diodes come into play, to prevent it. When the signals get too high for comfort, they cut them. Thus, one of the essential uses of diodes in circuits is signal clipping. The main uses of it are for protection and safety.

9. **Freewheeling:** A diode is used to stop current flow when there are loads where there is a possibility of current flowing back into the circuit. Freewheeling diodes are the diodes used for these kinds of applications. Our circuits may become damaged or cease to function due to back flow current. This typically occurs when using motors or when there are inductive loads.
10. **Voltage Regulations:** A constant output voltage, independent of changes in load or current flow, is known as voltage regulation. It is a crucial feature to consider when purchasing any kind of voltage source. if the output of your voltage source fluctuates over a certain load range.

Any device that depends on your bad voltage source for operation will experience a decrease in performance. Fortunately, the primary purpose of a diode in power supply circuits is to regulate voltage. The kind of diode used for this kind of operation is known as a Zener diode.

11. **Signal Peak Detection:** As the title implies. A diode is what we use to find the peaks in the signal. The most crucial piece of information to search for is a signal's peak. Designing components or circuit segments with those input signal peaks in mind is helpful.

12. **Envelope Detector**: As you may know, most of the time we spend using these communications devices, they are electronic devices that operate on radio frequencies or occasionally millimeter waves. These signals now transmit data in the form of envelopes. When I refer to an envelope, I mean a message signal that has been carrier signal modulated. Additionally, diodes are essential parts that enable us to identify the message envelopes transmitted by the signal transceiver.

13. **RF Variable Attenuator:** You are aware that the RF resistance can be altered by adjusting the bias current passing through a PN diode. The diode behaves like a resistor at high frequencies, with its resistance being inversely proportional to its forward current. Furthermore, diodes can be employed in output leveling (automatic gain control) circuits and amplitude modulators in various variable attenuator designs.

CASE STUDY

APPLICATION IN POWER RECTIFICATION

- **Scenario:** A company, "PowerTech Electronics," designs power supply units (PSUs) for consumer electronics. They aim to improve the efficiency of their PSU designs to meet stringent energy-saving standards. The engineers at PowerTech decided to replace conventional silicon diodes with Schottky diodes in the rectifier stage of their PSUs.

Implementation

1. *Component Selection*: The engineers chose 1N5819 Schottky diodes, which are known for their low forward voltage drop and adequate current handling capacity.
2. *Circuit Design:* The rectifier circuit was redesigned to incorporate Schottky diodes. This involved minimal changes as the pin configurations were compatible with the existing design.
3. *Testing and Validation*: The new design was tested under various load conditions to measure efficiency improvements and thermal performance.

Results

- *Efficiency Improvement*: The efficiency of the PSU increased by approximately 5%, mainly due to the reduced forward voltage drop and lower power dissipation in the rectifier stage.
- *Thermal Performance*: The reduced power loss resulted in lower operating temperatures, enhancing the reliability and longevity of the PSU components.
- *Compliance*: The improved efficiency helped PowerTech meet the required energy-saving standards, allowing them to market their products as eco-friendly.

Analysis

- **Advantages of Schottky Diodes:**
  - *Low Forward Voltage Drop*: Reduces power loss and improves efficiency.
  - *Fast Switching Speed*: Ideal for high-frequency applications, minimizing signal distortion and improving performance.
  - *Reduced Heat Generation*: Enhances the thermal management of electronic circuits.

Limitations

- *Reverse Leakage Current*: Schottky diodes exhibit higher reverse leakage current compared to silicon diodes, which can be a drawback in high-precision applications.

- *Voltage Rating*: They typically have lower maximum reverse voltage ratings, limiting their use in high-voltage applications.

Conclusion

Schottky diodes play a critical role in enhancing the efficiency and performance of electronic circuits, especially in power rectification and high-frequency applications. The case study of PowerTech Electronics demonstrates the practical benefits of using Schottky diodes, including improved efficiency, better thermal performance, and compliance with energy-saving standards. However, engineers must consider their limitations, such as higher reverse leakage current and lower voltage ratings, when designing circuits. Overall, Schottky diodes are valuable components in modern electronics, contributing significantly to the development of efficient and high-performance devices.

CLASS ACTIVITY

To draw the characteristic curve of a Zener diode and to determine its reverse breaks down voltage. To identify a diode, an LED, a resistor, and a capacitor from a mixed collection of such items. Use of a multimeter to see the unidirectional flow of current in the case of a diode and an LED and check whether a given electronic component (e.g., diode) is in working order.

SUMMARY

- A diode is a specialized electronic component that acts as a one-way switch. It conducts electric current in only one direction and restricts current from the opposite direction.
- The voltage applied to the anode is positive with respect to the cathode. Also, the voltage in the diode is higher than the threshold voltage, so it acts as a short circuit and allows current to flow.
- A blocking diode is used in some circuits for protection in case there is an accidental reverse connection problem like connecting the DC supply in a wrong way or reversing its polarities.
- A voltage regulator is used to step down the input voltage to the required level and will keep it the same in spite of fluctuations in the supply.
- The current flowing through the Zener diode is reduced in favor of the load current when the load resistor is connected in parallel to the Zener diode.
- A pn junction is an interface between two types of semiconductor materials, p-type and n-type, maintaining their orderly crystal lattice arrangements.
- When the pn diode consisting of a p+ and an n^- layer conducts, holes are injected from the p^+ layer into the n^- layer.
- A Schottky barrier diode is a type of diode formed by the metal-to-semiconductor junction that behaves as a rectifier.
- Silicon carbide (SiC) consists of silicon (Si) and carbon (C) atoms. Each atom is surrounded by four different atoms in the form of a regular tetrahedron.
- Zener diodes differ from typical rectifier and switching diodes in that they are used in the breakdown region in which the reverse characteristics of the pn junction are stable.
- The half-wave rectifier is the simplest diode circuit, and it is used to convert the AC signal into DC signal.

REVIEW QUESTIONS

1. Give an overview of diode.
2. Write the classification of diodes according to their applications and characteristics.
3. What are Schottky barrier diodes?
4. Differentiate Zener breakdown and avalanche breakdown.
5. Explain the diode functions in circuit.

REFERENCES

1. Braun, F. (1874). Ueber die Stromleitung durch Schwefelmetalle (On current conduction in metal sulphides). *Annalen der Physik und Chemie, 153*, 556–563.

2. Classification of components. (2009). Digikey.com.
3. Component Construction. (2009). Digikey.com.
4. Cornelius, E. C. (1946). Germanium crystal diodes. *Electronics, February*, 118.
5. Crecraft, F. M., & Gergely, S. (2002). *Analog Electronics: Circuits, Systems and Signal Processing*. Butterworth-Heinemann.
6. Current regulator diodes. (2009). Digikey.com.
7. Dowdey, J. E., & Travis, C. M. (1964). An analysis of steady-state nuclear radiation damage of tunnel diodes. *IEEE Transactions on Nuclear Science, 11*(5), 55. https://doi.org/10.1109/TNS2.1964.4315475.
8. EDAboard.com. (2010). Introduction dates of common transistors and diodes? Archived from the original on October 11, 2007. Retrieved 2010-08-06.
9. Edison, T. A. (1884). Electrical meter (U.S. Patent No. 307,030).
10. Elhami Khorasani, A., Griswold, M., & Alford, T. L. (2014). Gate-controlled reverse recovery for characterization of LDMOS body diode. *IEEE Electron Device Letters, 35*(11), 1079. https://doi.org/10.1109/LED.2014.2353301.
11. Giacoletto, L. J. (1977). *Electronics Designers' Handbook*. New York: McGraw-Hill. pp. 24–138.
12. Guarnieri, M. (2011). Trailblazers in solid-state electronics. *IEEE Ind. Electron. M., 5*(4), 46–47. https://doi.org/10.1109/MIE.2011.943016.
13. Guthrie, F. (1873). On a relation between heat and static electricity. *The London, Edinburgh, and Dublin Philosophical Magazine and Journal of Science, 4th series, 46*, 257–266.
14. Horowitz, P., & Hill, W. (1989). *The Art of Electronics* (2nd ed.). London: Cambridge University Press.
15. Jonscher, A. K. (1961). The physics of the tunnel diode. *British Journal of Applied Physics, 12*(12), 654. https://doi.org/10.1088/0508-3443/12/12/304.
16. North, H. Q. (1955). Asymmetrically conductive device (U.S. Patent No. 2,704,818).
17. Pickard, G. W. (1906). Means for receiving intelligence communicated by electric waves (U.S. Patent No. 836,531).
18. Preece, W. H. (1886). Multiplex telegraphy. *The Telegraphic Journal and Electrical Review, 19*, 252.
19. Protecting low current loads in harsh electrical environments. (2009). Digikey.com.
20. Redhead, P. A. (1998). The birth of electronics: Thermionic emission and vacuum. *Journal of Vacuum Science & Technology A: Vacuum, Surfaces, and Films, 16*(3), 1394–1401. https://doi.org/10.1116/1.581157.
21. Richardson, O. W. (1928). Thermionic phenomena and the laws which govern them. Nobel Lecture, December 12, 1929.
22. Roddick, R. G. (1962). Tunnel diode circuit analysis. doi: 10.2172/4715062.
23. Sarkar, T. K. (2006). *History of Wireless*. John Wiley and Sons. pp. 94, 291–308. ISBN 0-471-71814-9.

24. Scaff, J. H., & Ohl, R. S. (1947). Development of silicon crystal rectifiers for microwave radar receivers. *The Bell System Technical Journal, 24*(1), 1–30.
25. Sylvania Electric Products Co. (1949). *40 Uses for Germanium Diodes*. p. 9.
26. Sylvania. (1949). Data book page. Archived from the original on May 25, 2018.
27. Sze, S. M. (1998). *Modern Semiconductor Device Physics*. Wiley Interscience. ISBN 0-471-15237-4.
28. Tooley, M. (2013). *Electronic Circuits: Fundamentals and Applications* (3rd ed.). Routledge. p. 81. ISBN 978-1-136-40731-4.
29. Torrey, H. C., & Whitmer, C. A. (1948). *Crystal Rectifiers*. New York: McGraw-Hill.
30. Turner, L. W. (2015). *Electronics Engineer's Reference Book* (4th ed.). Butterworth-Heinemann. pp. 8.14–8.22. ISBN 978-1483161273.
31. U. S. Navy Center for Surface Combat Systems. (2012). *Navy Electricity and Electronics Training Series, Module 11*, pp. 2–81–2–83.
32. Wu, H., Wang, Y., Xu, Y., Sivakumar, P. K., Pasco, C., Filippozzi, U., & McQueen, T. (2022). The field-free Josephson diode in a van der Waals heterostructure. *Nature, 604*(7907), 653–656. https://doi.org/10.1038/s41586-022-04504-8.

CHAPTER 3

BJT Fundamentals

LEARNING OBJECTIVES

After studying this chapter, you will be able to:

- Learn about bipolar junction transistor;
- Understand the physical principles and operation laws of BJT;
- Know about terminals of BJT;
- Learn the physical meaning of active, saturation, and cutoff regions of operation;
- Learn transistor test circuits and the corresponding v–i dependencies;
- Obtain initial exposure to early effect;
- Discuss about BJT configuration and characteristics;
- Learn and apply the large-signal circuit model of a BJT.

KEY TERMS FROM THIS CHAPTER

Amplification
Integrated circuits
npn-junction transistor
Emitter-base junction
Base-emitter voltage
Collector-emitter voltage
Common-base configuration
Semiconductor
Transfer resistor
pnp junction transistor
Collector-base junction
Base-collector voltage
Common-emitter configuration
Common-collector configuration

INTRODUCTION

A bipolar junction transistor (BJT) is a type of transistor in which the charge carriers are electrons and electron holes. On the other hand, a single type of charge carrier is utilized by a unipolar transistor, like the field-effect transistor (FET). A bipolar transistor is able to control a much larger current flowing between its terminals with a small current injected at one of its terminals, allowing the device to be used for switching or amplification.

BJTs use two p–n junctions, which are areas within a single crystal of material, between two semiconductor types: n-type and p-type. A variety of techniques can be used to create the junctions, including doping the semiconductor material as it grows, forming alloy junctions by depositing metal pellets, and diffusing n- and p-type doping substances into the crystal. The original point-contact transistor was swiftly replaced by junction transistors due to their superior predictability and performance. Integrated circuits for analog and digital functions consist of diffused transistors and other components. It is very cost-effective to make hundreds of bipolar junction transistors in a single circuit.

A generation of mainframes and minicomputers used integrated circuits with bipolar transistors as their primary active components; however, many computer systems today use complementary metal-oxide-semiconductor (CMOS) integrated circuits, which rely on the field-effect transistor (FET). Bipolar transistors are still utilized in mixed-signal integrated circuits employing BiCMOS for signal amplification and switching. Specialized types include high voltage switches, radio-frequency (RF) amplifiers, and high current switches.

3.1. BIPOLAR JUNCTION TRANSISTOR

William Shockley, Brattain, and John Bardeen invented BJT in 1948, and it has completely changed not just the electronics industry but also our daily lives. Both electrons and holes are used as charge carriers in bipolar junction transistors. Unipolar transistors, like field-effect transistors, only employ one type of charge carrier, which is called indifference. BJT uses n-type and p-type semiconductors between two junctions for operational purposes. A BJT's primary purpose is to increase current. This characteristic makes BJTs useful for a variety of electrical devices, such as radio transmitters, televisions, industrial controls, and cell phones.

An example of a solid-state device is the bipolar junction transistor, which has three terminals: the base terminal controls the base terminal's current flow, while the emitter and collector terminals handle the device's current flow. Compared to the other kind of transistor, it is distinct. The input voltage regulates the output current of a field-effect transistor. Below is the basic symbol for the n-type and p-type BJTs (Figure 3.1):

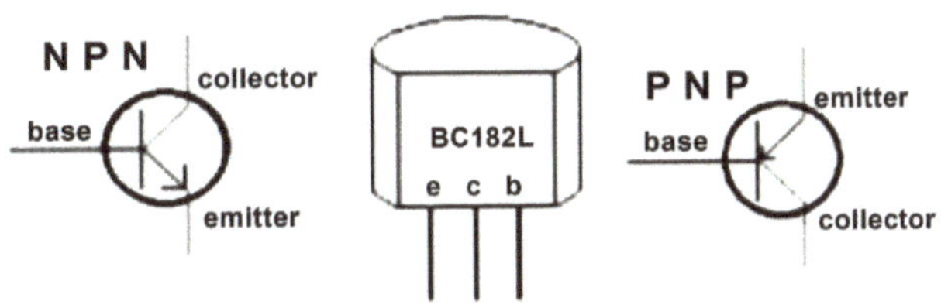

Figure 3.1. *BJT or bipolar junction transistor symbol.*

Source: https://www.elprocus.com/bipolar-junction-transistor-working/.

3.1.1. Why BJT Is Called Bipolar?

A BJT is a three-terminal semiconductor device. As the name implies, it is a bipolar transistor, meaning that current flows from two types of semiconductor materials, namely P-type (positive type) and n-type (negative type), in this type of transistor. Silicon is typically used in these transistors.

3.1.2. Construction of Bipolar Junction Transistor

The way BJT is constructed will dictate how it functions. Therefore, three doped semiconductor sections that are divided by two PN-junctions can be used to construct a BJT. The base, emitter, and collector are the three layers that make up this transistor. The physical structure of these transistors, which come in PNP and NPN varieties, is depicted below (Figure 3.2):

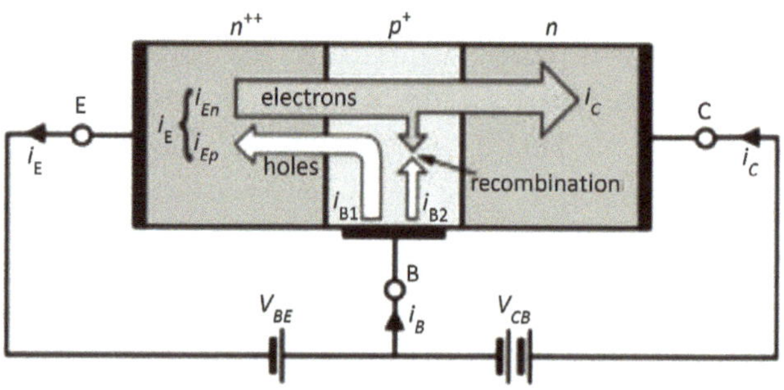

Figure 3.2. *Construction of bipolar junction transistor.*

Source: https://www.elprocus.com/bipolar-junction-transistor-working/.

It consists of two N-regions in NPN and one P-region separating them, while in PNP it consists of two P-regions and one N-region separating them. The use of both charge carriers, such as electrons and holes, within the structure is referred to as bipolar in BJT.

BE-junction is what's called when a PN-junction joins the emitter and base regions. Likewise, the point at which the collector region and base are joined by the pn-junction is referred to as the BC-junction. A wire lead connects to each area, and these leads are labeled E, B, and C for the emitter, base, and collector, respectively.

Compared to the heavily doped emitter terminal and the somewhat doped collector region, the base (B) region is much thinner and slightly doped. The following are the schematic symbols for the NPN and PNP BJTs.

3.1.3. Types of Bipolar Junction Transistor

We can refer to a transistor as a semiconductor device because, as we've seen, they provide low resistance in one direction and high resistance in another. There are two varieties of transistors that make up a bipolar junction transistor. BJT, including:

- Point contact
- Junction transistor

In contrast to point -contact transistors, junction transistors are used more frequently. Additionally, there are two types of junction transistors, which are listed below. Each junction transistor has three electrodes: the emitter, collector, and base.

- **PNP Junction Transistor:** The emitter in PNP transistors has a more positive relationship with both the base and the collector. A semiconductor device with three terminals is known as a PNP transistor. The transistor has three terminals: collector, base, and emitter. It is utilized in amplifying and switching applications. The PNP transistor's operation is depicted below.

In most cases, a resistor is used to connect the emitter or collector circuit to the negative supply, and the collector terminal is connected to the positive terminal. Applying a voltage to the base terminal causes the transistor to switch between an ON and OFF state. The transistor is in the OFF state when the emitter and base voltages are equal. When there is a decrease in base voltage relative to emitter, the transistor is in the ON state. This characteristic allows the transistor to function as a switch or amplifier. The PNP transistor's basic schematic is displayed below.

- **NPN Junction Transistor:** The PNP and NPN transistors are fundamentally opposed to each another. The emitter, collector, and base terminals of an NPN transistor are the same as those of a PNP transistor. The NPN transistor's operation is depicted below.

Typically, a resistor is used to connect the negative supply to the emitter terminal and the positive supply to the collector terminal, depending on whether the circuit is for the emitter or collector. Applying a voltage to the base terminal causes the transistor to operate in the ONN and OFF states. If the emitter and base voltages are equal, the transistor is in the off state. Transistor switches to ON when there is an increase in base voltage relative to emitter. This condition allows the transistor to function as both a switch and an amplifier. The basic symbol and the NPN configuration diagram are displayed below.

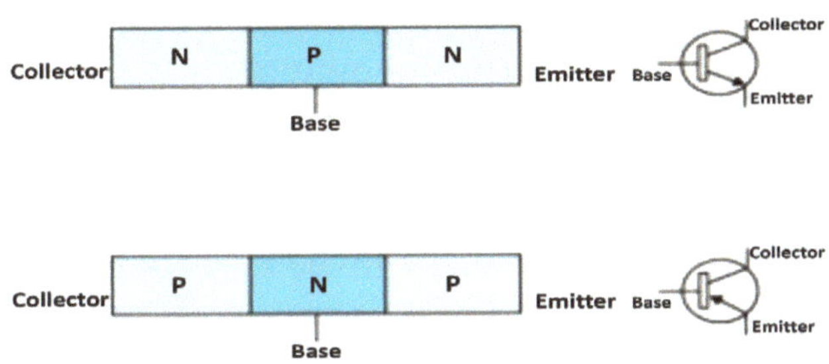

Figure 3.3. *PNP & NPN bipolar junction transistor.*

Source: https://www.elprocus.com/bipolar-junction-transistor-working/.

- **Hetero Bipolar Junction:** An additional variety of bipolar junction transistor is the hetero bipolar junction transistor. It creates heterojunction by using various semiconductor materials for the base and emitter regions. The HBT is typically used in radiofrequency and ultrafast circuits because it can handle singles at very high frequencies of several hundred GHz. RF power amplifiers and cell phones are two devices that use it.

3.1.4. Working Principle of BJT

Whereas the CB junction is reverse biased, the BE junction has forward bias. Compared to the BE junction, the CB junction's depletion region is wider. Electrons flow from the emitter to the base as a result of the forward bias at the BE junction, which lowers the barrier potential.

The base has very few holes and is only slightly doped. About 2% of the electrons from the emitter recombine in the base region with holes before exiting through the base terminal. The combination of electrons and holes causes the base current to flow as a result. To start the collector current, a significant portion of the remaining electrons will pass through the reverse bias collector junction. Using Kirchhoff's Current Law (KCL), we can observe the mathematical equation:

$$I_E = I_B + I_C$$

The base current is very less as compared to emitter and collector current:

$$I_E \sim I_C$$

The PNP transistor functions identically to the NPN transistor in this instance; the sole distinction is the absence of electrons in favor of holes. The PNP transistor for the active mode region is depicted in the diagram below (Figure 3.4).

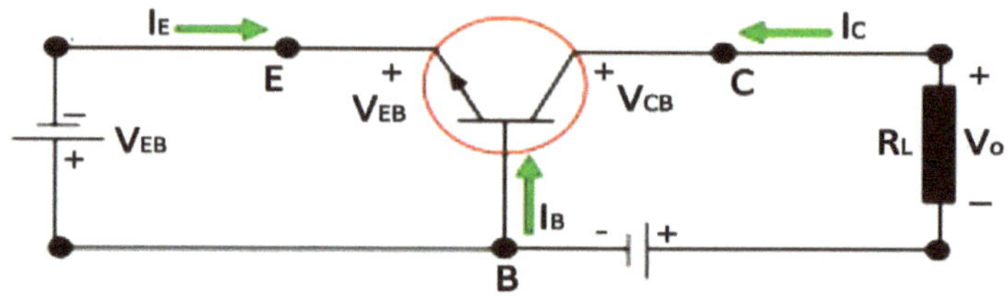

Figure 3.4. *Working principle of BJT.*

Source: https://www.elprocus.com/bipolar-junction-transistor-working/.

3.1.5. Functions of BJT

The most well- known functions of BJT are:

- **Amplification:** Because Junction Bipolar Transistors amplify signals by controlling a greater current flow between the collector and emitter based on a lesser input current at the base, they are essential for boosting electrical signals in devices like audio amplifiers.
- **Switching:** By controlling the current flow between the emitter and collector, BJTs operate as electronic switches. The BJT is a crucial component of digital circuits and logic gates. It acts as a switch by applying a signal to its base, allowing or restricting the flow of current.
- **Signal Modulation:** BJTs are used in modulation circuits to alter the characteristics of the carrier signals. These transistors are crucial parts of information-transmission communication systems because they can modify the base current, which can change signal properties like amplitude, frequency, or phase.

3.1.6. Advantages of BJT

The main advantages of bipolar junction transistors include the following:

- High driving capability
- High-frequency operation
- The digital logic family has an emitter-coupled logic used in BJTs as a digital switch
- It has a high gain bandwidth
- It gives good performance at high frequency
- Voltage gain is good
- It operates in low power or high-power applications
- It includes maximum current density
- Forward voltage drop is low

3.1.7. *Disadvantages of BJT*

The main disadvantages of bipolar junction transistors include the following:

- Thermal stability is less.
- It generates more noise.
- Less switching frequency.
- Base control is complex so needs skillful handling.
- The time taken for switching is not fast as compared to a high flashing frequency of voltage & current.

3.1.8. *Applications of BJT*

Here are some of the applications of Bipolar Junction Transistor:

- **Converters:** BJTs can be used in vast majority of converters. These converters can be of different types such as inverters, buck converters, boost converters or any **DC-DC**, **DC-AC**, **AC-DC** or **AC-AC**
- **Temperature Sensors:** Sensing temperature is one of the other applications of BJT. Where this can be found by two voltages at two different levels in a known ratio are subtracted
- **High Driving Capability**: It has High driving capability. For high voltage or current-handling capability, devices are connected in series and in parallel correspondingly. But the driving capability of individual devices is always considered.
- **High-Frequency Operation:** BJTs can be operated on a very high frequency. The frequency of the BJT for small signal is much higher than its switching frequency, principally due to storage delay. The storage time of 2N2222 is 310ns, thus the maximum switching frequency is about 3MHZ.
- **Digital Switch**: The digital logic family has emitter-coupled logic used in BJTs as a digital switch.
- **Oscillation Circuit**: These are preferred in oscillation circuits.
- **Clippers:** BJTs can be used in clipping circuits for changing the shape of waves. This can be used as a simple diode for clipping purpose but the problem with diode is that diode is uncontrolled.
- **Demodulator and Modulator:** BJTs can be used in demodulation and modulation circuits. BJTs are still using in the very old known modulation technique known as "**Amplitude modulation.**"
- **Detection Circuits**: BJTs can be used in detector circuits. BJT could be a new semiconductor sensor type for measuring ionization radiation dose.
- **Amplifiers:** One of the most important applications of BJTs is amplification where it is used in amplifier's circuit to amplify small signals. such as in audio amplifiers, these tiny components amplify very low audio signal to hearable range.
- **Electronic Switches:** It can be used as an electronic switch. BJTS are used in inverter for changing the direction of DC current and become AC current.
- **Automatic Switch:** It can be used instead of manual switch in an electrical circuit. the output signal of sensors is sometimes useless in electrical circuits because these signals are very low. However, these signals will become useful if they drive BJTs. As BJT operates on low signals. Then these BJT switches can run heavy loads including motors.

3.2. PHYSICAL PRINCIPLES AND OPERATION LAWS OF BJT

3.2.1. Physical Structure: Terminal Voltages and Currents

A semiconductor device with three terminals is called a bipolar junction transistor (BJT). As seen in Figure 3.5, the transistor's interior is divided into three separate doping regions. A heavily doped n^+ -type emitter, a p-type base, and an n-type collector make up an npn-junction transistor.

The transistor terminals are referred to as collector (C), base (B), and emitter (E) in standard terminology. Two semiconductor pn-junctions are represented by the doping structure in Figure 3.5. The emitter-base junction (EBJ) and collector-base junction (CBJ) are two types of pn-junctions. You must exercise caution because these junctions are not perfectly symmetrical. If you switch the transistor's collector and emitter, it will malfunction. As a three-terminal device, the transistor is characterized by three terminal voltages; this is also shown in Figure 3.5:

- Base-emitter (bias) voltage υ_{BE}
- Base-collector (bias) voltage υ_{BC} (or collector-base voltage $\upsilon_{CB} = -\upsilon_{BC}$)
- Collector-emitter (bias) voltage υ_{CE}

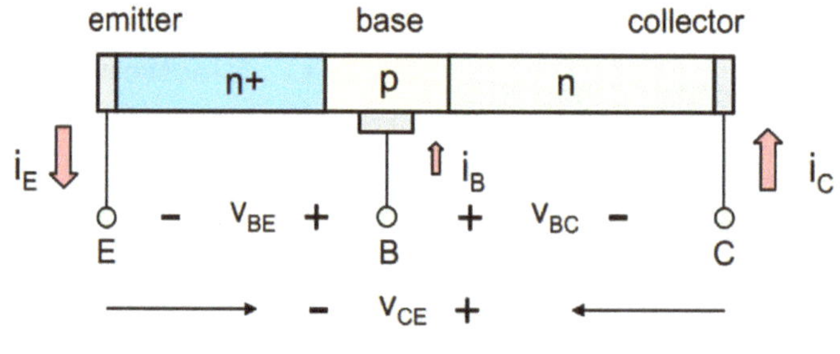

Figure 3.5. *Simplified physical structure of the npn-junction transistor.*

Source: Sergey N. Makarov, Reinhold Ludwig, and Stephen J. Bitar. Practical Electrical Engineering. Springer. DOI 10.1007/978-3-319-21173-2.

Only two of them are independent since KVL relates all three voltages to each other. The voltages υ_{BE} and υ_{CE} are chosen as independent variables. Then,

$$v_{BC} = v_{BE} - v_{CE} \tag{3.1}$$

Similarly, the transistor is characterized by three terminal currents shown in the same figure:

- Collector current i_C
- Emitter current i_E
- Base current i_B

Again, only two of them are independent since KCL relates all three currents to each other. The currents i_C and i_B are typically chosen as the primary parameters. Then,

$$i_E = i_C + i_B \tag{3.2}$$

Figure 3.6 (a) depicts the npn-junction transistor's circuit symbol; Figure 3.6 (b) shows the terminal currents; and Figure 3.6 (c) shows the terminal voltages. Transfer resistor is shortened to "transistor" in English. It describes a resistor with two terminals connected to it, whose resistance is controllable by a third terminal.

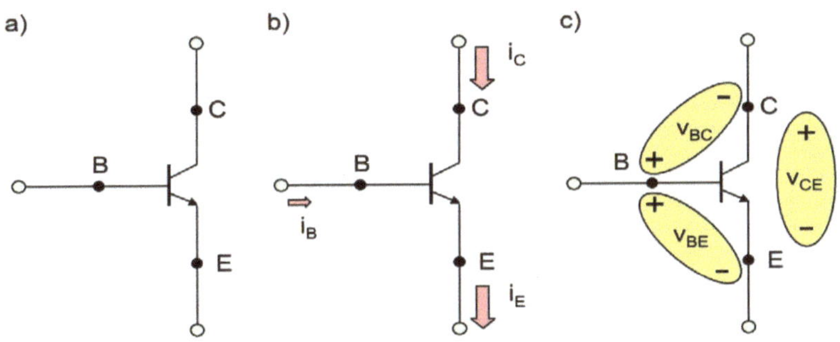

Figure 3.6. *Circuit symbol for the npn-junction transistor, terminal currents, and terminal voltages.*

Source: Sergey N. Makarov, Reinhold Ludwig, and Stephen J. Bitar. Practical Electrical Engineering. Springer. DOI 10.1007/978-3-319-21173-2.

3.2.2. Principle of Operation

Using a numerical simulation example for a specific n+ pn junction with a total length of R = 5μm and exponential (n^+ and n) donor doping profiles with terminal concentrations of 10^{18} cm^{-3} and 0.5×10^{15} cm^{-3}, respectively, we demonstrate how a junction transistor works. Acceptor doping is used in the central p-region; its distribution is cosine-shaped, with a maximum concentration of 0.3×10^{16} cm^{-3}. On top of Figure 3.7, doping profiles are displayed. The base-to-emitter pn-junction has a built-in voltage of about 0.80 V, and the collector-to-base pn-junction has a built-in voltage of about 0.6 V. These are typical npn-junction transistor values. At zero volts, the collector-base voltage (υ_{CB}) remains constant. In Figure 3.7, the transistor's emitter is grounded or a common node. This arrangement is called the common-emitter configuration. As an alternative, we may ground the collector (common-collector configuration) or the base (common-base configuration).

Along the junction, the built-in electric potential $\phi(x)$ is displayed in volts in Figure 3.7 (a–c). Along the junction, the intrinsic Fermi energy level of electrons, $E_i(x) = -q\phi(x)$, is shown in electron volts in Figure 3.7d–f. The base-emitter voltage curves are provided for three different values: υ_{BE} = 0, 0.5, and 0.7V. Since υ_{BE} and the collector-emitter voltage coincide, υ_{BC}, the base-collector voltage, is always fixed at zero volts. The device operation can be explained in three steps using simple mechanical analogies:

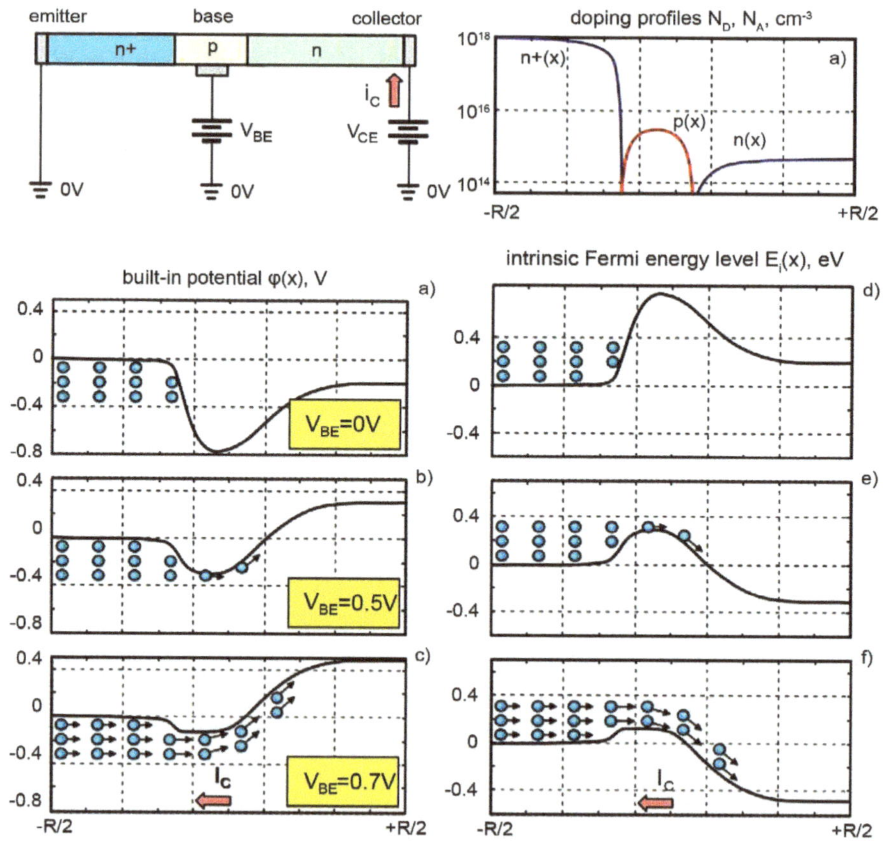

Figure 3.7. *Operation of the npn-junction transistor for three values of base-emitter voltage υ_{BE}. The collector-base voltage υ_{CB} is fixed at zero volts.*

Source: Sergey N. Makarov, Reinhold Ludwig, and Stephen J. Bitar. Practical Electrical Engineering. Springer. DOI 10.1007/978-3-319-21173-2.

1. The electric potential has a depth that extends deep into the p-region at υ_{BE} = 0V. As seen in Figure 3.7(a), the electrons in equilibrium try to gather in an area with a high electric potential. They resemble air bubbles in water that are floating upward toward the "potential" boundary when viewed from the perspective of the electric potential. They resemble water droplets in the air that decelerate into lower energy states when viewed from the perspective of electron energy. Since almost no electrons cross the barrier, there is very little electric current.
2. The potential depth (energy barrier) decreases with the bias voltage υ_{BE} = 0.5V. As shown in Figure 3.7(e), some electrons may now be able to diffuse across the p-type base, pass through the weaker potential barrier, and be swept into the n-type collector as they slide down the energy hill. The collection of electrons (air bubbles) beneath the collector's higher potential hill in Figure 3.7b offers an alternative interpretation.
3. Carrier diffusion into the base is fully enabled when the bias voltage reaches υ_{BE} = 0.7V. As a result, a sizable electric current (i_C) that is opposite to the motion of the electrons passes through the collector. To prevent any major recombination, the transistor's base, or p-region, is purposefully made narrow.

The npn transistor functions primarily as an electron-controlling device, in contrast to the pn-junction diode. In Figure 3.7, the potential/energy behavior to the right of the base region is determined by the base-collector voltage υ_{BC}. All the electrons diffusing across the base are swept into the n-type collector, and no electrons reach the collector as long as the potential (or energy) has a high enough elevation there. This indicates that when υ_{BC} is small or negative, the transistor operation is independent.

3.3. TERMINALS OF BJT

The base, emitter, and collector terminals—three components of the BJT—are briefly covered. The symbol representation of NPN and PNP transistors can be achieved in the following way: The emitter terminal always has an arrow symbol, and because of the charge carriers, the arrow direction indicates the direction of current flow.

3.3.1. Emitter Terminal

Charge carriers, such as electrons or holes, are released from emitter terminal and sent to the other two terminals. In order to generate large majority carriers, the base terminal is always reverse biased with regard to other terminals, such as emitter. The emitter is the region in BJT that is most heavily doped. In PNP and NPN transistors, the emitter-base junction needs to be connected in forward bias. In an NPN transistor, the emitter terminal supplies electrons to the emitter-base-junction, while in a PNP transistor, it supplies charge carriers such as holes to the same junction.

3.3.2. Collector Terminal

The component known as a collector, located on the opposite side of the emitter terminal is responsible for gathering the released charge carriers. This terminal is therefore highly doped, but its doping level is comparable to that of the base terminal, which is an emitter terminal that is both heavily and moderately doped. Reverse biasing of the CB-junction is required. This biasing is primarily done to remove charge carriers at the CB-junction. Whereas the PNP transistor's collector terminal gathers holes that are emitted through the emitter terminal, the NPN transistor's collector terminal is used to collect electrons that are emitted through the emitter terminal.

3.3.3. Base Terminal

The base terminal, which creates two PN junctions between the emitter and collector terminals, is in the middle. The transistor's base terminal is the section with the least amount of doping. Thus, its central position within the BJT will enable it to control the flow of charge carriers between the emitter and collector terminals. Because it can be connected in reverse bias, the BE junction exhibits high resistance.

3.4. OPERATING REGIONS OF BJT

Table 3.1 lists the three operating regions of the junction transistor. The conditions of the two pn-junction diodes dictate these regions, i.e., by the matching bias voltages υ_{BC} for the collector-base junction and υ_{BE} for the base-emitter junction. Table 3.1 contains a list of their generic values. The corresponding transistor test circuits in the common-emitter configuration are displayed in Figure 3.8. Using the two voltage sources, υ_{BE} and υ_{CE} are maintained at constant values in Figure 3.8(a). i_B is fixed in Figure 3.8, as opposed to υ_{BE}.

Table 3.1. *Operating Regions of an NPN-Junction Transistor*

Region	Junctions	
	EBJ	CBJ
Active	Forward bias $v_{BE} > 0.5$ V	Reverse-bias or small positive forward bias $v_B c < 0.4V$
Saturation	Forward bias $v_{BE} > 0.5$ V	Forward bias $v_B c > 0.4V$
Cutoff	Reverse-bias or small positive forward bias $v_{BE} < 0.5$ V	Reverse-bias or small positive forward bias $v_{Bc} < 0.4V$

Source: Sergey N. Makarov, Reinhold Ludwig, and Stephen J. Bitar. Practical Electrical Engineering. Springer. DOI 10.1007/978-3-319-21173-2.

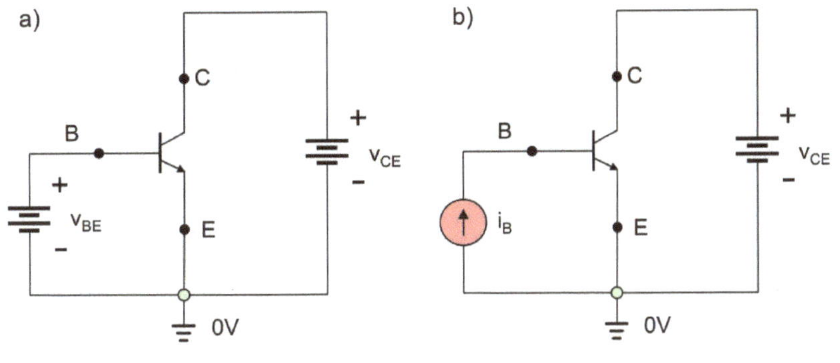

Figure 3.8. *Conceptual test circuits for a npn transistor in common-emitter configuration.*

Source: Sergey N. Makarov, Reinhold Ludwig, and Stephen J. Bitar. Practical Electrical Engineering. Springer. DOI 10.1007/978-3-319-21173-2.

3.4.1. Active Region

In this area, there are two junctions: one with a forward bias and the other with a reverse bias. Here, the amount of Ic (collector current) is managed by the base current (Ib). Therefore, the active region can be primarily utilized for amplification whenever this transistor functions as an amplifier by using a gain of "β" as shown by the following equation:

$Ic = \beta \times Ib$

The region that lies between the cutoff and saturation is known as the active region, or linear region. This region is where this transistor typically operates.

- **Collector Current: Qualitative Description:** As shown in Figure 3.9, the collector current is the diffusion current of minority electron carriers through the electrically neutral p-doped base. It is possible to calculate this current; the ideal-diode Shockley equation was used for the analysis. One major simplification applies to the junction transistor: The BJT's base region is extremely narrow, making it possible to overlook the minority carriers' recombination there. This results in an altered saturation current expression but the same Shockley equation.

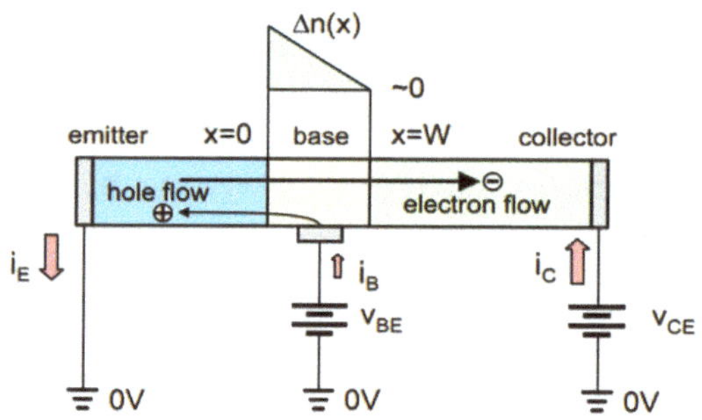

Figure 3.9. *Transistor test circuit of the common-emitter configuration in the active region.*

Source: Sergey N. Makarov, Reinhold Ludwig, and Stephen J. Bitar. Practical Electrical Engineering. Springer. DOI 10.1007/978-3-319-21173-2.

- **Collector Current: Quantitative Derivation of Shockley Equation:** With reference to Figure 3.9, boundary conditions for the excess minority carrier concentration Δn(x) are:

$$\Delta n(x=0) = \frac{n_i^2}{N_{A0}}\left(\exp\left(\frac{\upsilon_{BE}}{V_T}\right) - 1\right),$$
$$\Delta n(x=W) = \frac{n_i^2}{N_{A0}}\left(\exp\left(\frac{\upsilon_{BC}}{V_T}\right) - 1\right) \approx 0 \qquad (3.3)$$

The CBJ is thought to be either slightly positive biased ($\upsilon_{BC} \leq 0.4V$) or reverse biased. The thermal voltage in this case is given by $V_T = kT/q$, the intrinsic concentration of free carriers in the base is given by n_i, the (terminal) acceptor doping concentration in the base is given by N_{A0}, and the base width is given by W. The diffusion equation is:

$$\frac{d^2\Delta n}{dx^2} = 0 \qquad (3.4)$$

The solution of Equations (3.3) and (3.4) is a simple linear function:

$$\Delta n = \frac{n_i^2}{N_{A0}}\left(\exp\left(\frac{v_{BE}}{V_T}\right) - 1\right)(1 - x/W) \qquad (3.5)$$

While the concentration does drop with increasing distance along the base, this does not mean that the minority carriers vanish. They just move more quickly toward the collector. The diffusion current of excess minority carriers in the base is the transistor collector current. The formula for calculating the current density per unit area is $J_C = qD_n d\Delta n/dx$, where D_n represents the minority carriers' base diffusion constant. The current density multiplied by the junction area A yields the total collector current. Using Equation (3.5) yields:

$$i_C = I_S\left[\exp\left(\frac{v_{BE}}{V_T}\right) - 1\right],\ I_S = A\frac{qn_i^2 D_n}{N_{A0}W} \qquad (3.6)$$

The ideal-diode Shockley equation is (3.6). The fact that the saturation current I_S in Equation (3.6) scales linearly with the emitter-base junction area A is highlighted by its other name, the scale current. Typical values are in the range $I_S = 10^{-12} - 10^{-15}$ A.

- **Base Current:** A certain number of holes are injected from the p-type base into the emitter when the emitter-base junction is forward biased. The base current i_B represented in Figure 3.7 is made up of these holes. One unavoidable consequence of the junction transistor is the base current. The diffusion equation method is applicable in exactly the same way. The final result has the form:

$$i_B = I_{SB}\left[\exp\left(\frac{v_{BE}}{V_T}\right) - 1\right] \qquad (3.7)$$

The saturation current I_{SB} has the form of second Equation (3.6) related to the emitter region.

- **Relation Between Transistor Currents: Common-Emitter Current Gain:** According to Equations (3.6) and (3.7), transistor currents are directly proportional to each other:

$$i_C = \beta i_B,\ i_E = (\beta + 1)\, i_B \qquad (3.8)$$

where the dimensionless constant β is the (forward) common-emitter current gain of the transistor. The uncertainty in the manufacturing process makes it impossible to precisely control the current gain. Although much higher values can be obtained, the typical values are β = 20 – 200. The junction transistor's most crucial DC parameter is the current gain. Examine a general-purpose small-signal npn 2N3904 Si transistor, which is low-power. According to the device's datasheet, the current gain can be as low as 30–100 (depending on the collector current) or as high as 300 (at room temperature). Along with β, another parameter α is of interest, called the common-base current gain:

$$i_C = \alpha i_E,\ \alpha = \frac{\beta}{\beta + 1} \qquad (3.9)$$

In summary, in the active region the junction transistor operates as a current amplifier with terminal currents given by:

$$i_C = I_S\left[\exp\left(\frac{v_{BE}}{V_T}\right) - 1\right],\ i_B = \frac{I_S}{\beta}\left[\exp\left(\frac{v_{BE}}{V_T}\right) - 1\right] \qquad (3.10)$$

Equation (3.7) is the accurate nonlinear exponential model of the transistor in the active region. For practical purposes the factor 1 in the square brackets may be ignored.

3.4.2. Saturation Region

In this region, the transistors both the junctions are connected in forwarding bias. So, this region is mainly used for the ON-state for a switch wherever;

Ic = Isat

When this transistor is connected in the saturation region, "Isat," or the saturation current, is the maximum amount of current flowing through both the emitter and collector terminals. Because of the forwarding bias between these junctions, the transistor functions like a short circuit.

- **Qualitative Description of the Saturation Region:** As saturation occurs, both pn-junctions exhibit forward bias. Consequently, there is an opposing directed diffusion electron motion from the collector in addition to the diffusion electron motion from the emitter depicted in Figure 3.7. Currently, the contribution of two diffusion currents from two pn-junctions that are operating simultaneously determines the net collector (diffusion) current. In an effort to reduce the total collector current, these two currents try to cancel each other. Since it now serves two forward-biased pn-junctions rather than just one, the base current, on the other hand, increases. Equation (3.8) must be replaced by:

$$i_C < \beta i_B \tag{3.11}$$

- **Quantitative Description of the Saturation Region:** Let us assume that both bias voltages, υ_{BE} and υ_{BC}, have an impact on the collector current and examine two scenarios. For υ_{BE} and υ_{BC} = 0V, the first scenario corresponds to a forward bias. Equation (3.10) expresses the collector and base currents. The second scenario has a forward bias for υ_{BC} and υ_{BE} = 0V. In this instance, the two pn-junctions' roles—that of the emitter and the collector—are reversed; the emitter acts as the collector and vice versa. This situation could arise in a lab setting where the junction transistor in the active region has a reverse connection. With reference to Figure 3.10, the emitter and base currents are given by

$$i_E = I_S\left[\exp\left(\frac{\upsilon_{BC}}{V_T}\right) - 1\right], i_B = \frac{I_S}{\beta_R}\left[\exp\left(\frac{\upsilon_{BC}}{V_T}\right) - 1\right] \tag{3.12}$$

Since the saturation current in Equation (3.6) is only influenced by the base parameters and remains unchanged if the collector and emitter are switched, the emitter current in Equation (3.12) has the same form as the collector current in Equation (3.10). Similar in nature is the base current, but it has a different constant called the reverse common-emitter current gain, β_R. The reverse current gain is negligible, with $\beta_R \geq 1$, while $\beta \gg 1$.

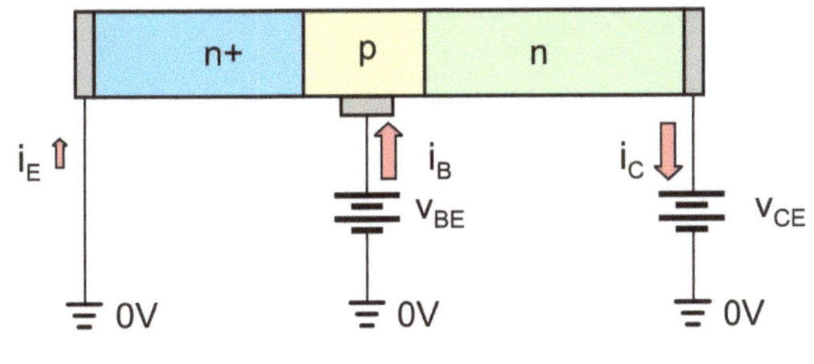

Figure 3.10. *Current flow in the transistor when EBJ is at zero volts and CBJ is forward biased.*

Source: Sergey N. Makarov, Reinhold Ludwig, and Stephen J. Bitar. Practical Electrical Engineering. Springer. DOI 10.1007/978-3-319-21173-2.

When both bias voltages υ_{BE} and υ_{BC} are present, the collector and emitter currents are the superposition of two solutions given by Equations (3.10) and (3.12), respectively,

$$i_C = I_S\left[\exp\left(\frac{\upsilon_{BE}}{V_T}\right) - 1\right] - I_S\left(1 + \frac{1}{\beta_R}\right)\left[\exp\left(\frac{\upsilon_{BC}}{V_T}\right) - 1\right] \quad (3.13a)$$

$$i_B = \frac{I_S}{\beta}\left[\exp\left(\frac{\upsilon_{BE}}{V_T}\right) - 1\right] + \frac{I_S}{\beta_R}\left[\exp\left(\frac{\upsilon_{BC}}{V_T}\right) - 1\right] \quad (3.13b)$$

The general analytical Ebers-Moll model of the junction transistor, which is applicable for all values of the bias voltages, is provided by equations (3.13). Circuit simulators such as SPICE employ it. For the active region, its simplification is given by equation (3.10). The saturation region may be also described in terms of the "forced" common-emitter current gain (the forced beta):

$$\beta_{\text{forced}} \equiv i_C/i_B \quad (3.13c)$$

- *Example 3.1:* Using the Ebers-Moll model and assuming room temperature of 25°C, estimate the forced common-emitter current gain $\beta_{forced} \equiv i_C/i_B$ of the BJT when $\beta = 100$, $\beta_R = 0.02$, $I_S = 10^{-12}$ A for two sets of bias voltages:

A. $\upsilon_{BE} = 0.5$ V, $\upsilon_{BC} = 0.1$, 0.2, 0.3V

B. $\upsilon_{BE} = 0.6$V, $\upsilon_{BC} = 0.2$, 0.3, 0.4 V

- *Solution:* We plug the above data into the Ebers-Moll model and obtain:

A. $\beta_{forced} = 99.9$, 95.3, 29.8

B. $\beta_{forced} = 99.9$, 95.3, 29.8

The forced current gain indeed decreases when the base-collector voltage increases. Remarkably, both sets of β forced are identical to each other. This means that the current gain degradation in the saturation region is solely determined by the collector-emitter voltage, which is the $\upsilon_{CE} = \upsilon_{BE} - \upsilon_{BC}$.

3.4.3. Cutoff Region

In this region, both the transistor junctions are connected in reverse bias. Here this transistor works like the off condition of a switch wherever

Ic = 0

In contrast to the saturation region, the operation is completely reversed in the cut-off region. Therefore, no external supplies are linked. There won't be any emitter current if there is no collector current.

By lowering the base voltage to below the voltage of the emitter and collector, the transistor can function as an off-state of the switch in this manner.

Vbe < 0.7

EBJ and CBJ are both reverse biased in the cutoff region. Transistor currents are all at zero. The corresponding inequalities of Table 3.1 must be satisfied by transistor voltages.

3.5. TRANSISTOR V–I DEPENDENCIES

The circuits depicted in Figure 3.8 are used to measure the transistor v-i curves experimentally. In the initial scenario, we adjust υ_{BE} in Figure 3.8a while maintaining a constant υ_{CE}. The relationship between the collector current (i_C) and base-emitter voltage (υ_{BE}) is depicted in Figure 3.11. The corresponding dependence varies with temperature and is of the Shockley type.

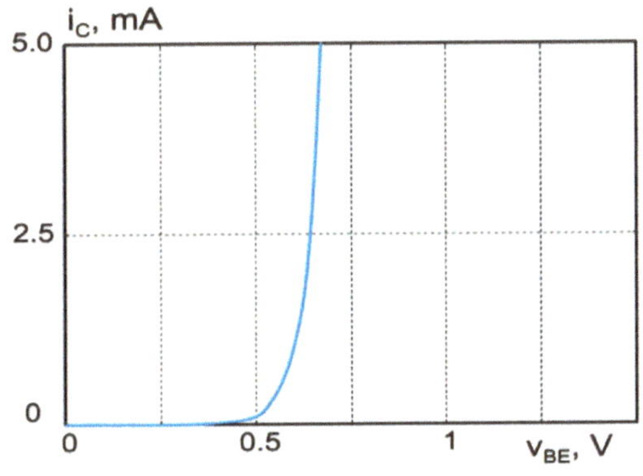

Figure 3.11. *Typical dependence of the collector current on υ_{BE}.*

Source: Sergey N. Makarov, Reinhold Ludwig, and Stephen J. Bitar. Practical Electrical Engineering. Springer. DOI 10.1007/978-3-319-21173-2.

The collector current as a function of the collector-emitter voltage, or υ_{CE}, is a new and significant plot that is specifically related to junction transistors. Here, we change υ_{CE} in Figure 3.8 while maintaining a constant υ_{BE} or i_B.

This plot enables us to observe the transistor equivalent "resistance" between the collector and emitter terminals if we treat the junction transistor as a "transfer resistor". An idealized plot of this kind at three different base current values is shown in Figure 3.12 (a). Figure 3.8 (b) shows the test circuit.

The EBJ is thought to be forward biased. In the saturation region, the collector current nonlinearly decreases to zero at small υ_{CE} (less than approximately 0.3 V). The transistor enters the active region at larger υ_{CE}. As long as i_B (and hence υ_{BE}) is a fixed number, the collector current on υ_{CE} (and on υ_{BC}) stays independent—refer to Equation (3.1). The transistor collector-emitter resistance is infinitely high in the active region but relatively small in the saturation region, as shown in Figure 3.12a.

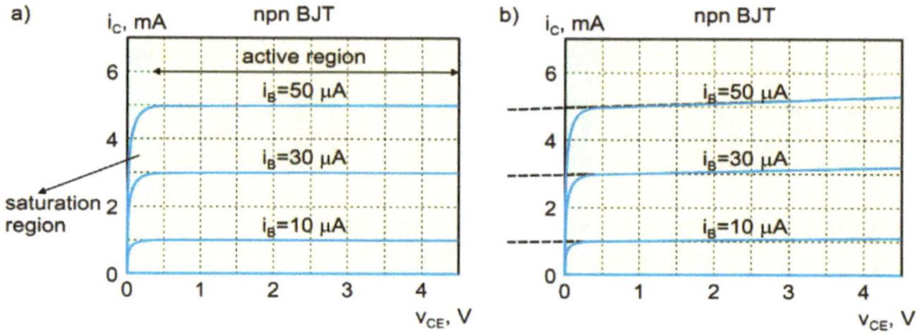

Figure 3.12. *Typical common-emitter characteristics of the npn BJT. The collector current is independent of the collector-to-emitter voltage in the active region in (a). (b) shows a more accurate representation which includes the early effect.*

Source: Sergey N. Makarov, Reinhold Ludwig, and Stephen J. Bitar. Practical Electrical Engineering. Springer. DOI 10.1007/978-3-319-21173-2.

3.6. EARLY EFFECT

A more realistic scenario is depicted in Figure 3.12(b), where the collector current increases slightly—in fact, linearly—as υ_{CE} increases. We refer to this as the Early effect. The CBJ is significantly reverse biased at large positive υ_{CE} (large negative υ_{BC}), which causes the neutral base length W in Equation (3.6) to shorten and the corresponding depletion region to widen. As a result, when the diffusion current gains prominence, the saturation current in Equation (3.6) increases. For this reason, the base-width modulation effect is another name for the Early effect. It leads to a finite output resistance of the transistor, r_o,

$$r_o = \frac{V_A}{i_C} \tag{3.14}$$

where i_C is the collector current given by Equation (3.10) and shown in Figure 3.12 with the Early effect ignored and V_A is a transistor constant known as the Early voltage (a typical V_A is 50 V). The Early voltage is the common intersection point of all dashed asymptotes in Figure 3.12(b) with the υ_{CE}-axis.

3.7. BJT CONFIGURATION AND CHARACTERISTICS

A bipolar junction transistor has three terminals, allowing it to be connected to a circuit in three different ways. One of the terminals is common among the others, meaning that it is common between the input and the output. All connections merely react to the input signal differently.

3.7.1. Common Base Configuration

The base terminal of the transistor is shared by the input and output signals in its CB configuration. In this case, the output can be obtained between the two terminals, base and collector, while the input signal can be given between the two terminals, base and emitter. In comparison to the input signal at the emitter terminal, the output signal at the collector terminal is lower.

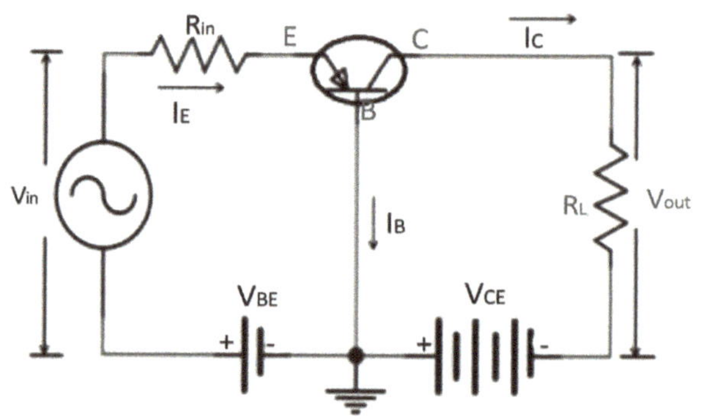

Figure 3.13. *CB configuration.*

Source: https://www.elprocus.com/bipolar-junction-transistor-working/.

Since its gain is less than 1, the signal is attenuated. Because this configuration's output is non-inverting, the input and output signals will both be in phase. Due to its high voltage gain, this common- base configuration is not utilized very often. This type of arrangement is limited to single-phase amplifiers due to its exceptionally high-frequency response. These single-phase amplifiers can be applied as microphone pre-amps and RF amplifiers.

Gains of CB Configuration:

- Voltage Gain = Av = Vout/Vin = Ic x RL/IE X RIN
- Current Gain = Ic/ie
- Resistance Gain = RL/Rin

3.7.2. Common Emitter Configuration

The emitter terminal in this configuration is shared by the input and output. While the output can be obtained between the collector and emitter terminals, the input is provided between the base and emitter terminals.

Input and output are obtained from the base and collector terminals, respectively, when the emitter terminal is connected to GND. The maximum current and power gain across all configuration types are included in the CE configuration.

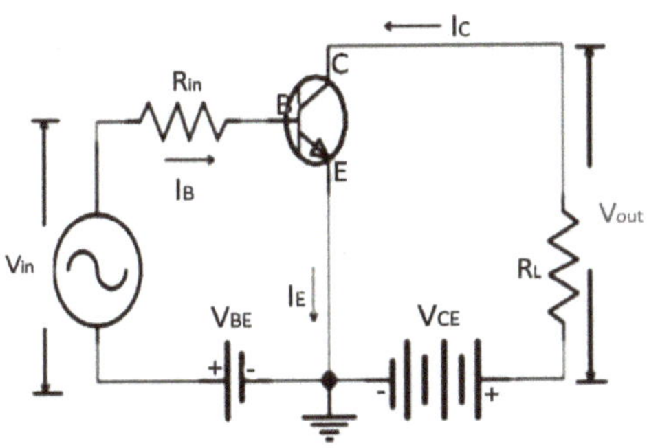

Figure 3.14. *CE configuration.*

Source: https://www.elprocus.com/bipolar-junction-transistor-working/.

The primary cause is that the output can be received from the reverse bias junction, but the input impedance is very low at the forwarding bias junction. Its output impedance is very high as a result. The emitter current in a CE configuration equals the sum of the base and collector currents. So, the equation is:

Ie= Ic + Ib

The emitter current is denoted by "Ie" in the above formula. Due to the ability to connect the load resistance in series through the collector terminal, this configuration has the maximum current gain, similar to Ic/Ib systems. From the equation above, even a slight increase in base current will result in an extremely high current in the output region. When there is a complete polarity difference between the input and output signals, the CE configuration functions similarly to an inverting amplifier. As a result, there is a 180° shift in the output signal relative to the input.

3.7.3. Common Collector Configuration

The CC configuration, which has a grounded collector, is also known as an emitter follower or voltage follower. The collector terminal in a CC configuration can be connected to ground in the direction of the supply.

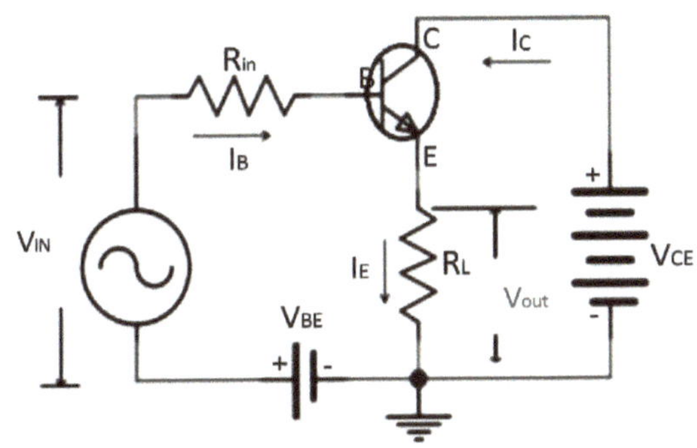

Figure 3.15. *CC configuration.*

Source: https://www.elprocus.com/bipolar-junction-transistor-working/.

The collector terminal is shared by both input and output. The input is sent directly to the base terminal, while the output is obtained from the emitter through a load connected in series. Both the input and output impedances are high in this configuration. It can function similarly to an impedance matcher because of this. This type of arrangement is, therefore, very helpful when using the impedance matching technique.

3.7.4. BJT Characteristics

When it comes to input and output impedances and gains, such as voltage, power, and current, the behavior of the bipolar transistor varies greatly depending on the circuit configuration. Three primary categories can be used to categorize a BJT's fixed characteristics; these are described below:

- **Input Characteristics**
  - o Common Base (CB) = $\Delta V_{EB}/\Delta I_E$
  - o Common Emitter (CE) = $\Delta V_{BE}/\Delta I_B$
- **Output Characteristics**
  - o Common Base (CB) = $\Delta V_C/\Delta I_C$
  - o Common Emitter (CE) = $\Delta V_{C/}\Delta I_C$
- **Transfer Characteristics**

o Common Base (CB) = $\Delta I_C/\Delta I_E$

o Common Emitter (CE) = $\Delta I_C/\Delta I_B$

The different transistor configuration characteristics are given below:

Table 3.2. *The Different Transistor Configuration Characteristics*

Characteristics	CB	CE	CC
Input Impedance	Low	Medium	High
Output Impedance	Very High	Low	Low
Phase Angle	0°	180°	0°
Voltage Gain	High	Medium	Low
Current Gain	Low	Medium	High
Power Gain	Low	Very High	Medium

3.8. LARGE-SIGNAL CIRCUIT MODEL OF A BJT

Dependent sources and diodes are two examples of simpler circuit blocks that must be combined in a BJT's equivalent circuit model to explain how it operates. A BJT's corresponding first-order circuit can be of the π- or T-type (or Y-type). First-order implies that there are, in fact, more accurate models. The first-order π-type npn BJT model in the active region of operation is depicted in Figure 3.16 (b) and 3.1(c). The circuit model for the large-signal BJT is shown in Figure 3.16 (b) and 3.16(c). This indicates that it applies equally well to all conceivable υ_{BE} values.

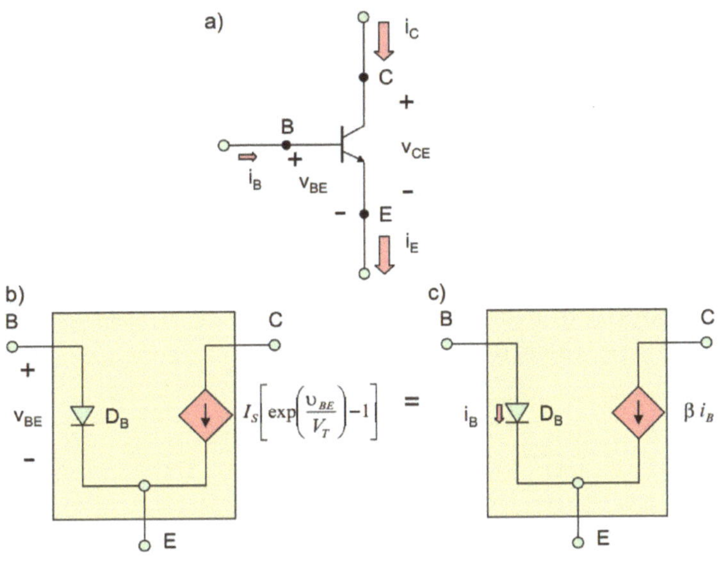

Figure 3.16. *First-order π-type BJT model in the active region of operation.*

Source: Sergey N. Makarov, Reinhold Ludwig, and Stephen J. Bitar. Practical Electrical Engineering. Springer. DOI 10.1007/978-3-319-21173-2.

The EBJ Shockley diode and a nonlinear voltage-controlled current source take the place of the transistor in Figure 3.16(b). An alternative current source that is controlled by a linear current is shown in Figure 3.16(c). In Figure 3.16(b) and 3.16(c), both circuits represent the BJT operation in the active region, as described by Equation (3.10), which is repeated here for convenience:

$$i_C = I_S\left[\exp\left(\frac{\upsilon_{BE}}{V_T}\right) - 1\right], i_B = \frac{I_S}{\beta}\left[\exp\left(\frac{\upsilon_{BE}}{V_T}\right) - 1\right], i_C = \beta i_B \quad (3.15)$$

- *Example 3.2:* Solve the base-bias transistor circuit in Figure 3.17(a) (which is essentially the transistor test circuit in Figure 3.8(b) at 25°C by determining voltages υ_{BE}, υ, and υ_{CE} using the large-signal model. Assume $I_S = 10^{-14}$ A for the EBJ Shockley diode.
- *Solution:* First, we need to replace the transistor by an equivalent circuit model. We assume the active region of operation. Since the base current is given, the model of Figure 3.16(b) is more appropriate. The transistor circuit is transformed as shown in Figure 3.17(b). Then, we solve this equivalent circuit. We find the base-emitter voltage υ_{BE} from the Shockley diode model:

$$i_B = \frac{I_S}{\beta}\left[\exp\left(\frac{\upsilon_{BE}}{V_T}\right) - 1\right] \Rightarrow \upsilon_{BE} = V_T \ln\left(\frac{\beta i_B}{I_S} + 1\right) = 0.68\,\text{V} \quad (3.16a)$$

The next step is to find the collector current. From the circuit in Figure 3.17(b),

$$i_C = \beta i_B = 1.93\text{mA} \quad (3.16b)$$

Resistor voltage υ is obtained in the form

$$\upsilon = R_C i_C = 9.07\text{V} \quad (3.16c)$$

Therefore, the collector-emitter voltage is given by

$$\upsilon_{CE} = 20 - 9.07 = R_C i_C = 10.93\text{V} \quad (3.16d)$$

After the solution is complete, we check if the transistor is really in the active operation region following Table 3.1. Since υ_{BE} = 0:68 V, we obtain $\upsilon_{BC} = \upsilon_{BE} - \upsilon_{CE}$ = −10.25V so that both conditions of the active region (the EBJ is forward biased and the CBJ is reverse biased) are satisfied.

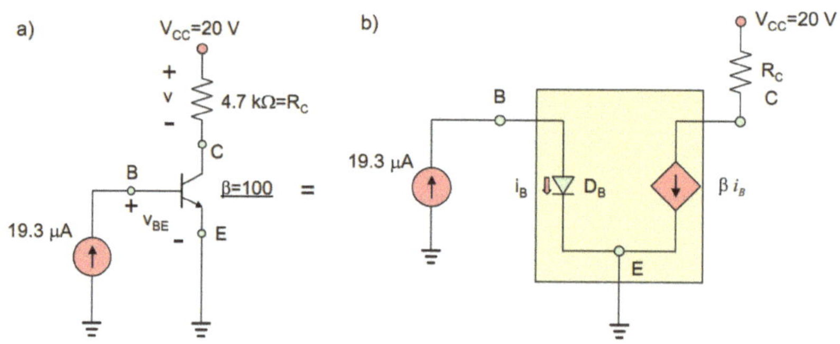

Figure 3.17. *A base-bias transistor circuit and its equivalent representation.*

Source: Sergey N. Makarov, Reinhold Ludwig, and Stephen J. Bitar. Practical Electrical Engineering. Springer. DOI 10.1007/978-3-319-21173-2.

Using the exponential large signal model, other (more complex) transistor circuits can be solved using the procedure shown in Example 3.2. The Ebers-Moll formalism and the second diode can be added to the model to extend it to the saturation region. A transcendental equation may need to be solved in many real-world situations. This can be done iteratively or graphically using the load-line method.

3.8.1. Large-Signal DC Circuit Model of a BJT

The circuit parameters are fixed in the DC steady state. Turn-on voltages have the potential to substitute for the BJT's steep exponential dependencies. The idea is described in Figure 3.18. This produces a large-signal DC BJT circuit model that is simplified, and enables us to estimate the main circuit parameters as well as identify the operation region (active, saturation, or cutoff). This model performs admirably when determining a transistor amplifier circuit's DC operating point and when qualitatively analyzing transistor power circuits.

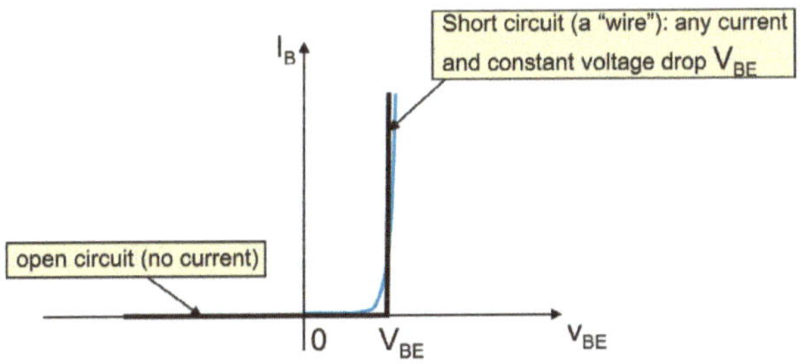

Figure 3.18. *Large-signal DC circuit model of the EBJ.*

Source: Sergey N. Makarov, Reinhold Ludwig, and Stephen J. Bitar. Practical Electrical Engineering. Springer. DOI 10.1007/978-3-319-21173-2.

Figure 3.19 shows the DC circuit model for three different transistor regions:

1. In the active region, $V_{BE} \approx 0.7V$ (turn-on voltage of the EBJ), and the CBJ is reverse biased or has a small positive bias below its turn-on voltage of $V_{BC} \approx 0.5V$ (turn-on voltage of the CBJ is smaller due to more shallow collector doping). Therefore, $V_{CE} = V_{BE} - V_{BC} > 0.2V$. This data falls within the wider active region of Table 3.1. Indeed, $I_C = \beta I_B$.
2. In the saturation region, $V_{BE} \approx 0.7V$ (turn-on voltage of the EBJ) and $V_{BC} \approx 0.5V$ (turn-on voltage of the CBJ). Therefore, $V_{CE} = V_{BE} - V_{BC} \approx 0.2V$. This data again falls within the wider saturation region of Table 3.1. In saturation, $I_C < \beta I_B$. The value of the collector current is determined by the rest of the circuit. The transistor is most efficient as a switch (has the lowest relative power loss) in the saturation region.
3. In the cutoff region, both junctions are reverse biased or have a small positive bias so that $V_{BE} \leq 0.5V$, $V_{BC} \leq 0.4V$ according to Table 3.1. We round the last value to 0.5 V to better memorize it. All terminal currents are zeros; the transistor is an open circuit.

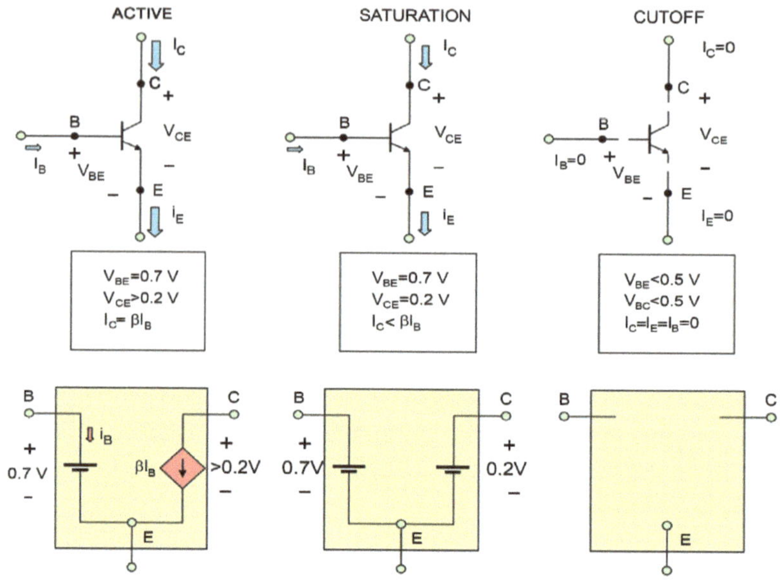

Figure 3.19. *Large-signal DC circuit model of a npn BJT transistor for three regions of operation.*

Source: Sergey N. Makarov, Reinhold Ludwig, and Stephen J. Bitar. Practical Electrical Engineering. Springer. DOI 10.1007/978-3-319-21173-2.

- *Example 3.3:* Create the analog of the chart in Figure 3.19 for a pnp transistor.
- *Solution:* The pnp BJT model is reduced to npn model after switching voltage polarities, i.e., after substitutions $V_{BE} \rightarrow V_{EB}$, $V_{BC} \rightarrow V_{CB}$, and $V_{CE} \rightarrow V_{EC}$. Therefore, the chart of Figure 3.19 is straightforwardly extended to the pnp case as shown in Figure 3.20.

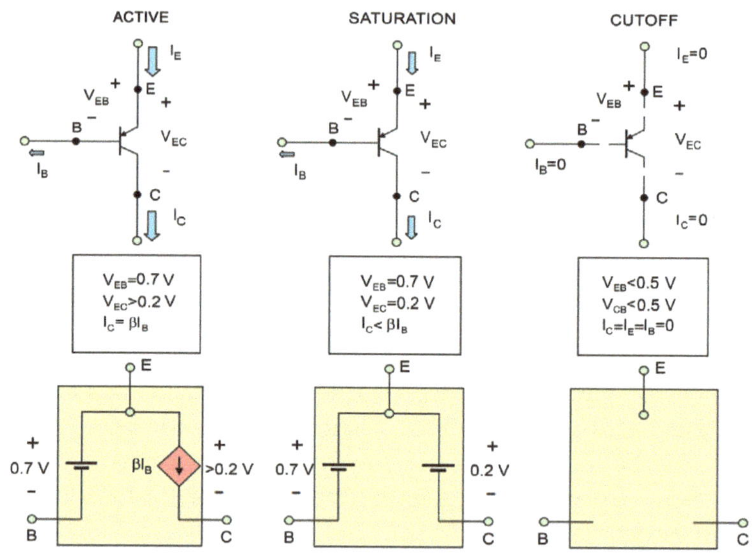

Figure 3.20. *Large-signal DC circuit model of a pnp BJT transistor.*

Source: Sergey N. Makarov, Reinhold Ludwig, and Stephen J. Bitar. Practical Electrical Engineering. Springer. DOI 10.1007/978-3-319-21173-2.

In practice, for different junction transistors and under different operating conditions, the voltage values shown in Figure 3.18 and 3.19 may slightly vary. We will ignore these variations.

3.8.2. Method of Assumed States

When applying the large-signal DC circuit model of a transistor is applied, the circuit is solved using the method of assumed states. Its concept is as follows:

1. We begin by guessing the transistor state (active, saturation, or cutoff).
2. We replace the transistor by the large-signal DC circuit model for the corresponding region
3. We solve the resulting circuit. Not all model conditions are necessary for this purpose.
4. Subsequently, we verify if the remaining model conditions are satisfied.
5. If this is not the case, the model for another region is selected.

The large-signal DC circuit model always provides a unique solution.

3.8.3. Transistor Circuit Analysis Using the Method of Assumed States

The transistor test circuit displayed in Figure 3.8(a) serves as our starting point. As seen in Figure 3.21, we will add collector and base resistances to make it more realistic. Notice the power supply indexing has changed from Figure 3.8(a). This is necessary because the base-emitter and collector-emitter supply voltages are no longer the source voltages. First, we will use the large-signal DC model and the method of assumed states to solve the resulting transistor circuit. Subsequently, we will use the large-signal exponential model to solve the same circuit. Finally, we will compare and contrast the two solutions.

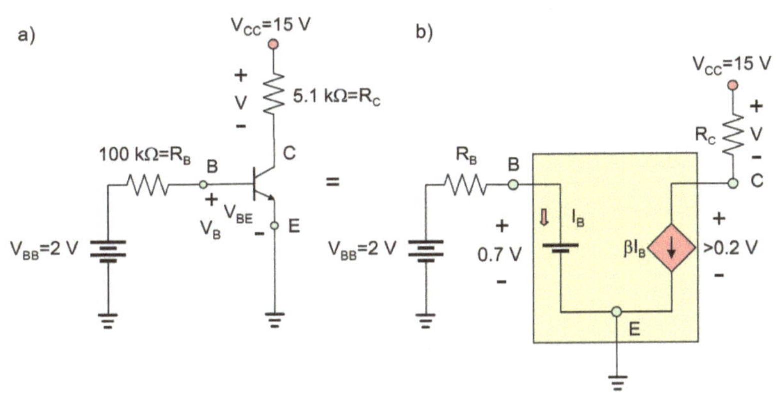

Figure 3.21. *Transistor circuit solved using the method of assumed states.*

Source: Sergey N. Makarov, Reinhold Ludwig, and Stephen J. Bitar. Practical Electrical Engineering. Springer. DOI 10.1007/978-3-319-21173-2.

- *Example 3.4:* Solve the transistor circuit shown in Figure 3.21(a) by determining unknown voltages V_{BE} and V_{CE} using the large-signal DC BJT model and the method of assumed states. Assume $\beta = 100$.
- *Solution:* We need to choose one of the regions first; let us choose the active region of operation. This guess leads to the circuit shown in Figure 3.21(b) with:

$V_{BE} = 0.7V$ (3.17a)

The base current is found using KVL for the base-emitter loop:

$$I_B = \frac{V_{BB} - 0.7V}{R_B} = 13\,\mu A \quad (3.17b)$$

In the active region, $I_C = \beta I_B = 1.3mA$. From KVL for the base-to-collector branch:

$V_{CE} = V_{CC} - V = 15 - R_C I_C = 15 - 6.63 = 8.37V$ (3.17c)

The circuit is thus "tentatively" solved. As a last step, we must check if our initial guess was correct. The only remaining condition to check in the active region is that of the collector-to-emitter voltage:

$V_{CE} = 8.37V > 0.2V$ (3.17d)

Thus, every requirement for the active region is satisfied. It is not necessary to check the saturation and cutoff regions of operation; doing so will inevitably produce unfavorable outcomes because a solution obtained using the method of assumed states is always unique.

- *Example 3.5:* Solve the transistor circuit in Figure 3.22(a) at room temperature of 25°C by determining unknown voltages υ_{BE} and υ_{CE} using the large-signal BJT model. Assume $I_S = 10^{-14}A$ for the EBJ Shockley diode and $\beta = 100$.
- *Solution:* The current-controlled current source model from Figure 3.16c is selected. This model is somewhat more convenient for the present problem. The rightmost base-to-emitter branch is solved either using the load-line method or iteratively. Using KVL and the definition of the Shockley diode, one has:

$$2 - R_B i_B - \upsilon_{BE} = 0 \Rightarrow i_B = \frac{2 - \upsilon_{BE}}{R_B} \Rightarrow i_B = \frac{2 - V_T \ln\left(\frac{\beta i_B}{I_S} + 1\right)}{R_B} \quad (3.18a)$$

We will use the iterative solution for i_B with the starting guess of $i_B = 0$. The convergence process is very fast. We present the final result, which is:

$i_B = 13.3\mu A$ (3.18b)

The base-emitter voltage is obtained as:

$$\upsilon_{BE} = V_T \ln\left(\frac{\beta i_B}{I_S} + 1\right) = 0.67V \quad (3.18c)$$

The collector current is found from the equivalent circuit, $i_C = \beta i_B = 1.33mA$. Voltage across the collector resistance is obtained in the form $\upsilon = R_C i_C = 6.71V$. Therefore, the collector-emitter voltage is given by:

$$\upsilon_{CE} = 15 - \upsilon = 8.29V \qquad (3.18d)$$

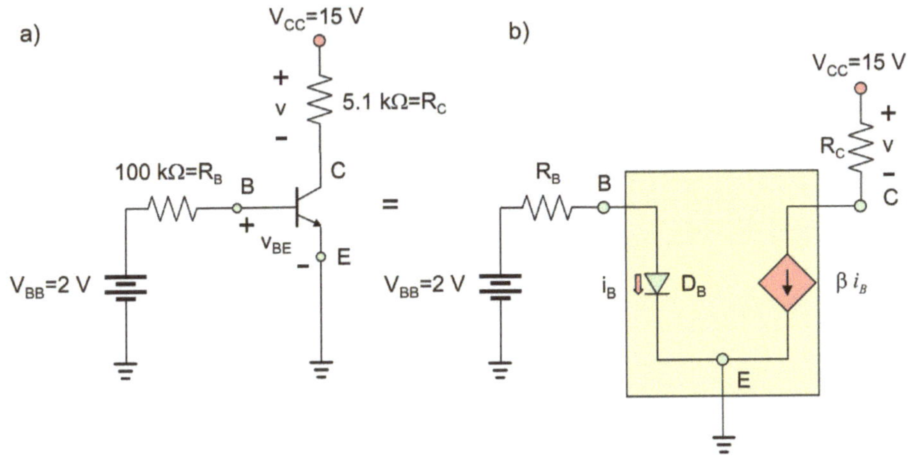

Figure 3.22. *Transistor circuit solved using the large-signal exponential model.*

Source: Sergey N. Makarov, Reinhold Ludwig, and Stephen J. Bitar. Practical Electrical Engineering. Springer. DOI 10.1007/978-3-319-21173-2.

Examples 3.4 and 3.5 show results that are nearly identical, with errors of no more than 5% for each parameter. If we change the saturation current, for example, or other key exponential model parameters, we also get a good agreement. This indicates that a practical yet basic tool for transistor circuit analysis is the large-signal DC model.

3.8.4. DC Transistor Bias Circuits

The analysis of BJT DC transistor bias circuits makes use of the large-signal DC circuit model and the method of assumed states. All transistor amplifiers, whether discrete or integrated, are designed using these circuits as a foundation. Typically, bias circuits consist of a transistor, several resistors surrounding it, and/or one or more sources of voltage or current. The transistor must function in the active region, which is ensured by the bias circuits. Controlling the corresponding currents and base, collector, and emitter voltages (V_B, V_C, V_E) is the primary objective of the bias circuit. We will be able to determine amplifier gain, input and output resistances, and other relevant parameters in this way. The bias circuit acts as the "puppet master" if we (not very seriously) swap out the transistor for the marionette depicted in Figure 3.23 and give its arms and legs specific voltage positions.

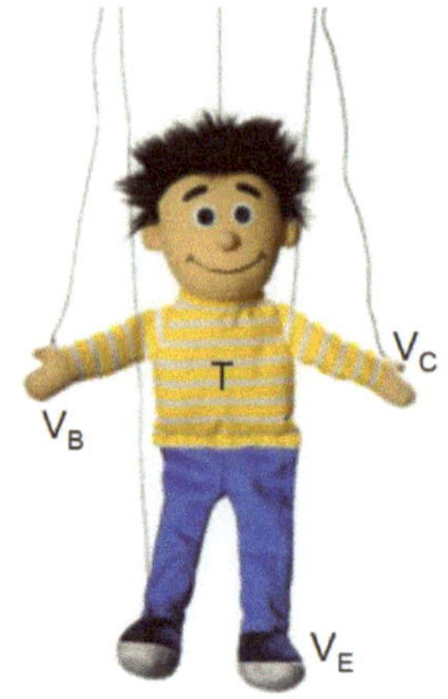

Figure 3.23. *A "puppet master" analogy of the transistor bias circuit.*

Source: Sergey N. Makarov, Reinhold Ludwig, and Stephen J. Bitar. Practical Electrical Engineering. Springer. DOI 10.1007/978-3-319-21173-2.

Our goal now is to design a simple active-region bias circuit that is typical for transistor amplifiers. The following conditions should be met:

1. Provide a constant DC collector current I_C that remains nearly the same at large β variations (and temperature variations).
2. Make sure that the transistor has enough output voltage swing (a sufficiently wide dynamic range) in the active region.

ROLE MODEL

JOHN BARDEEN

John Bardeen (May 23, 1908 – January 30, 1991) was an American physicist and electrical engineer. He is the only person to be awarded the Nobel Prize in Physics twice: first in 1956 with William Shockley and Walter Brattain for the invention of the transistor; and again in 1972 with Leon N. Cooper and John Robert Schrieffer for a fundamental theory of conventional superconductivity known as the BCS theory (Hoddeson, Lillian and Vicki Daitch (2002)).

The transistor revolutionized the electronics industry, making possible the development of almost every modern electronic device, from telephones to computers, and ushering in the Information Age. Bardeen's developments in superconductivity—for which he was awarded his second Nobel Prize—are used in nuclear magnetic resonance spectroscopy (NMR), medical magnetic resonance imaging (MRI), and superconducting quantum circuits.

Born and raised in Wisconsin, Bardeen received a Ph.D. in physics from Princeton University. After serving in World War II, he was a researcher at Bell Labs and a professor at the University of Illinois. In 1990, Bardeen appeared on *Life* magazine's list of "100 Most Influential Americans of the Century."

Bardeen is the first of only three people to have won multiple Nobel Prizes in the same category (the others being Frederick Sanger and Karl Barry Sharpless in chemistry), and one of five persons with two Nobel Prizes.

Education and Early Life

Bardeen was born in Madison, Wisconsin, on May 23, 1908. He was the son of Charles Bardeen, the first dean of the University of Wisconsin Medical School.

Bardeen attended University of Wisconsin High School in Madison. He graduated from the school in 1923 at age 15. He could have graduated several years earlier, but this was postponed because he took courses at another high school and because of his mother's death. Bardeen entered the University of Wisconsin in 1923. While in college, he joined the Zeta Psi fraternity. He raised a part of the needed membership fees by playing billiards. Bardeen was initiated as a member of Tau Beta Pi engineering honor society. Not wanting to be an academic like his father, Bardeen chose engineering. He also felt that engineering had good job prospects.

Bardeen received his Bachelor of Science degree in electrical engineering in 1928 from the University of Wisconsin. Despite taking a year off to work in Chicago, he graduated in 1928. Taking all the graduate courses in physics and mathematics that had interested him, Bardeen graduated in five years instead of the usual four. This allowed him time to complete his master's thesis, supervised by Leo J. Peters. He received his Master of Science degree in electrical engineering in 1929 from Wisconsin.

Bardeen furthered his studies by staying on at Wisconsin, but he eventually went to work for Gulf Research Laboratories, the research arm of the Gulf Oil Corporation that was based in Pittsburgh. From 1930 to 1933, Bardeen worked there on the development of methods for the interpretation of magnetic and gravitational surveys. He worked as a geophysicist. After the work failed to keep his interest, he applied and was accepted to the graduate program in mathematics at Princeton University.

As a graduate student, Bardeen studied mathematics and physics. Under physicist Eugene Wigner, he wrote his thesis on a problem in solid-state physics. Before completing his thesis, he was offered a position as junior fellow of the Society of Fellows at Harvard University in 1935. He spent the next three years there, from 1935 to 1938, working with to-be Nobel laureates in physics John Hasbrouck van Vleck and Percy Williams Bridgman on problems in cohesion and electrical conduction in metals,and also did some work on level density of nuclei. He received his Ph.D. in mathematical physics from Princeton in 1936.

- **Career and Research**

a. *World War II Service*

From 1941 to 1944, Bardeen headed the group working on magnetic mines and torpedoes and mine and torpedo countermeasures at the Naval Ordnance Laboratory. During this period, his wife Jane gave birth to a son (Bill, born in 1941) and a daughter (Betsy, born in 1944).

b. *Bell Labs*

In October 1945, Bardeen began work at Bell Labs as a member of a solid-state physics group led by William Shockley and chemist Stanley Morgan. Other personnel working in the group were Walter Brattain, physicist Gerald Pearson, chemist Robert Gibney, electronics expert Hilbert Moore and several technicians. He moved his family to Summit, New Jersey (Daitch, Vicki; Hoddeson, Lillian (2002).).

The assignment of the group was to seek a solid-state alternative to fragile glass vacuum tube amplifiers. Their first attempts were based on Shockley's ideas about using an external electrical field on a semiconductor to affect its conductivity. These experiments mysteriously failed every time in all sorts of configurations and materials. The group was at a standstill until Bardeen suggested a theory that invoked surface states that prevented the field from penetrating the semiconductor. The group changed its focus to study these surface states, meeting almost daily to discuss the work. The rapport of the group was excellent, and ideas were freely exchanged. By the winter of 1946, they had enough results that Bardeen submitted a paper on the surface states to *Physical Review*. Brattain started experiments to study the surface states through observations made while shining a bright light on the semiconductor's surface. This led to several more papers (one of them co-authored with Shockley), which estimated the density of the surface states to be more than enough to account for their failed experiments. The pace of the work picked up significantly when they started to surround point contacts between the semiconductor and the conducting wires with electrolytes. Moore built a circuit that allowed them to vary the frequency of the input signal easily and suggested that they use *glycol borate* (gu), a viscous chemical that did not evaporate. Finally, they began to get some evidence of power amplification when Pearson, acting on a suggestion by Shockley, put a voltage on a droplet of gallium placed across a p–n junction.

c. Invention of the Transistor

On December 23, 1947, Bardeen and Brattain were working without Shockley when they succeeded in creating a point-contact transistor that achieved amplification. By the next month, Bell Labs' patent attorneys started to work on the patent applications.

Bell Labs' attorneys soon discovered that Shockley's field effect principle had been anticipated and patented in 1930 by Julius Lilienfeld, who filed his MESFET-like patent in Canada on October 22, 1925.

Shockley publicly took the lion's share of the credit for the invention of the transistor; this led to a deterioration of Bardeen's relationship with him. Bell Labs management, however, consistently presented all three inventors as a team. Shockley eventually infuriated and alienated Bardeen and Brattain, essentially blocking the two from working on the junction transistor. Bardeen began pursuing a theory for superconductivity and left Bell Labs in 1951. Brattain refused to work with Shockley further and was assigned to another group. Neither Bardeen nor Brattain had much to do with the development of the transistor beyond the first year after its invention.

The "transistor" (a portmanteau of "transconductance" and "resistor") was 1/50 the size of the vacuum tubes it replaced in televisions and radios, used far less power, was far more reliable, and it allowed electrical devices to become more compact.

d. University of Illinois at Urbana–Champaign

By 1951, Bardeen was looking for a new job. Fred Seitz, a friend of Bardeen, convinced the University of Illinois at Urbana–Champaign to make Bardeen an offer of $10,000 a year. Bardeen accepted the offer and left Bell Labs, joining the engineering and physics faculties at Illinois in 1951, where he was professor of electrical engineering and of physics.

At Illinois, he established two major research programs, one in the electrical engineering department and one in the physics department. The research program in the electrical engineering department dealt with both experimental and theoretical aspects of semiconductors, and the research program in the physics department dealt with theoretical aspects of macroscopic quantum systems, particularly superconductivity and quantum liquids.

He was an active professor at Illinois from 1951 to 1975 and then became *professor emeritus*. In his later life, Bardeen remained active in academic research, during which time he focused on understanding the flow of electrons in charge density waves (CDWs) through metallic linear chain compounds. His proposals that CDW electron transport is a collective quantum phenomenon were initially greeted with skepticism. However, experiments reported in 2012 show oscillations in CDW current versus magnetic flux through tantalum trisulfide rings, like the behavior of superconducting quantum interference devices, lending credence to the idea that collective CDW electron transport is fundamentally quantum in nature. Bardeen continued his research throughout the 1980s, and published articles in *Physical Review Letters* and *Physics Today* less than a year before he died.

A collection of Bardeen's personal papers are held by the University of Illinois Archives.

e. Nobel Prize in Physics in 1956

In 1956, John Bardeen shared the Nobel Prize in Physics with William Shockley of Semiconductor Laboratory of Beckman Instruments and Walter Brattain of Bell Telephone Laboratories "*for their research on semiconductors and their discovery of the transistor effect*".

At the Nobel Prize ceremony in Stockholm, Brattain and Shockley received their awards that night from King Gustaf VI Adolf. Bardeen brought only one of his three children to the Nobel Prize ceremony. King Gustav chided Bardeen because of this, and Bardeen assured the King that the next time he would bring all his children to the ceremony. He kept his promise.

f. BCS Theory

In 1957, Bardeen, in collaboration with Leon Cooper and his doctoral student John Robert Schrieffer, proposed the standard theory of superconductivity known as the BCS theory (named for their initials).

g. Nobel Prize in Physics in 1972

In 1972, Bardeen shared the Nobel Prize in Physics with Leon N Cooper of Brown University and John Robert Schrieffer of the University of Pennsylvania "for their jointly developed theory of superconductivity, usually called the BCS-theory". This was Bardeen's second Nobel Prize in Physics. He became the first person to win two Nobel Prizes in the same field.

Bardeen brought his three children to the Nobel Prize ceremony in Stockholm. Bardeen gave much of his Nobel Prize money to fund the Fritz London Memorial Lectures at Duke University.

In the late 1960s, Bardeen felt that Cooper and Schrieffer deserved the Nobel prize for BCS. He was concerned that they might not be awarded because of the Nobel Committee's reticence to award the same person twice, which would be his case as a co-author of the theory. Bardeen nominated scientists who worked on superconducting tunneling effects such as the Josephson effect for the prize in 1967: Leo Esaki, Ivar Giaever and Brian Josephson. He recognized that because the tunneling developments depended on superconductivity, it would increase the chances that BCS itself would be awarded first. He also reasoned that the Nobel Committee had a predilection for multinational teams, which was the case for his tunneling nominees, each being from a different country. Bardeen renewed the nominations in 1971, 1972, when BCS received the prize, and finally 1973, when tunneling was awarded.

He is the only double laureate in physics, and one of three double laureates of the same prize; the others are Frederick Sanger who won the 1958 and 1980 Prizes in Chemistry and Karl Barry Sharpless who won the 2001 and 2022 Prizes in chemistry.

h. Other Awards

In addition to being awarded the Nobel prize twice, Bardeen has numerous other awards including:

- 1952 Franklin Institute's Stuart Ballantine Medal.
- 1954 elected a member of the United States National Academy of Sciences
- 1958 elected to the American Philosophical Society
- 1959 elected a Fellow of the American Academy of Arts and Sciences
- 1965 National Medal of Science.
- 1971 IEEE Medal of Honor for "his profound contributions to the understanding of the conductivity of solids, to the invention of the transistor, and to the microscopic theory of superconductivity."
- Elected a Foreign Member of the Royal Society (ForMemRS) in 1973.
- 1975 Franklin Medal.
- On January 10, 1977, John Bardeen was presented with the Presidential Medal of Freedom by President Gerald Ford. He was represented at the ceremony by his son, William Bardeen.
- Bardeen was one of 11 recipients given the Third Century Award from President George H. W. Bush in 1990 for "exceptional contributions to American society" and was granted a gold medal from the Soviet Academy of Sciences in 1988.
- 1987 Golden Plate Award of the American Academy of Achievement

Personal Life

Bardeen married Jane Maxwell on July 18, 1938. While at Princeton, he met Jane during a visit to his old friends in Pittsburgh.

Bardeen was a scientist with a very unassuming personality. While he served as a professor for almost 40 years at the University of Illinois, he was best remembered by neighbors for hosting cookouts where he would prepare food for his friends, many of whom were unaware of his accomplishments at the university. He would always ask his guests if they liked the hamburger bun toasted (since he liked his that way). He enjoyed playing golf and going on picnics with his family. Lillian Hoddeson said that because he "differed radically from the popular stereotype of 'genius' and was uninterested in appearing other than ordinary, the public and the media often overlooked him."

When Bardeen was asked about his beliefs during a 1988 interview, he responded: "I am not a religious person, and so do not think about it very much". However, he has also said: "I feel that science cannot provide an answer to the ultimate questions about the meaning and purpose of life." Bardeen did believe in a code of moral values and behavior. John Bardeen's children were taken to church by his wife, who taught Sunday school and was a church elder. Despite this, he and his wife made it clear that they did not have faith in an afterlife and other religious ideas. He was the father of James M. Bardeen, William A. Bardeen, and daughter Elizabeth.

Death

Bardeen died of heart disease at age 82 at Brigham and Women's Hospital in Boston, Massachusetts, on January 30, 1991. Although he lived in Champaign-Urbana, he had come to Boston for medical consultation. Bardeen and his wife Jane (1907–1997) are buried in Forest Hill Cemetery, Madison, Wisconsin. They were survived by three children, James, William and Elizabeth Bardeen Greytak, and six grandchildren.

Legacy

In honor of Bardeen, the engineering quadrangle at the University of Illinois at Urbana-Champaign is named the Bardeen Quad.

Also in honor of Bardeen, Sony Corporation endowed a $3 million John Bardeen professorial chair at the University of Illinois at Urbana-Champaign, beginning in 1990. Sony Corporation owed much of its success to commercializing Bardeen's transistors in portable TVs and radios and had worked with Illinois researchers. As of 2022, the John Bardeen Professor is Yurii Vlasov.

At the time of Bardeen's death, then-University of Illinois chancellor Morton Weir said, "It is a rare person whose work changes the life of every American; John's did."

Bardeen was honored on a March 6, 2008, United States postage stamp as part of the "American Scientists" series designed by artist Victor Stabin. The $0.41 stamp was unveiled in a ceremony at the University of Illinois. His citation reads: "Theoretical physicist John Bardeen (1908–1991) shared the Nobel Prize in Physics twice—in 1956, as co-inventor of the transistor and in 1972, for the explanation of superconductivity. The transistor paved the way for all modern electronics, from computers to microchips. Diverse applications of superconductivity include infrared sensors and medical imaging systems." The other scientists on the "American Scientists" sheet include biochemist Gerty Cori, chemist Linus Pauling and astronomer Edwin Hubble.

CLASS ACTIVITY

Students draw the energy variation curve in the conduction band for an open circuited n–p–n transistor. How is the curve modified when the transistor is operating in the active region?

SUMMARY

- A bipolar junction transistor (BJT) is a type of transistor that uses both electrons and electron holes as charge carriers.
- BJTs use two p–n junctions between two semiconductor types, n-type and p-type, which are regions in a single crystal of material. The junctions can be made in several different ways, such as changing the doping of the semiconductor material as it is grown, by depositing metal pellets to form alloy junctions, or by such methods as diffusion of n-type and p-type doping substances into the crystal.
- The Bipolar junction transistor is a solid-state device and, in these transistors, the current flow in two terminals are emitter and collector, and the flow of current controlled by the third terminal is the base terminal. It is different from the other type of transistor, i.e., Field-effect transistor which is the output current is controlled by the input voltage.
- A BJT is a 3-terminal semiconductor device, as the name suggests, the term bipolar is taken from the truth that this kind of transistor includes two kinds of semiconductor materials like P-type (positive type) and n-type (negative type) where the current flows from these regions Usually, these transistors include Silicon.
- The PNP transistor is a three-terminal device that is made from semiconductor material. The three terminals are collector, base, and emitter, and the transistor is used for switching and amplifying applications.
- The NPN transistor contains three terminals which are the same as the PNP transistor which are emitter, collector, and base.
- The emitter terminal emits the charge carriers like holes or electrons to the other two terminals. The base terminal is always reverse biased with respect to other terminals like emitter so that it can produce huge majority carriers. In BJT, the emitter is the most heavily doped region.

- The part on the reverse side of the Emitter terminal will collect the emitted charge carriers called a collector. So, this terminal is heavily doped however the doping level of this terminal is among the base terminal which is a lightly doped & heavily doped emitter terminal. CB-junction must be reversely biased in both the transistors.
- The base terminal is the center part among collector & emitter terminals which forms two PN junctions among them. The base terminal is the most lightly doped segment in the transistor. So, being the central part of the BJT will allow it to manage the charge carriers flow among emitter & collector terminals. The BE junction shows high resistance as this junction can be connected in reversed bias.
- In the DC steady state, the circuit parameters are fixed. The steep exponential dependencies of the BJT transistor may be replaced by turn-on voltages.

REVIEW QUESTIONS

1. What is a Bipolar Junction Transistor (BJT)?
2. Who invented bipolar junction transistor?
3. Name the two main types of BJTs.
4. Which are three operating region of a bipolar junction transistor?
5. List the advantages of BJT.

REFERENCES

1. Antognetti, P., & Massobrio, G. (1993). *Semiconductor Device Modeling with Spice*. McGraw–Hill Professional. ISBN 978-0-07-134955-0.
2. Ashcroft, N., & Mermin, N. D. (1976). *Solid State Physics* (1st ed.). Holt, Rinehart, and Winston. pp. 299–302. ISBN 978-0030839931.
3. Daitch, V., & Hoddeson, L. (2002). *True Genius: The Life and Science of John Bardeen*. Joseph Henry Press. ISBN 978-0309084086.
4. Gildenblat, G. (2010). *Compact Modeling: Principles, Techniques and Applications*. Springer Science & Business Media. Part II: Compact Models of Bipolar Junction Transistors, pp. 167–267 cover Mextram and HiCuM in-depth. ISBN 978-90-481-8614-3.
5. Hoddeson, L., & Daitch, V. (2002). *True Genius: The Life and Science of John Bardeen*. National Academy Press. ISBN 0-309-08408-3.
6. Horowitz, P., & Hill, W. (1989). *The Art of Electronics* (2nd ed.). Cambridge University Press. ISBN 978-0-521-37095-0.
7. Liou, J. J., & Yuan, J. S. (1998). *Semiconductor Device Physics and Simulation*. Springer. ISBN 978-0-306-45724-1.

8. Morgan, D. V., & Williams, R. H. (1991). *Physics and Technology of Heterojunction Devices*. London: Institution of Electrical Engineers (Peter Peregrinus Ltd.). ISBN 978-0-86341-204-2.

9. Morris, P. R. (1990). A History of the World Semiconductor Industry. In D. V. Morgan & R. H. Williams (Eds.), *IEE History of Technology Series 12*. London: Peter Peregrinus Ltd. p. 29. ISBN 978-0-86341-227-1.

10. Schröter, M. (2010). *Compact Hierarchical Bipolar Transistor Modeling with HiCUM*. World Scientific. ISBN 978-981-4273-21-3.

11. Sedra, A. S., & Smith, K. C. (2004). *Microelectronic Circuits* (5th ed.). New York: Oxford. ISBN 978-0-19-514251-8.

CHAPTER 4

Field Effect Transistors

LEARNING OBJECTIVES

After studying this chapter, you will be able to:

- Understand the field effect transistors (FETS) types and working principles;
- Discuss on basics, working principle and applications of MOSFET;
- Explain the MOSFET dc circuit analysis;
- Describe the metal oxide semiconductor field effect transistor (MOSFET).

KEY TERMS FROM THIS CHAPTER

Bipolar junction transistors (BJT)
Channel length modulation
Field-effect transistor (FET)
Gate capacitance
Insulated-gate FET (IGFET)
Junction FET (JFET)
Organic field-effect transistors (OFET)
Subthreshold region
Body diode
Depletion mode
Flat-band voltage
Gate leakage
Integrated circuit (IC)
N-Channel FET
P-channel FET

INTRODUCTION

A Field-Effect Transistor (FET) is a commonly used type of transistor for amplifying weak signals, such as wireless ones. The FET can amplify both analog and digital signals, switch DC, or function as an oscillator. In the FET, current moves through a semiconductor path known as the channel, with electrodes called the source at one end and the drain at the other end. By applying voltage to a control electrode called the gate, the electrical diameter of the channel can be adjusted. The conductivity of the FET depends on the electrical diameter of the channel at any given time. Even a slight change in gate voltage can result in a significant change in current from the source to the drain, allowing for signal amplification. Field-Effect Transistors come in two main classifications: the junction FET (JFET) and the metal-oxide-semiconductor FET (MOSFET).

The junction field-effect transistor (JFET) features a channel composed of either N-type or P-type semiconductor material. The gate is composed of the opposite semiconductor type. Electric charges are primarily carried in P-type material as electron deficiencies known as holes. Most of the charge carriers in N-type material are electrons. The junction in a JFET is the line dividing the channel from the gate. This P-N junction is normally reverse-biased, meaning that a DC voltage is applied to it, preventing any current from passing between the gate and the channel. However, in certain situations, a tiny current flows through the junction for a portion of the input signal cycle.

The channel of a MOSFET may consist of a P-type or N-type semiconductor. An oxidized metal piece serves as the gate electrode. The oxide layer separates the gate from the channel electrically. Because of this, the MOSFET was formerly referred to as the insulated-gate FET (IGFET), though this nomenclature is no longer frequently used. There is practically no current flowing between the gate and the channel at any point during the signal cycle because the oxide layer functions as a dielectric. The MOSFET has a very high input impedance as a result. The MOSFET is vulnerable to electrostatic charge destruction due to the incredibly thin oxide layer. MOS devices must be handled and transported with extra caution. Compared to the bipolar transistor, the FET has both benefits and drawbacks. When working with weak signals, such as in broadcast receivers and wireless communications, field-effect transistors are the preferred choice. They are also favored in high- impedance circuits and systems. In general, the FET is not utilized for high-power amplification, which is necessary in broadcast transmitters and large wireless communications systems. The fabrication of field-effect transistors is done on silicon integrated circuits (IC) chips. Many thousands of FETs can be found in a single integrated circuit (IC), along with additional parts like resistors, capacitors, and diodes.

4.1. FIELD EFFECT TRANSISTORS (FETS) TYPES AND WORKING PRINCIPLES

An electric field can control the output current of a transistor known as a field effect transistor (FET). Because a FET operates using a single carrier type, it is sometimes referred to as a uni-polar transistor. The fundamental types of BJT and FET transistors are entirely different. FETs are semiconductor devices with three terminals: source, drain, and gate. Electrons or holes, which carry charge, move through an active channel from the source to the drain. A voltage across the source and gate terminals regulates the flow of electrons from the source to the drain. The electronics industry now uses FET very frequently. It came into the mainstream of electronics later than bipolar transistors, but despite this, it is still more widely used than any other active device due to its wide range of applications.

Discrete small signal devices and power electronics both use a wide variety of FET types. Nonetheless, the field effect transistor has become increasingly important due to its application in integrated circuits. Because of its ability to use significantly less current, it has made it possible to achieve far higher degrees of integration than would have been feasible with any other method. The foundation of the FET is the idea that charges in a semiconductor channel can be drawn to one another by charges on an adjacent object. In this sense, the FET's gate regulates the movement of carriers—electrons or holes—from the source to the drain. By adjusting the conductive channel's dimensions and form, it achieves this. There are two types of semiconductor channels where current flow can occur: P-type and N-type. The result is the P-Channel and N-Channel FETs, two types or categories of FET.

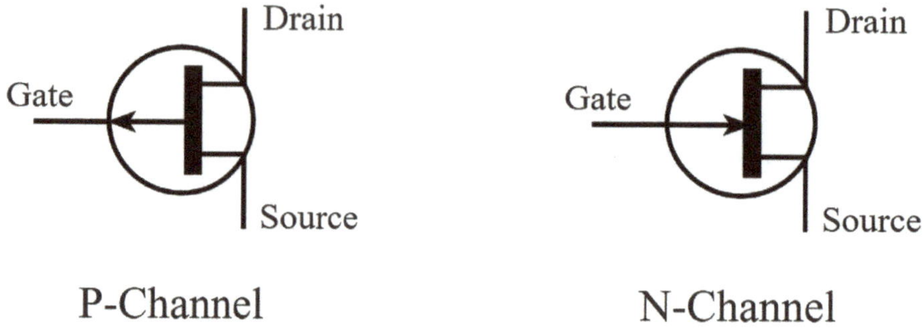

Figure 4.1. *Junction FET circuit symbol.*

Source: https://www.electronics-notes.com/images/circuit-symbols-jfet-n-and-p-channel.svg.

A gate, or third electrode, receives the electric field used to regulate the current. Since the device's high input impedance—typically many megaohms—and solely the electric field regulate the current flowing in the channel, it is referred to as voltage operated. Compared to the current-operated, significantly lower input impedance bipolar transistor, this can be a clear advantage. An FET is referred to as an enhancement mode FET when its external field enhances the carriers in the channel, or depletes the channel of carriers, as in the case of a depletion mode FET.

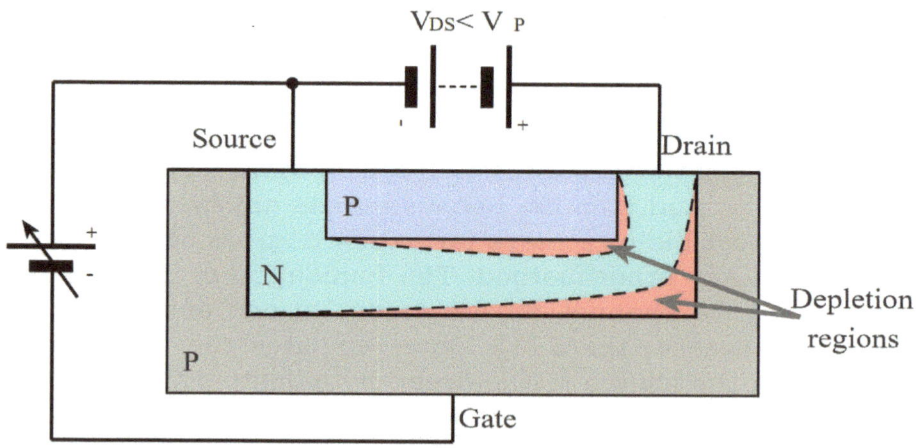

Figure 4.2. *Junction FET working below saturation.*

Source: https://www.electronics-notes.com/images/fet-junction-jfet-operation-below-saturation-01.svg.

4.1.1. Field Effect Transistor (FET) History

The FET concept has long been understood. Some of its earliest roots can be found in a proposal by Lilien Field from 1926 and in another Heil paper from 1935. Subsequently, Bell

Laboratories established a semiconductor research team in the 1940s. A gadget that would modify the current flowing in a semiconductor channel by applying an electric field nearby was one of the many topics they investigated in relation to semiconductors and semiconductor technology. The researchers' initial attempts at implementing the idea proved unsuccessful, so they moved on to another concept and eventually created the bipolar transistor, a different type of semiconductor electronics component. Following this, a significant portion of semiconductor research was devoted to enhancing the bipolar transistor, and the concept of a field effect transistor was not thoroughly explored for a while. These days, FETs are incredibly popular and are often used as the primary active component in integrated circuits. The state of electronics technology today would be drastically different without them (Grundmann, Marius (2010)).

An electronic component that is frequently utilized in integrated circuits is the field effect transistor (FET). They are a special kind of transistor that changes its output voltage in response to the input. Bipolar junction transistors (BJT) are made to have on and off states based on the flow of current, in contrast to this. Because it provides greater speed with less energy consumption than BJTs, the Metal-Oxide-Semiconductor Field-Effect Transistor (MOSFET), the most widely used type of FET, is widely used in computer memory design. For the circuits they are intended for, transistors perform a wide range of features and tasks. The organic layer substrate, typically a polymer, serves as the foundation for organic field-effect transistors (OFET). These transistors, which are biodegradable and flexible, are used to create solar cell sheets and plastic-based video displays. The junction field-effect transistor (JFET) is a different kind of FET variation that functions in a circuit like a diode by conducting current only when the voltage is reversed.

An experimental type of field-effect transistor known as a carbon nanotube field-effect transistor (CNTFET) is constructed using individual carbon nanotubes rather than the more common silicon substrate. They can therefore be produced in a size that is roughly 20 times smaller than the smallest transistors possible using traditional thin-film technology. They are promising to provide higher memory capacities and significantly faster computer processing speeds at a reduced price. Although they have been successfully demonstrated since 1998, they have remained experimental due to issues like the nanotubes' deterioration in the presence of oxygen and their long-term dependability under stresses from temperature or electrical fields.

4.1.2. Field Effect Transistor (FET) Types

The various kinds of FETs that are available can be defined in a variety of ways. Though there are many ways to classify them, the tree diagram below covers some of the most common types of FET.

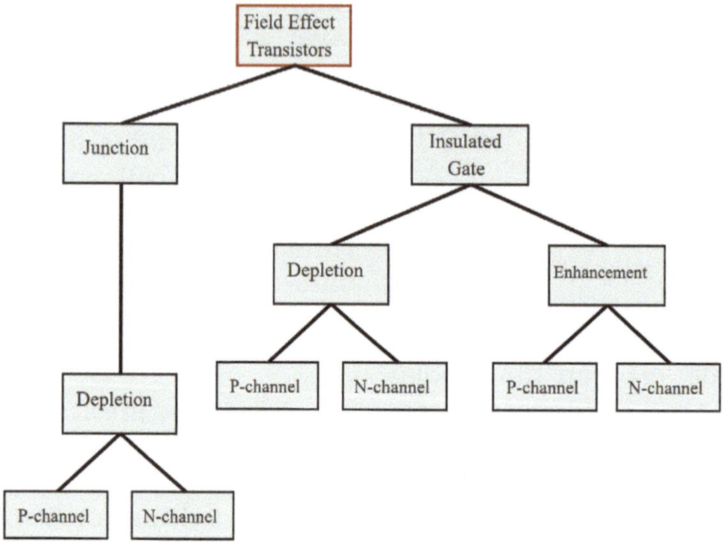

Figure 4.3. *Types of FET.*

Source: https://www.electronics-notes.com/images/fet-types-tree-01.svg.

There are many different types of FET on the market for which there are various names. Some of the major categories are delayed below:

- **Junction FET, JFET**: As the name suggests, the isolation from the channel is provided by this type of FET is using a reverse biased diode junction. It is the earliest type of FET and the most fundamental one. Nonetheless, it continues to offer top-notch support in a variety of electronic areas.
- **Insulated Gate FET/Metal Oxide Semiconductor FET MOSFET**: An insulated later is used by this kind of FET between the gate and the channel. This is usually formed by the semiconductor's oxide layer the most prevalent kind of IGFET. The Metal Oxide Semiconductor FET, or MOSFET. The gate in this instance is composed of a metal layer that is placed on top of silicon oxide, which is situated on the silicon channel.
- **Dual Gate MOSFET:** This is a Speciale form of MOSFET that has two gates in series along the channel. This enables some considerable performance improvements to be made, especially at RF, when compared to single gate devices. Read more about the Dual Gate MOSFET
- **MESFET**: The Metal Semiconductor FET is normally fabricated using gallium Arsenide and is often referred to as a GaAs FET. It provides very high performance, but in view of the gate structure, it is very sensitive to ESD. Read more about the MESFET/GaAsFET
- **HEMT/PHEMT**: The High Electron Mobility Transistor and Pseudomorphic High Electron Mobility Transistor are developments of the basic FET concept but developed to enable very high frequency operation. Read more about the HEMT/PHEMT

- **FinFET**: FinFET technology is now being used within integrated circuits to enable higher levels of integration to be achieved by allowing smaller feature sizes. Read more about the FinFET
- **VMOS**: VMOS standard for vertical MOS. It is a type of FET that uses a vertical current flow to improve the switching and current carrying performance. VMOS FETs are widely used for power applications. Read more about the VMOS

1. *Junction Field Effect Transistor:* Field effect transistors that can be used as electrically controlled switches include the Junction FET transistor type. Between sources and drain terminals, electrical energy travels via an active channel. The channel is strained, and the electric current is totally cut off by applying a reverse bias voltage to the gate terminal.

The junction FET transistor is available in two polarities which are:

a. **N- Channel JFET:** An n-type bar with two doped p-type layers on either side makes up an N-channel JFET. The device's N- channel is made up of the electron channel. The N-channel device has two ohmic contacts on both ends, which are joined to form the gate terminal. The other two sides of the bar are where the source and drain terminals are located. The terms "Vsd" and "Vgs" refer to the potential differences between the source and drain terminals and the gate and source terminal, respectively. Electrons move from the source to the drain, which causes the charge to flow. Electrons flow from the source (S) to the drain (D) terminal whenever a positive voltage is applied across the drain and source terminals, whereas conventional drain current (Id) flows through the drain to the source. The device is in the on state when current is flowing through it. A depletion region is formed in the channel when a voltage with negative polarity is applied to the gate terminal. The channel resistance between the source and drain rises as a result of the narrowing of the channel. The device is in the off condition because the gate-source junction is reverse biased and there is no current flowing through it. Essentially, less current will flow from the source to the drain if the gate terminal voltage is raised. Compared to the P-channel JFET, the N-channel JFET is more conductive. Thus, when compared to P-channel JFETs, N-channel JFETs are more efficient conductors.

b. **P-Channel JFET:** The P-channel JFET is made up of two doped n-type layers on either side of a P-type bar. Connecting the ohmic contacts on both sides creates the gate terminal. As in an N-channel JFET, the drain and source terminals are removed from the remaining two bar sides. Between the drain terminal and the source, a P-type channel is created with holes acting as charge carriers. The device operates in the ohmic region and the flow of current from source to drain terminal is ensured by applying a negative voltage to the drain and source terminals. The channel resistance rises as a result of the channel width being reduced when a positive voltage is applied to the gate terminal. Less current passes through the device when the gate voltage is higher.

2. *MOSFET Transistor:* A metal oxide layer is deposited on the surface of a MOSFET transistor, which is a p-type (n-type) semiconductor bar with holes removed to form source and drain terminals. The transistor has two heavily doped n-type regions diffused into it, as suggested by its name. The oxide layer is covered with a layer of metal to create the gate terminal. Switching a MOSFET is one of the fundamental uses of a field effect transistor. The source, drain, and gate terminals of this type of FET transistor are its three terminals. Current flow from source to drain is regulated by the voltage applied to the gate terminal. The device has a high input impedance because of the metal oxide insulating layer (Julius Edgar (1926-10-08)).

a. **Types of MOSFET Transistor Based on Operation Modes:** The type of field effect transistor that is most frequently used is the MOSFET. Two modes of operation are available for MOSFETs, and these modes determine the classification of MOSFET transistors. A channel gradually forms during MOSFET operation in enhancement mode, whereas a channel that has already diffused occurs during depletion mode MOSFET operation. CMOS is a sophisticated MOSFET application.

i. **Enhancement MOSFET Transistor:** The MOSFET's gate terminal experiences a negative voltage when it is applied, which causes the carriers or holes that carry positive charges to accumulate closer to the oxide layer. From the source to the drain terminal, a channel forms. Current flows from source to drain terminal as channel width widens and voltage becomes more negative. This device is known as an Enhancement type MOSFET because the current flow "enhances" when the gate voltage is applied.

ii. **Depletion Mode MOSFET Transistor:** A channel that is diffused between the drain and source terminals makes up a depletion mode MOSFET. Because of the channel, current flows from source to drain when there is no gate voltage. Positive charges build up in the channel when this gate voltage is made negative. This stops the current from flowing through the channel and creates a depletion region, or region of immobile charges. This device is called a depletion mode MOSFET because the formation of a depletion region affects the flow of current.

b. **Applications involving MOSFET as a Switch**

i. **Controlling speed of BLDC Motor:** A DC motor can be operated by a MOSFET acting as a switch. In this setup, the MOSFET is triggered by a transistor. A microcontroller uses PWM signals to control the on-off state of the transistor. The OPTO coupler functions when the microcontroller pin generates a logic low signal, producing a high logic signal at the output. When the PNP transistor is turned off, the MOSFET is triggered and turns on as a result. By shorting the source and drain terminals, current flows to the motor windings, causing the motor to begin rotating. PWM signals ensure the motor's speed control.

ii. **Driving an array of LEDs:** When a MOSFET is used as a switch, an array of LEDs' intensities is applied and controlled. Here, the MOSFET is driven by a transistor that is powered by signals from an external source, such as a microcontroller. The MOSFET receives the supply and turns on when the transistor is turned off, giving the LED array the correct biasing.

iii. **Switching Lamp using MOSFET:** A MOSFET can be used as a switch to regulate how lights are turned on and off. A transistor switch is also used in this instance to activate the MOSFET. The conduction of the transistor is controlled by PWM signals from an external source, such as a microcontroller. The MOSFET then turns on or off, controlling the lamp's switching.

4.2. BASICS, WORKING PRINCIPLE, AND APPLICATIONS OF MOSFET

The conductivity of a metal-oxide-semiconductor field-effect transistor, also known as a MOSFET, MOS-FET, or MOS FET, is determined by the voltage applied to it. It is applied to signal switching and amplification. Electronic signals can be switched or amplified by using the conductivity's to change with applied voltage. In both digital and analog circuits, MOSFETs are even more prevalent than BJTs (bipolar junction transistors).

A memory chip or microprocessor may contain hundreds of thousands or even millions of MOSFETs, making them the most prevalent type of transistor in digital circuits by far. Complementary pairs of MOS transistors can be used to create switching circuits with extremely low power consumption in the form of CMOS logic because they can be made with either p-type or n-type semiconductors.

MOSFETs have a nearly infinite input impedance, which makes them especially helpful in amplifiers because it lets the amplifier absorb almost all the incoming signal. The primary benefit is that, in comparison to bipolar transistors, it requires nearly no input current to control the load current. MOSFETs are available in two basic forms:

- **Depletion Type:** The transistor requires the Gate-Source voltage (V_{GS}) to switch the device "OFF". The depletion mode MOSFET is equivalent to a "Normally Closed" switch.
- **Enhancement Type:** The transistor requires a Gate-Source voltage (V_{GS}) to switch the device "ON". The enhancement mode MOSFET is equivalent to a "Normally Open" switch.

A semiconductor device is called a MOSFET (Metal Oxide Semiconductor Field Effect Transistor). The most common application for a MOSFET is in power electronics. Manufactured materials that behave neither like conductors nor like insulators make up semiconductors. A dry piece of wood or any other naturally occurring material that doesn't conduct electricity is an insulator. Any natural substance that passes or conducts electricity is called a conductor. The most prevalent types of conductors are metals. Devices like a MOSFET are made of semiconductor material, which possesses both conduction and insulation-like qualities. The ability to control the conduction or insulation properties is perhaps the most significant feature of semiconductor design.

The transistor is perhaps the best-known semiconductor device. Early transistors use a technology referred to as bi-polar material. Pure silicon can be doctored or "corrupted" --a process that is referred to as "doping". It is possible to make either p-type (positive) material or n-type (negative) material depending upon material used to "dope" or corrupt the pure silicon. If you combine p-type material and n-type material, you have a bipolar device. The transistor is a basic example of a bipolar device. The transistor has three terminals, the collector, the emitter, and the base. The current in the base terminal is used to control the flow of current between the emitter and the collector.

The fact that a MOSFET has a positive temperature coefficient over bipolar technology is another important improvement. This indicates that the device's propensity to conduct current reduces as its temperature rises. The system's capacity can be easily increased by the designer by utilizing this feature in parallel. The opposite is true with a bipolar device. Devices connected in parallel by MOSFET technology will automatically share current. When a device tries to conduct more than it should, it will heat up and lose its propensity to conduct current, thereby reduce the amount of current flowing through it until all the devices are once more sharing equally. Conversely, bipolar devices connected in parallel will heat up if one begins to conduct more current. This implies that more current will be switched to this device, raising both its temperature and current simultaneously. This runaway condition that causes the device to break quickly. Because of this, connecting bipolar devices in parallel is far more challenging, and MOSFET devices are currently the most widely used power semiconductor type of transistor.

4.2.1. Two-Terminal MOS Structure

The metal-oxide-semiconductor capacitor (MOSFET) at its core is depicted in Figure 4.4. The metal could be aluminum or another kind of metal entirely. A high-conductivity polycrystalline silicon layer deposited on the oxide frequently takes the place of the metal. Nonetheless, MOSFETs are still frequently referred to as "metals". The oxide thickness, t_{ox}, and the oxide permittivity, $\boxtimes_{ox}$, are represented in the Figure.

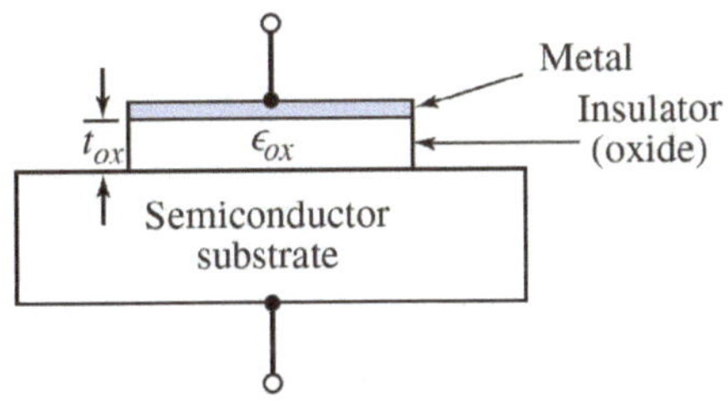

Figure 4.4. *The basic MOS capacitor structure.*

Source: http://www.mhhe.com/engcs/electrical/neamen01/etext/ch05.pdf.

A basic parallel-plate capacitor can be used to explain the physics of the MOS structure. A parallel-plate capacitor with the top plate at a negative voltage in relative to the bottom plate is depicted in Figure 4.5(a). The two plates are separated by an insulator substance. As can be seen, this bias results in an electric field being induced between the plates, a negative charge on the top plate, and a positive charge on the bottom plate. Figure 4.5(b) depicts a MOS capacitor with a p-type semiconductor substrate. Regarding the semiconductor substrate, the top metal terminal, also known as the gate, is negatively biased. We can observe that an electric field will be induced in one direction and that the top metal plate of the parallel-plate capacitor will have a negative charge. The p-type semiconductor's holes will feel a force directed toward the oxide-semiconductor interface if the electric field enters the semiconductor.

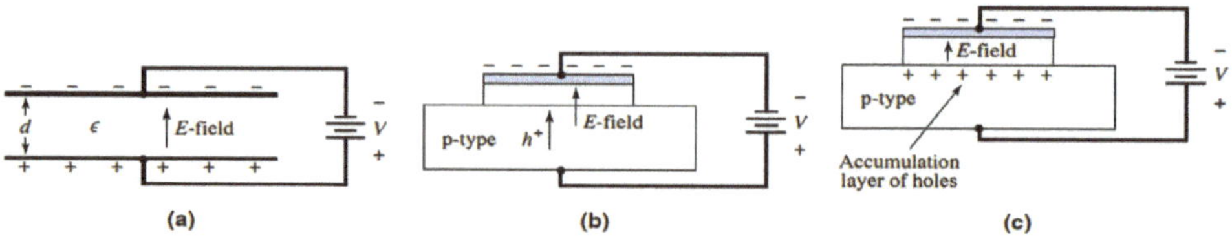

Figure 4.5. *(a) A parallel plate capacitor, showing the electric field and conductor charges, (b) a corresponding MOS capacitor with a negative gate bias, showing the electric field and charge flow, and (c) the MOS capacitor with an accumulation layer of holes.*

Source: http://www.mhhe.com/engcs/electrical/neamen01/etext/ch05.pdf.

Using this specific applied voltage, Figure 4.5(c) displays the charge equilibrium distribution in the MOS capacitor. As the MOS capacitor's bottom plate becomes positively charged, an accumulation layer of positively charged holes forms at the oxide-semiconductor junction. Reversing the polarity of the applied voltage, Figure 4.6(a) displays the identical MOS capacitor. The induced electric field is pointing in the opposite direction and that the top metal plate is now positively charged. P-type material holes in this scenario will feel a force away from the oxide-semiconductor interface if the electric field enters the semiconductor. A negative space-charge region is generated as the holes are pushed away from the interface because of the fixed acceptor impurity atoms. The MOS capacitor's bottom plate has a negative charge that is equal to that of the induced depletion region itself. With this applied voltage, Figure 4.6(b) depicts the MOS capacitor's equilibrium charge distribution.

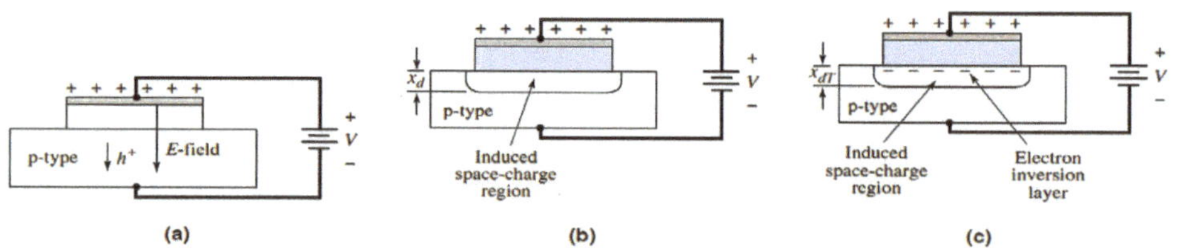

Figure 4.6. *The MOS capacitor with p-type substrate: (a) effect of positive gate bias, showing the electric field and charge flow, (b) the MOS capacitor with an induced space-charge region due to a moderate gate bias, and (c) the MOS capacitor with an induced space-charge region and electron inversion layer due to a larger gate bias.*

Source: http://www.mhhe.com/engcs/electrical/neamen01/etext/ch05.pdf.

The induced electric field's magnitude rises with increasing positive voltage applied to the gate. As seen in Figure 4.6(c), minority carrier electrons are drawn to the oxide-semiconductor interface. Forming an electron inversion layer in this region. The applied gate voltage determines how much charge is present in the inversion layer. When using an n-type semiconductor substrate in a MOS capacitor, the same fundamental charge distributions can be achieved. This MOS capacitor structure is depicted in Figure 4.7(a), where the top gate terminal is positively biased. The top gate becomes positively charged, inducing an electric field in the indicated direction. In this case, the n-type semiconductor experiences the induction of an electron accumulation layer. The situation where a negative voltage is applied to the gate terminal is depicted in Figure 4.7(b).

The induced electric field causes the n-type substrate to induce a positive space-charge region. As seen in Figure 4.7(c), a region of positive charge is produced at the oxide-semiconductor interface when a greater negative voltage is applied. A hole inversion layer is the term used to describe this area of minority carrier holes. The applied gate voltage determines how much positive charge is present in the inversion layer.

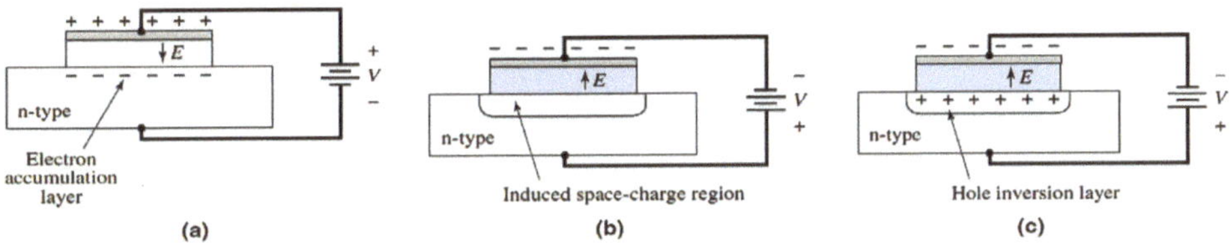

Figure 4.7. *The MOS capacitor with n-type substrate for: (a) a positive gate bias, (b) a moderate negative bias, and (c) a larger negative bias.*

Source: http://www.mhhe.com/engcs/electrical/neamen01/etext/ch05.pdf.

4.2.2. MOSFET Operation

The MOS capacitor is essential to a MOSFET's operation. The primary component of a MOSFET is the MOS capacitor. The semiconductor surface is situated between the source and drain terminals at the bottom of the oxide layer. By applying either positive or negative gate voltages, it can be inverted from p-type to n-type. The holes beneath the oxide layer have a repulsive force when a positive gate voltage is applied, and they are pushed downward along with the substrate. The bound negative charges connected to the acceptor atoms populate the depletion region. A channel for electrons to reach is created. Additionally, electrons from the n+ source and drain regions are drawn into the channel by the positive voltage. The current now flows freely between the source and drain and the gate voltage regulates the electrons in the channel if a voltage is applied between the drain and source. A hole channel will form beneath the oxide layer if we apply a negative voltage (Shichman, H. & Hodges, D. A. (1968)).

1. **P-Channel MOSFET***:* The substrate is in the n-type, and the drain and source are heavily doped in the p+ region. The positively charged hole flow, also referred to as p-channel MOSFET, is the cause of the current flow. Applying a negative gate voltage causes the electrons under the oxide layer to be pushed into the substrate by a repulsive force. The bound positive charges connected to the donor atoms populate the depletion region. Additionally, holes from the

p+ source and drain region are drawn into the channel region by the negative gate voltage.

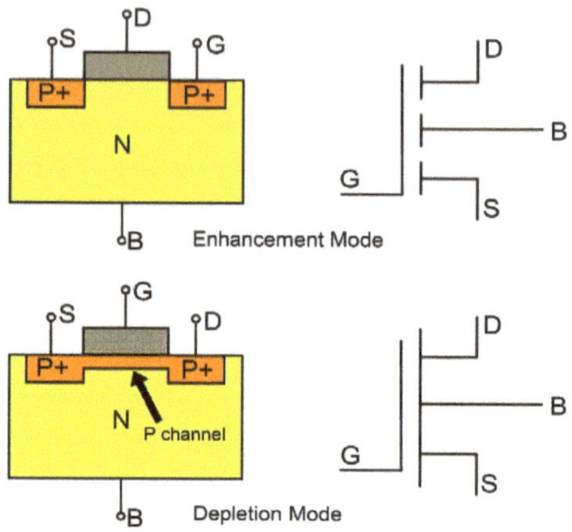

Figure 4.8. *P-channel MOSFET.*

Source: https://electronicsforu.com/wp-contents/uploads/2017/07/p-channel-mosfet.jpg.

2. **N-channel MOSFET:** The substrate is p-type, and the drain and source are heavily doped in the n+ region. Also referred to as an n-channel MOSFET, the current flows because of negatively charged electrons flowing. The holes under the oxide layer encounter a repulsive force when the positive gate voltage is applied, and they are forced downward into the bound negative charges connected to the acceptor atoms. An electron reach channel is created when the positive gate voltage draws electrons into the channel from the n+ source and drain regions.

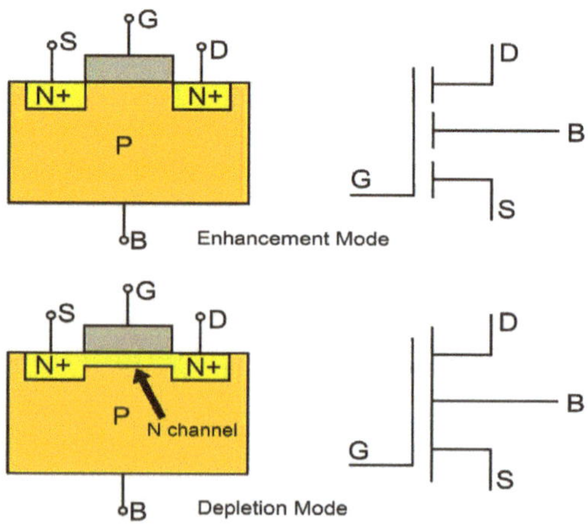

Figure 4.9. *N-channel MOSFET.*

Source: https://electronicsforu.com/wp-contents/uploads/2017/07/p-channel-mosfet.jpg.

4.2.3. n-Channel Enhancement-Mode MOSFET

We will now apply the concepts of an inversion layer charge in a MOS capacitor to create a transistor.

- **Transistor Structure:** A simplified cross-section of a MOS field-effect transistor is displayed in Figure 4.10(a). The gate, oxide, and p-type substrate regions of the MOS capacitor are identical. Furthermore, the source and drain terminals are two distinct n-regions. A MOSFET's current is produced by the charge flow in the channel region, or inversion layer, which is located next to the oxide-semiconductor interface. The figure defines the channel width (W) and length (L). MOSFETs are small devices because their typical integrated circuit has a channel length of less than 1 gm (10^{-6} m). Usually, the oxide thickness t_{ox} is 400 angstroms or smaller. The basic schematic of the transistor's structure is shown in Figure 4.10(a). A more thorough cross-section of a MOSFET assembled into an integrated circuit is displayed in Figure 4.10(b). Outside the region where the metal interconnect lines are formed, a thick oxide known as the field oxide is deposited. Heavily-doped polysilicon is typically used as the gate material. It is possible to develop the fundamental transistor characteristics using the simplified diagram, even though the actual structure of a MOSFET may be complex.

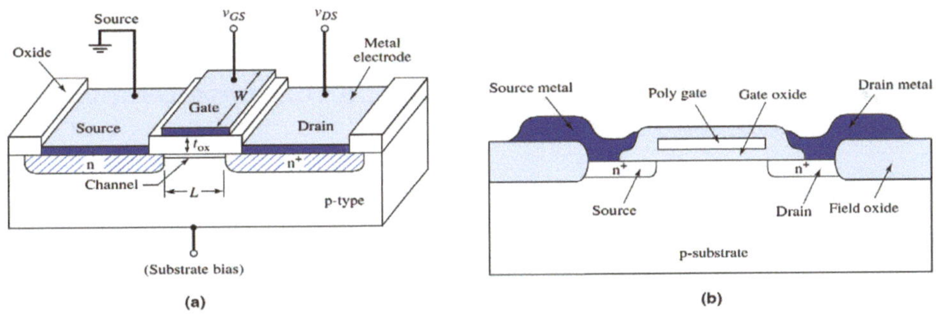

Figure 4.10. *(a) Schematic diagram of an n-channel enhancement mode MOSEFT and (b) an n-channel MOSFET, showing the field oxide and poly silicon gate.*

Source: http://www.mhhe.com/engcs/electrical/neamen01/etext/ch05.pdf.

- **Basic Transistor Operation:** The source and drain terminals are separated by the p-region when the gate is applied with zero bias, as seen in Figure 4.11(a). This is the same as two diodes placed back-to-back, as Figure 4.11(b) illustrates. In this instance, the current is practically zero. The n-source and n-drain are connected by an electron inversion layer that forms at the oxide semiconductor interface when a sufficiently high positive gate voltage is applied, as seen in Figure 4.11(c). The sources and drain terminals can then produce a current. This type of transistor is known as an enhancement-mode MOSFET because a voltage needs to be applied to the gate to generate the inversion charge. Also, this device is also known as an n-channel MOSFET because the carriers in the inversion layer are electrons.

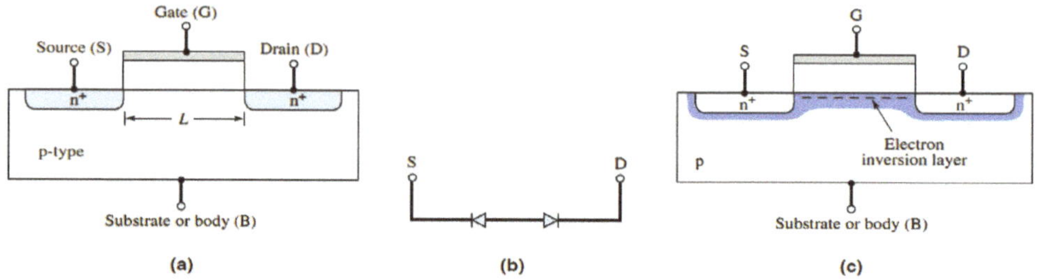

Figure 4.11. *(a) Cross section of the n-channel MOSFET prior to the formation of an electron inversion layer, (b) equivalent back-to-back diodes between sources and drain when the transistor is in cutoff, and (c) cross section after the formation of an electron inversion layer.*

Source: http://www.mhhe.com/engcs/electrical/neamen01/etext/ch05.pdf.

Carrier flow through the channel is supplied by the source terminal, and carrier outflow from the channel is permitted by the drain terminal. With an applied drain-to-source voltage, electrons move from the source to the drain of an n-channel MOSFET, also known as an NMOS transistor. This means that conventional current enters the drain and exits the source. The amount of charge in the inversion layer, which is a function of the applied gate voltage, determines the size of the current. There is no gate current because an oxide or insulator divides the gate terminal from the channel. Similarly, there is essentially no current through the substrate because a space-charge region separates the channel and substrate.

4.2.4. Additional MOSFET Structures and Circuit Symbols

In addition to the n-channel enhancement-mode device, there are several other MOSFET structures that must be taken into account before we begin analyzing MOSFET circuits.

- **N-Channel Depletion-Mode MOSFET:** The cross section of an n-channel depletion-mode MOSFET is displayed in Figure 4.12(a). Because of impurities introduced during device fabrication, for example, an n-channel region or inversion layer exists beneath the oxide when zero volts are applied to the gate. Even with zero gate voltage, a drain-to-source current may still be generated in the channel because an n-region connects the n-source and n-drain. The existence of a channel even at zero gate voltage is referred to as depletion mode. To turn off the n-channel depletion-mode MOSFET, a negative gate voltage needs to be applied. The n-channel depletion mode MOSFET with a negative applied gate-to-source voltage is depicted in Figure 4.12(b). By creating a space-charge region beneath the oxide, a negative gate voltage thins the n-channel region. The drain current is lowered as a result of the decreased channel conductance caused by the thinner layer.

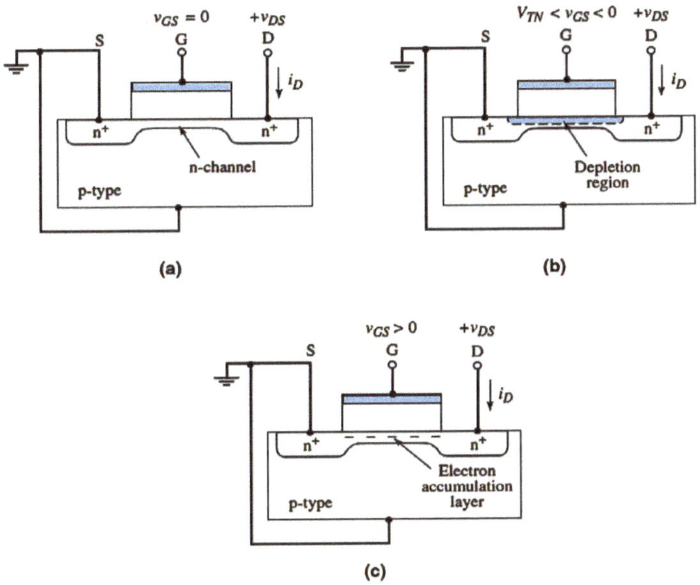

Figure 4.12. *Cross section of n-channel depletion mode MOSFET for (a) v_{GS}=0, (b) v_{GS}<0, (c) v_{GS}>0.*

Source: http://cdcpc.ce.ncu.edu.tw/classes/EEShortversion/Elect/Ch3%20The%20Field%20Effect%20 Transistor.pdf.

The induced space-charge region completely fills the n-channel region and the current drops to zero when the gate voltage reaches the threshold voltage, which in this case is negative. As seen in Figure 4.12(c), a positive gate voltage produces an electron accumulation layer that raises the drain current. Figure 13 displays the general i_D versus V_{DS} family of curves for the n-channel depletion-mode MOSFET.

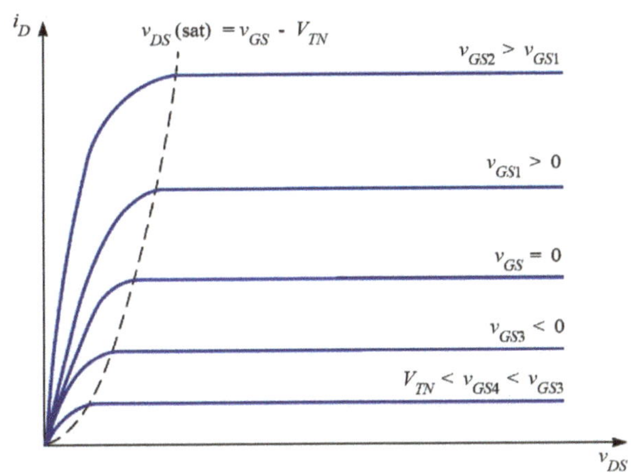

Figure 4.13. *Family of i_D versus v_{DS} curves for an n-channel depletion mode MOSFET.*

Source: http://www.mhhe.com/engcs/electrical/neamen01/etext/ch05.pdf.

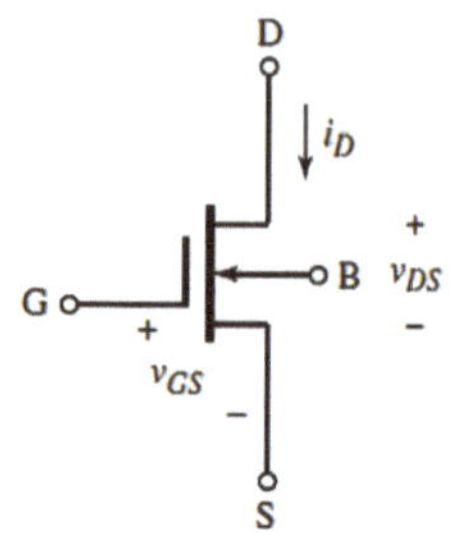

(a)

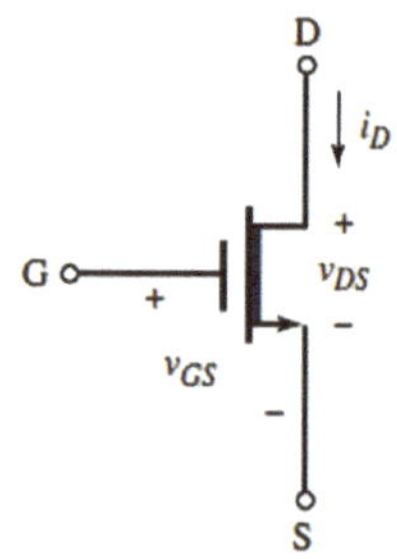

(b)

Figure 4.14. *The n-channel depletion mode MOSFET: (a) conventional circuit symbol and (b) simplified circuit symbol.*

Source: http://www.mhhe.com/engcs/electrical/neamen01/etext/ch05.pdf.

Equations define the current-voltage characteristics that are applicable to n-channel devices in both depletion and enhancement modes. The enhancement-mode MOSFET's threshold voltage (VTN) is positive, whereas the depletion-mode MOSFET's is negative. This is the only distinction. For the sake of clarity, different circuit symbols are used even though the current voltage characteristics of enhancement- and depletion-mode devices are described by the same equations.

Figure 4.14(a) displays the standard circuit symbol for the n-channel depletion-mode MOSFET. The device is in depletion mode as indicated by the vertical solid line that represents the channel. The broken versus the solid line denoting the channel is the only distinction between the enhancement- and depletion-mode symbols, as can be seen by comparing Figures 4.11(a) and 4.14(a). Figure 4.14(b) depicts a simplified symbol for the n-channel depletion-mode MOSFET. The arrowhead once again points to the source terminal, indicating that the current is flowing out of the source in the case of the n-channel device. The region of the depletion-mode channel is represented by the heavy solid line. Again, the circuit diagram's clarity is the only reason the depletion-mode device has a different circuit symbol than the enhancement-mode device.

- **P-Channel MOSFETs***:* Cross sections of a p-channel enhancement-mode and a p-channel depletion-mode MOSFET, along with the biasing configurations and current directions, are shown in Figures 4.15(a) and 4.15(b). In contrast to the n-channel device, the p-channel MOSFET, or PMOS transistor, has different types of impurity doping in the source, drain, and substrate regions. To form the inversion layer, or channel region, of holes in the p-channel enhancement-mode device, a negative gate-to-source voltage must be applied. This region connects the source and drain regions. For an enhancement-mode device, the threshold voltage, or VTP, is negative for the p-channel device. For a device operating in the p-channel depletion mode, the threshold voltage VTP is positive. Conventional current enters the source and exits the drain because holes flow from the source to the drain. The depletion-mode device has a p-channel region where the gate voltage is zero. Except for the conventional current direction and voltage polarities being inverted, the p-channel device operates similarly to the n-channel device. The hole is the charge carrier instead of the electron. For the p-channel device biased in the non-saturation region, the current is given by:

$$i_D = K_p[2(v_{SG} + V_{TP})v_{SD} - v_{SD}^2] \quad (1)$$

In the saturation region, the current is given by:

$$i_D = K_p(v_{SG} + V_{TP})^2 \quad (2)$$

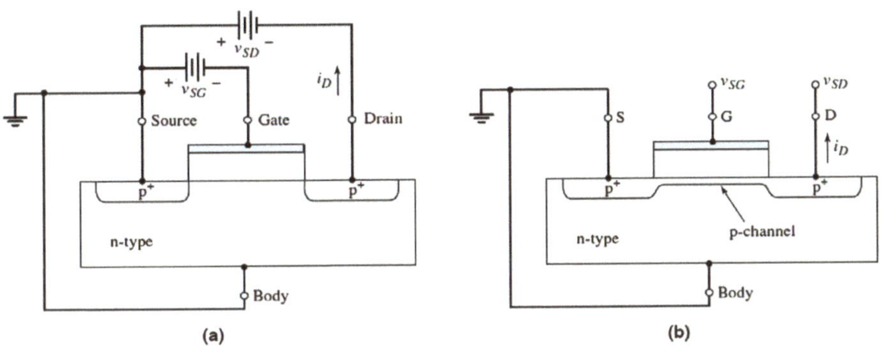

Figure 4.15. *Cross section of p-channel MOSFETs: (a) enhancement mode and (b) depletion mode*

Source: http://cdcpc.ce.ncu.edu.tw/classes/EEShortversion/Elect/Ch3%20The%20Field%20Effect%20 Transistor.pdf.

and the drain current exits the drain terminal. The parameter K, is the con-duction parameter for the p-channel device and is given by

$$K_p = \frac{W \mu_p C_{ox}}{2L} \quad (3)$$

where W, L, and Cox are the channel width, length, and oxide capacitance per unit area, as previously defined. The parameter p is the mobility of holes in the whole inversion layer. In general, the whole inversion layer mobility is less than the electron inversion layer mobility. We can also rewrite Equation (3) in the form

$$K_p = \frac{k_p'}{2} \cdot \frac{W}{L} \quad (4)$$

where k; = ApCox

For a p-channel MOSFET biased in the saturation region, we have

$$v_{SD} > v_{SD}(\text{sat}) = v_{SG} + V_{TP} \quad (5)$$

Figure 4.16(a) shows the p-channel enhancement-mode MOSFET's conventional circuit symbol. It should be noted that the substrate terminal's arrowhead direction is different from the n-channel enhancement-mode devices. Figure 4.16(b) depicts the simplified circuit symbol that we shall employ. The p-channel device's current flow is directed into the source by the arrowhead that is present on the source terminal.

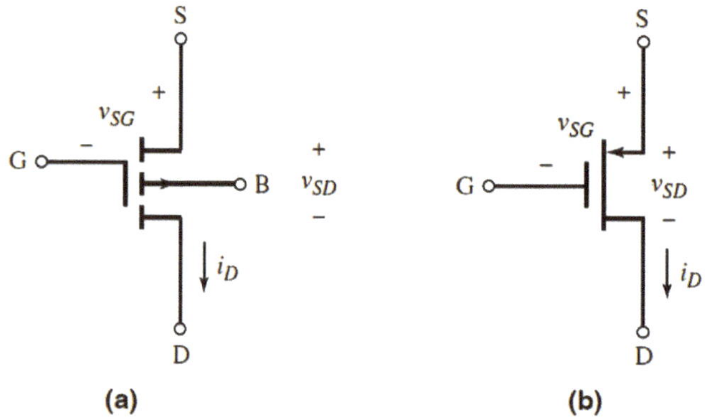

Figure 4.16. *The p-channel enhancement mode MOSFET: (a) conventional circuit symbol and (b) simplified circuit symbol.*

Source: http://www.mhhe.com/engcs/electrical/neamen01/etext/ch05.pdf.

Figure 4.17 (a & b) displays the p-channel depletion-mode device's conventional and simplified circuit symbols. In the simplified symbol, the depletion-mode device and the channel region are indicated by the heavy solid line. Once more, the source terminal has the arrowhead, which shows the direction in which it is pointing.

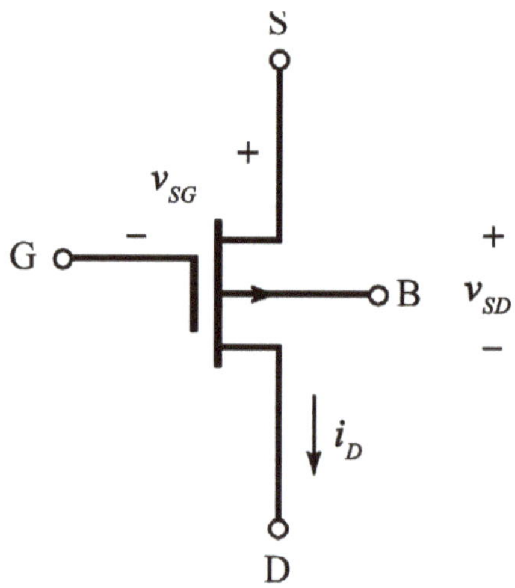

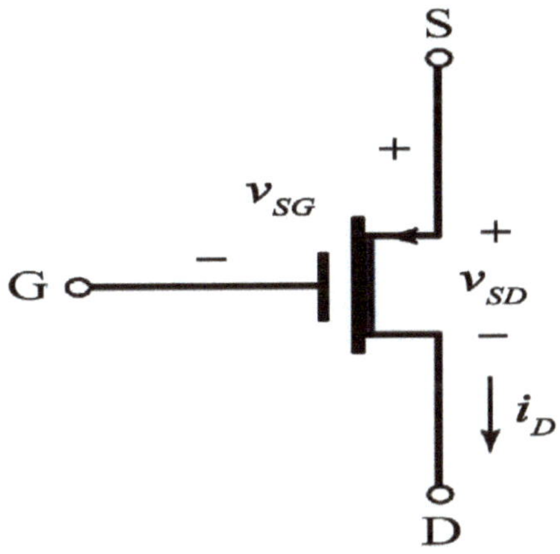

(b)

Figure 4.17. *The p-channel depletion mode MOSFET: (a) conventional circuit symbol and (b) simplified circuit symbol.*

Source: http://www.mhhe.com/engcs/electrical/neamen01/etext/ch05.pdf.

4.3. MOSFET DC CIRCUIT ANALYSIS

The world of electronic circuits is revolutionized by MOSFET, the three terminal device that is now the most widely used. These days, it seems impossible to design integrated circuits without MOSFET. These are fairly small, and they are made using a very straightforward process. Because of the properties of MOSFETs, integrated circuits for both analog and digital applications can be implemented successfully. MOSFET circuits can be analyzed using either the large-signal model or the small signal-model. The model of large-signals is nonlinear. It is employed to find the voltage and current DC values of the device. The linearization of the large signal model can be used to derive the small signal model. The three regions of operation of a MOSFET are the cut-off region, the triode region, and the saturation region. The device is in the cut off region when the gate to source voltage (V_{GS}) is less than the threshold voltage (V_{th}). A MOSFET operates in the saturation region when it is utilized as an amplifier. When utilized as a switch, it is in the triode or cut off region. We looked at the fundamental traits and attributes of MOSFETs. We now turn our attention to the analysis and design of MOS transistor circuits for dc biasing. Resistors are used in conjunction with MOS transistors in many of the circuits that are shown. However, in a real MOSFET integrated circuit, the resistors are typically swapped out for additional MOSFETs, making the circuit made up completely of MOS devices.

4.3.1. Common-Source Circuit

The common-source circuit is a fundamental MOSFET circuit configuration. One such circuit utilizing an n-channel enhancement-mode MOSFET is depicted in Figure 4.18. The source terminal, which is shared by the circuit's input and output sections, is at ground potential. The coupling capacitor Cc permits the signal voltage to be coupled to the MOSFET gate while functioning as an open circuit to DC. In Figure 4.19(a), the DC equivalent circuit is displayed. We once more employ the notation for DC and voltages in the DC analyses that follow. Since the gate current into the transistor is zero, the voltage at the gate is given by a voltage divider, which can be written as:

$$V_G = V_{GS} = \left(\frac{R_2}{R_1 + R_2}\right) V_{DD} \qquad (6)$$

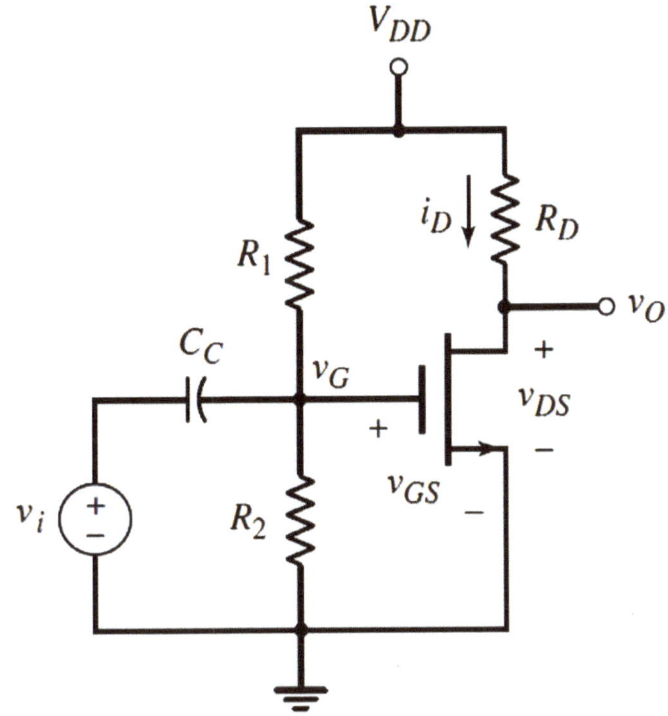

Figure 4.18. *An NMOS common source circuit.*

Source: http://www.mhhe.com/engcs/electrical/neamen01/etext/ch05.pdf.

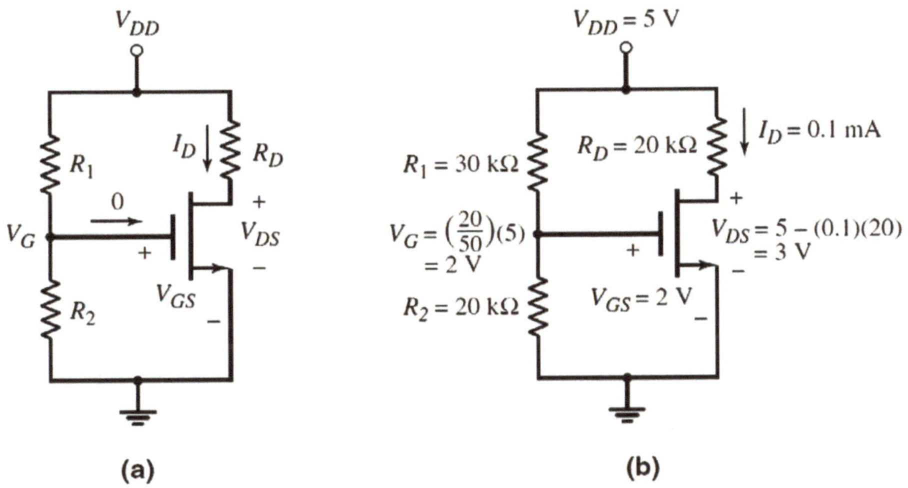

Figure 4.19. *(a) An NMOS common source circuit and (b) the NMOS circuit.*

Source: http://www.mhhe.com/engcs/electrical/neamen01/etext/ch05.pdf.

Assuming that the gate-to-source voltage given by Equation (6) is greater than V_{TN}, and that the transistor is biased in the saturation region, the drain current is:

$$I_D = K_n(V_{GS} - V_{TN})^2 \quad (7)$$

The drain to sources voltage is:

$$V_{DS} = V_{DD} - I_D R_D \quad (8)$$

If $V_{DS} > V_{DS}(sat) = V_{GS}$-$V_{TN}$, then the transistor is biased in the saturation region, as we initially assumed, and our analysis is correct. If $V_{DS} < V_{DS}(sat)$, then the transistor is biased in the non-saturation region, and the drain current is given by Equation (3).

Figure 20(a) shows a common-source circuit with a p-channel enhancement-mode MOSFET. The source terminal is tied to $+V_{DD}$, which becomes signal ground in the ac equivalent circuit. Thus, the terminology common-source applies to this circuit.

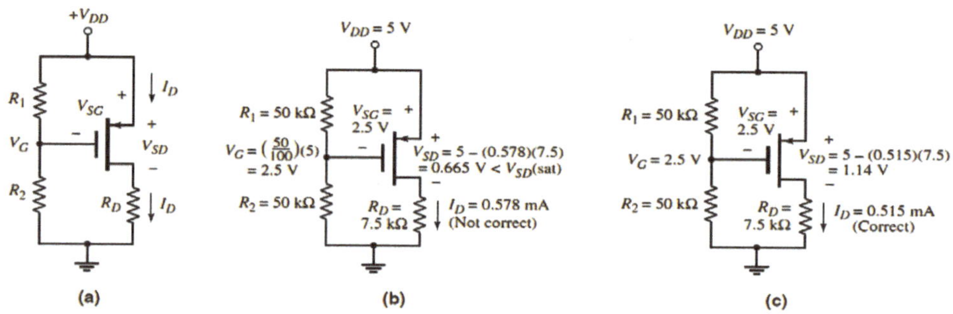

Figure 4.20. *(a) A PMOS common source circuit, (b) result when saturation region bias assumption in incorrect, and (c) result when non saturation region bias assumption is correct.*

Source: http://www.mhhe.com/engcs/electrical/neamen01/etext/ch05.pdf.

The dc analysis is essentially the same as for the n-channel MOSFET circuit. The gate voltage is:

$$V_G = \left(\frac{R_2}{R_1 + R_2}\right)(V_{DD}) \quad (9)$$

and the source-to-gate voltage is;

$$V_{SG} = V_{DD} - V_G \quad (10)$$

Assuming that $V_{GS} < V_{TP}$, or $V_{SG} > |V_{TP}|$, and that the device is biased in the saturation region, the drain current is given by:

$$I_D = K_p(V_{SG} + V_{TP})^2 \tag{11}$$

and the source-to-drain voltage is:

$$V_{SD} = V_{DD} - I_D R_D \tag{12}$$

If $V_{SD} > V_{SD}(\text{sat}) = V_{SG} + V_{TP}$, then the transistor is indeed biased in the saturation region, as we have assumed. However, if $V_{SD} < V_{SD}$ (Sat), the transistor is biased in the non-saturation region.

4.3.2. Load Line and Modes of Operation

The region where the MOSFET is biased can be easily seen using the load line. Examine the common-source circuit depicted in Figure 4.20(b) once more. Equation (8), the load line equation, is obtained by enclosing the drain—source loop in a Kirchhoff's voltage law equation. This equation illustrates a linear relationship between the drain current and drain-to-source voltage. The V_Ds(sat) characteristic for the transistor mentioned is displayed in Figure 4.21.

The load line is given by:

$$V_{DS} = V_{DD} - I_D R_D = 5 - I_D(20) \tag{13}$$

or

$$I_D = \frac{5}{20} - \frac{V_{DS}}{20} \text{ (mA)} \tag{14}$$

and is also plotted in the figure. The two end points of the load line are determined in the usual manner. If $I_D = 0$, then $V_{DS} = 5$ V; if $V_{DS} = 0$, then $I_D = 5/20 = 0.25$ mA. The Q-point of the transistor is given by the DC drain current and drain-to-source voltage, and it is always on the load line, as shown in the figure. A few transistor characteristics are also shown on the figure.

The transistor is in cutoff and the drain current is zero if the gate-to-source voltage is less than VTN. The transistor turns on and becomes biased in the saturation region as soon as the gate-to-source voltage is slightly higher than VT_N. The Q-point advances up the load line as V_{GS} rises. The transition point, defined as the point where $V_{DS} = V_{DS}$ (sat) $= V_{GS} - V_{TN}$, is the boundary between the saturation and non-saturation regions. In the non-saturation region, the transistor becomes biased as Vas rises above the transition point value.

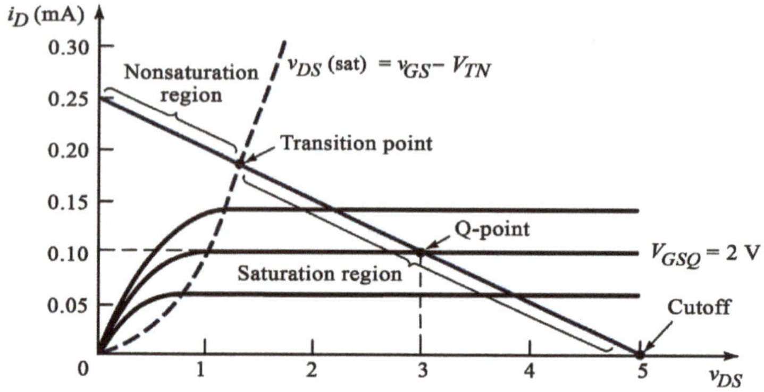

Figure 4.21. *Transistor characteristics, v_{DS} (sat) curve, load line, and Q-point for the NMOS common source circuit in Figure 18 (b).*

Source: http://www.mhhe.com/engcs/electrical/neamen01/etext/ch05.pdf.

4.3.3. Common MOSFET Configurations: DC Analysis

In addition to the fundamental common-source circuit, there are numerous other MOSFET circuit configurations. We keep working on the DC analysis and design of MOSFET circuits in order to improve our skills and knowledge of this kind of circuit analysis. We found that the addition of an emitter resistor to bipolar circuits tends to stabilize the Q-point. Similarly, the addition of a source resistor tends to stabilize the Q-point of MOSFET circuits against changes in transistor parameters.

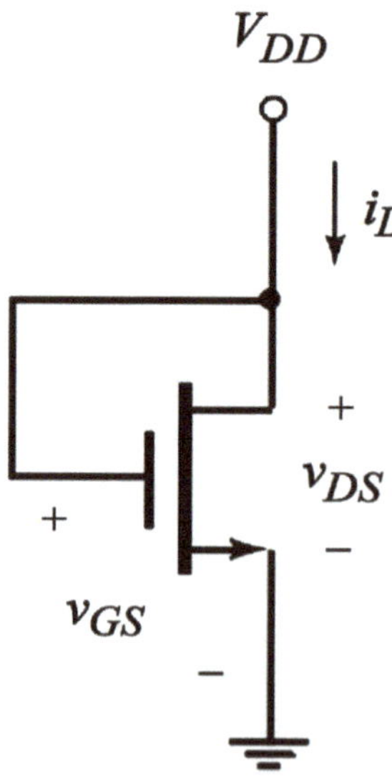

Figure 4.22. *Enhancement mode NMOS device with the gate connected to the drain.*

Source: http://www.mhhe.com/engcs/electrical/neamen01/etext/ch05.pdf.

Because of fabrication tolerances in oxide thickness, carrier mobility, channel width, and length, the transistor conduction parameters can differ from device to device. Additionally, different devices may have different threshold voltages. Changes in these device parameters will affect the Q-point in a given circuit; however, by adding a source resistor, the change will be mitigated. Furthermore, the source resistor is often replaced for a constant-current source in modern MOSFET circuits. This stabilizes the Q-point by biasing the transistor with a constant current regardless of the transistor's parameters.

A nonlinear resistor can be made from an enhancement-mode MOSFET connected in the manner that is depicted in Figure 4.22. An enhancement load device is a transistor having this connection. $V_{TN} > 0$ because the transistor is an enhancement mode device. Furthermore, the transistor in this circuit is always biased in the saturation region since $v_{DS} = v_{GS} > v_{DS}(sat) = v_{GS} - V_{TN}$.

The general i_D versus V_{DS} characteristics can then be written as

$$i_D = K_n(v_{GS} - V_{TN})^2 = K_n(v_{DS} - V_{TN})^2 \qquad (15)$$

Figure 4.23 shows a plot of Equation (14) for the case when $K_n = 1\ mA/V^2$ and $V_{TN} = 1V$. We will see how this transistor is used in place of a load resistor in an amplifier circuit.

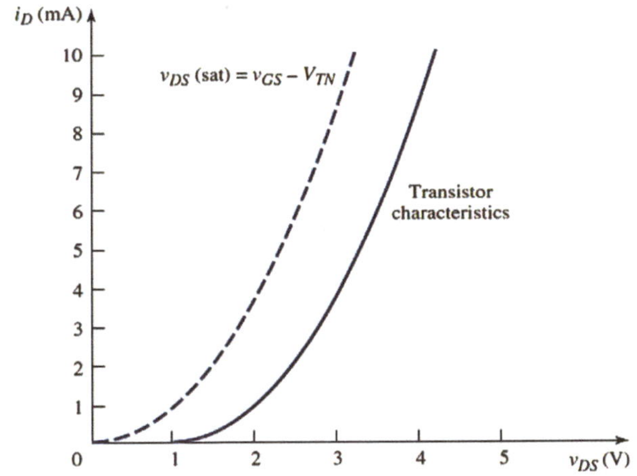

Figure 4.23. *Current voltage characteristics of an enhancement load device.*

Source: http://www.mhhe.com/engcs/electrical/neamen01/etext/ch05.pdf.

A circuit can be used as an amplifier or an inverter in a digital logic circuit by connecting an enhancement load device to another MOSFET in the configuration shown in Figure 4.24. The transistor M_D, also referred to as the driver transistor, can be biased in either the non-saturation or saturation region, contingent upon the input voltage value, while the load device M_L is always biased in the saturation region. The DC analysis of this circuit for DC input voltages to the M_D gate is covered in the following example.:

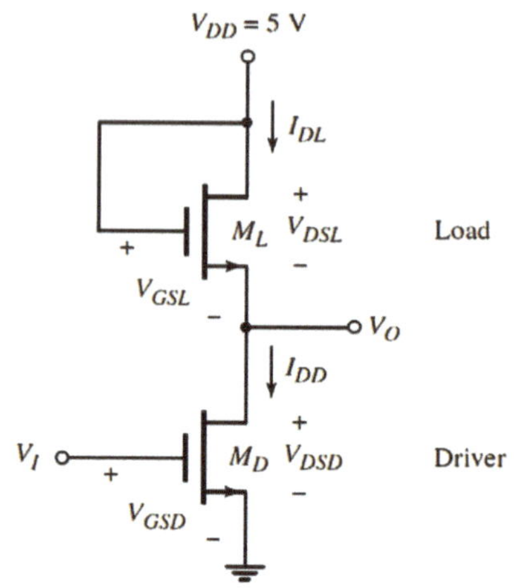

Figure 4.24. *Circuit with enhancement load device and NMOS driver.*

Source: https://www.chegg.com/homework-help/questions-and-answers/vtnd-vtnl-1v-knd-50-v-2-knl-10-v-2-nd-nl-0-d-driver-transistor-l-load-transistor-vi-5v-fin-q111485747.

We can identify the driver transistor's transition point in the circuit depicted in Figure 4.24, which divides the circuit into the saturation and non-saturation regions. The equation establishes the transition point.

$$V_{DSD}(\text{sat}) = V_{GSD} - V_{TND} \tag{16}$$

Again, the drain currents in the two transistors are equal. Using the saturation drain current relationship for the driver transistor, we have:

$$I_{DD} = I_{DL} \tag{15}$$

or

$$K_{nD}[V_{GSD} - V_{TND}]^2 = K_{nL}[V_{GSL} - V_{TNL}]^2 \tag{17}$$

Again, noting that $V_{GSD} = V_I$ and $V_{GSL} = V_{DSL} = V_{DD} - V_O$, and taking the square root, we have:

$$\sqrt{\frac{K_{nD}}{K_{nL}}}(V_I - V_{TND}) = (V_{DD} - V_O - V_{TNL}) \tag{18}$$

At the transition point, we can define the input voltage as $V_I = V_{It}$ and the output voltage as $V_{Ot} = V_{DSD}(\text{sat}) = V_{It} — V_{TND}$. Then, from Equation (29?), the input voltage at the transition point is:

$$V_{It} = \frac{V_{DD} - V_{TNL} + V_{TND}(1 + \sqrt{K_{nD}/K_{NL}})}{1 + \sqrt{K_{nD}/K_{nL}}} \tag{19}$$

Equation (19) can be applied to the previous example to demonstrate that our preliminary assumptions were accurate. Thus far, the only load device that we have taken into consideration is the n-channel enhancement-mode MOSFET. It is also possible to use an n-channel depletion-mode MOSFET. Examine the depletion-mode MOSFET depicted in Figure 4.25(a), which has its gate and source connected. Figure 4.25(b) displays the characteristics of the current voltage. Bias can occur in the saturation or non-saturation regions of the transistor. The plot additionally displays the transition point. Since the n-channel depletion-mode MOSFET has a negative threshold voltage, v_{DS}(sat) is positive.

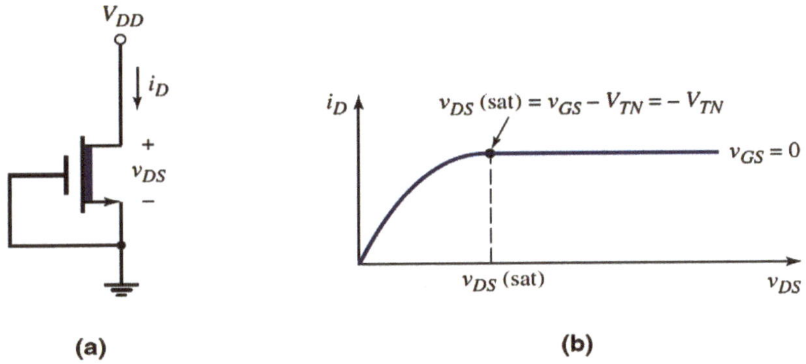

Figure 4.25. *(a) Depletion mode NMOS device with the gate connected to the source and (b) current voltage characteristics.*

Source: http://www.mhhe.com/engcs/electrical/neamen01/etext/ch05.pdf.

`Examine the circuit depicted in Figure 4.26, where the transistor serves as a depletion load component. Depending on the values of the transistor parameters, V_{DD}, and R_S, it could be biased in the non-saturation or saturation region.

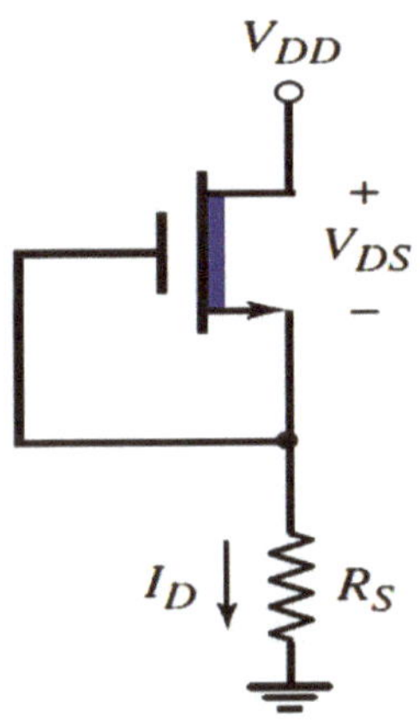

Figure 4.26. *Circuit containing a depletion load device.*

Source: http://www.mhhe.com/engcs/electrical/neamen01/etext/ch05.pdf.

A circuit that can be utilized as an amplifier or an inverter in a digital logic circuit can be constructed by combining a depletion load device with another MOSFET. Depending on the input voltage value, the driver transistor M_D and the load device M_L may be biased in the non-saturation or saturation region. For a specific DC input voltage to the driver transistor's gate, we will conduct the circuit's DC analysis. A CMOS inverter is illustrated in Figure 4.27. The load, also known as an active load, is the PMOS device, and the NMOS transistor serves as the amplifier or driver. Analog applications are primarily used for this setup. The two gates are connected to form the input in an alternative configuration.

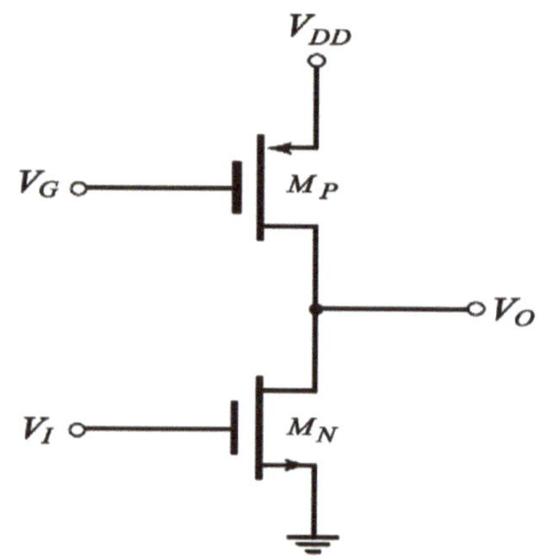

Figure 4.27. *Example of CMOS inverter.*

Source: http://www.mhhe.com/engcs/electrical/neamen01/etext/ch05.pdf.

The two transistors depicted in Figure 4.25 may be biased in the saturation or no saturation region, contingent upon the input voltage value, like the preceding two NMOS inverters. The voltage transfer characteristic is most easily determined from a P Spice analysis.

4.3.4. Constant-Current Source Biasing

It is possible to bias a MOSFET by applying a constant current source IQ. This circuit's benefit is that the drain current is unaffected by the characteristics of the transistor. MOSFETs can be used to implement the constant-current source, as seen in Figure 4.28. The current source is composed of transistors M_3, M_3, and M_4. Transistors M_3 and M_4 create a reference current by connecting to one another in a diode-like fashion. We observed that the transistor is implied to be always biased in the saturation region by this diode-type connection. As a result, transistors M_3 and M_4 are biased in the saturation region, and it is presumed that transistor M_3 is biased as well. The bias current (IQ) is determined by applying the gate-to-source voltage that results to M_3.

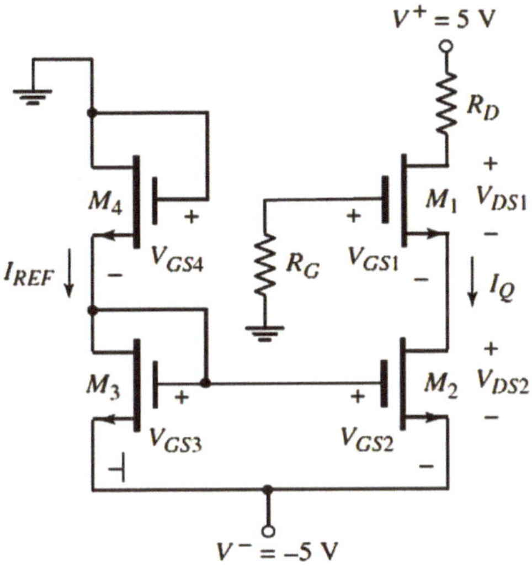

Figure 4.28. *Implementation of a MOSFET constant current source.*

Source: http://www.mhhe.com/engcs/electrical/neamen01/etext/ch05.pdf.

Since the reference current is the same in transistors M_3 and M_4, we can write:

$$K_{n3}(V_{GS3} - V_{TN3})^2 = K_{n4}(V_{GS4} - V_{TN4})^2 \quad (20)$$

We also know that,

$$V_{GS4} + V_{GS3} = (-V^-) \quad (21)$$

Solving Eq. (21) for V_{GS4} and substituting the result into Eq. (20) yields:

$$\frac{V_{GS3} = \sqrt{\frac{K_{n4}}{K_{n3}}}[(-V^-) - V_{TN4}] + V_{TN3}}{1 + \sqrt{\frac{K_{n4}}{K_{n3}}}} \quad (22)$$

Since $V_{GS3=}$ V_{GS2} the bias current is:

$$I_Q = K_{n2}(V_{GS3} - V_{TN2})^2 \quad (23)$$

4.3.5. Basic MOSFET Applications

MOSFETs can be used to amplify small time-varying signals, switch currents, voltages, and power, and carry out digital logic operations.

- **NMOS Inverter:** There are many different electronic applications where the MOSFET can be utilized as a switch. Reliability and speed are two areas where transistor switches excel over mechanical ones. The circuit for an n-channel enhancement-mode MOSFET inverter is shown in Figure 4.29. The transistor is in cutoff and $I_D = 0$ if $v_I < V_{TN}$. The output voltage is $v_O = V_{DD}$, and there is no voltage drop across R_D. Additionally, no power is lost in the transistor because $I_D = 0$.

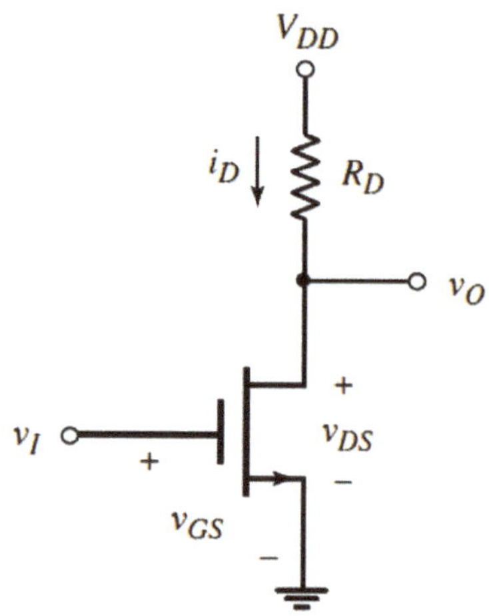

Figure 4.29. *NMOS inverter circuit.*

Source: http://www.mhhe.com/engcs/electrical/neamen01/etext/ch05.pdf.

The transistor is on and initially biased in the saturation region if $v_I > V_{TN}$ because $v_{DS} > v_{GS}$ − VTN at this point. The drain-to-source voltage drops with increasing input voltage, and the transistor eventually becomes biased in the no saturation region. A minimum value for V_O is reached, a maximum value for the drain current is reached, and the transistor is biased in the non-saturation region when $V_I = V_{DD}$.

Given by

$$i_D = K_n[2(v_I - V_{TN})v_O - v_O^2] \quad (24)$$

and,

$$v_O = v_{DD} - i_D R_D \quad (25)$$

where $v_O = v_{DS}$ and $v_I = v_{GS}$.

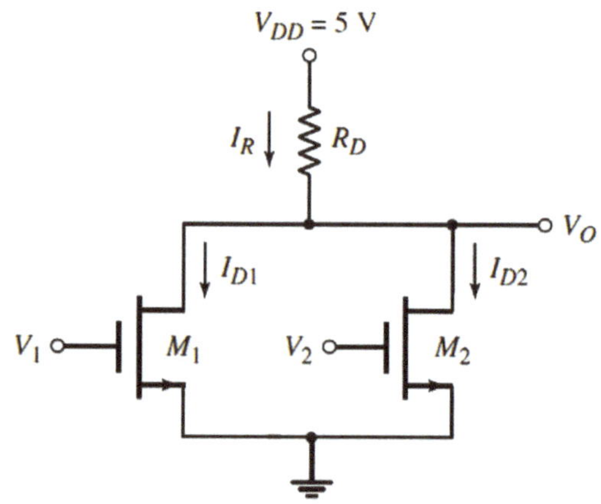

Figure 4.30. *A two input NMOS NOR logic gate.*

Source: http://www.mhhe.com/engcs/electrical/neamen01/etext/ch05.pdf.

- **Digital Logic Gate:** In the transistor inverter circuit shown in Figure 4.30, the transistor is turned off when the input falls below about zero volts, and the output rises to equal V_{DD}. The transistor is biased in the nonsaturation region when the input is high and equal to V_{DD}, resulting in a low output. We can analyze the circuit in terms of DC parameters, because the input voltages will either be high or low. Now let's look at the scenario where a second transistor is connected in parallel, as Figure 4.30 illustrates. M_1 and M_2 are both cut off and

V_O=5V if both inputs are zero. Transistor M_1 turns on when V_1 = 5V and V_2 = 0, but M_2 remains cut off. As V_O approaches a low value, transistor M_1 becomes biased in the non-saturation region. M1 is cut off and M_2 is biased in the nonsaturation region if the input voltages are reversed so that V_1 = 0 and V_2 = 5V. Once more, the value of V_O is low. In the event that both inputs are high, specifically at V_1, = V_2 = 5V, then V_O is low and both transistors are biased in the nonsaturation region.

Table 4.1 displays the different conditions of the circuit depicted in Figure 4.30. In a positive logic setting, the outcomes suggest that the circuit operates as a NOR logic function, thus classifying it as a two-input NOR logic circuit. In NMOS logic circuits, resistor RD is substituted with another NMOS transistor.

Table 4.1. *NMOS NOR Logic Gate Response*

V_1(V)	V_2(V)	V_o(V)
0	0	High
5	0	Low
0	5	Low
5	5	Low

4.4. METAL OXIDE SEMICONDUCTOR FIELD EFFECT TRANSISTOR (MOSFET)

A semiconductor device that is frequently used in electronic devices for switching and signal amplification is the MOSFET (Metal Oxide Semiconductor Field Effect Transistor). The development of a new MOS gate-controlled power device technology in the 1980s put the power electronics industry's reliance on bipolar devices to the test. The MOS integrated circuit technology gave rise to the power MOS field effect transistor (MOSFET). The new gadget promised incredibly low input power levels and no intrinsic switching speed limitations. Consequently, it created the opportunity to lower the size and weight of power electronic systems by raising the operating frequency of those systems. The necessity to build the gate drive circuit to account for the pulse currents required to charge and discharge the high input capacitance of these devices, however, tempered the initial claims of infinite current gain for the power MOSFET. The amount of gate drive power needed increases significantly at higher operation frequencies. Additionally, MOSFETs have a relatively higher on-state resistance per unit area of the device cross section, which rises as the device's blocking voltage rating does. As a result, MOSFET use is now limited to low voltage (less than 500 volts) applications where the ON-state resistance remains at reasonable levels. By using these devices' inherent fast switching speed, it is possible to raise the switching frequency above several hundred kHz.

From an operational principle perspective, a MOSFET is a majority carrier device that is controlled by voltage. As the name implies, a MOSFET's majority carrier movement is controlled by the voltage applied to the gate, or control electrode, which is separated from the bulk semiconductor body by a thin layer of metal oxide. Between the main current-carrying terminals known as the Drain (D) and the Source (S), the electric field created by the gate voltage modifies the conductivity of the semiconductor material. Similar to their integrated circuit counterpart, power MOSFETs come in two varieties: enhancement type (i) and depletion type (ii). Both of these, depending on the bulk semiconductor's characteristics, can be either n-channel or p-channel types. The circuit symbol and transfer characteristics (drain current vs. gate-source voltage) of these four types of MOSFETs are displayed in Figure 4.31(a).

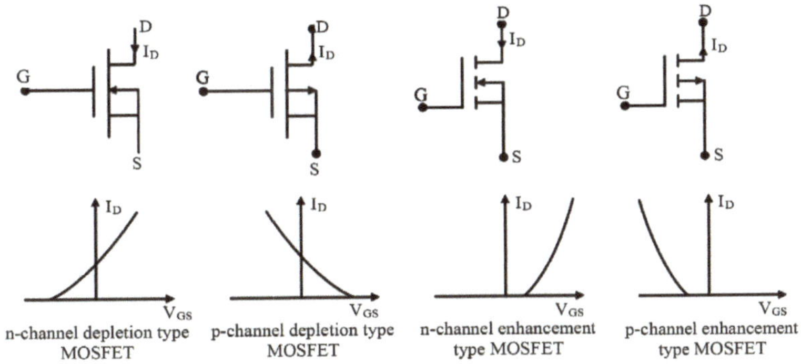

(a)

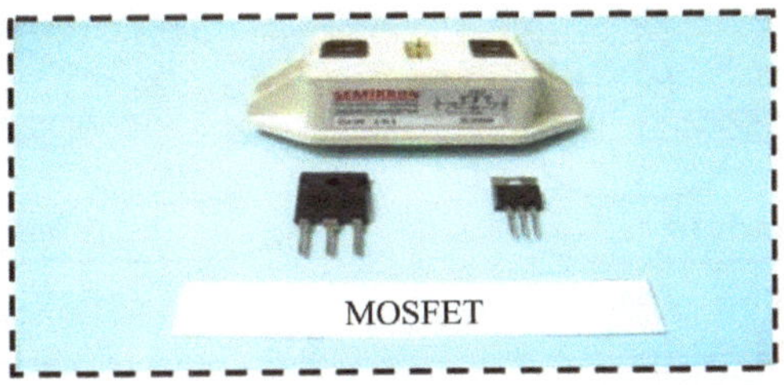

(b)

Figure 4.31. *Different types of power MOSFET. (a) Circuit symbols and transfer characteristics (b) Photograph of n-channel enhancement type MOSFET.*

Source: https://nptel.ac.in/courses/Webcourse-contents/IIT%20Kharagpur/Power%20Electronics/PDF/L-6(DK)(PE)%20((EE)NPTEL).pdf.

Depletion type MOSFETs are typically ON type switches, meaning that a nonzero drain current can flow in these devices when the gate terminal is open, as can be inferred from Figure 4.31(a). In many powers electronic applications, this is inconvenient. For power electronics applications, enhancement type MOSFETs—especially those with n channels—are therefore more common. A photo of some commercially available n-channel enhancement type Power MOSFETs can be found in Figure 4.31(b).

4.4.1. Constructional Features of a Power MOSFET

The MOS integrated circuit technology gave rise to the power MOSFET. Initially, lateral MOSFETs were redesigned to boost their voltage-blocking capacity in an effort to create high-voltage MOSFETs. Lateral double-diffused MOS (DMOS) is the name of the resulting technology. It was quickly discovered, nevertheless, that using a vertically oriented structure would allow for much higher breakdown voltage and current ratings. Since then, almost all manufacturers of Power MOSFETs have adopted the Vertical Double-diffused MOS (VDMOS) structure.

As illustrated in Figure 4.32(a), a schematic representation of a single MOSFET cell structure, a power MOSFET utilizing VDMOS technology has a vertically oriented three-layer structure of alternating p-type and n type semiconductors. A complete device is formed by connecting many of these cells in parallel, as shown in Figure 4.32(b).

The two n+ end layers labeled "Source" and "Drain" have similar levels of heavy doping. The body, also known as the substrate, is the middle layer of the p-type and has a moderate doping level, two to three orders of magnitude lower than the n+ regions on both sides. The lowest doping density is found in the n-drain drift region. The device's breakdown voltage is determined by the thickness of this area.

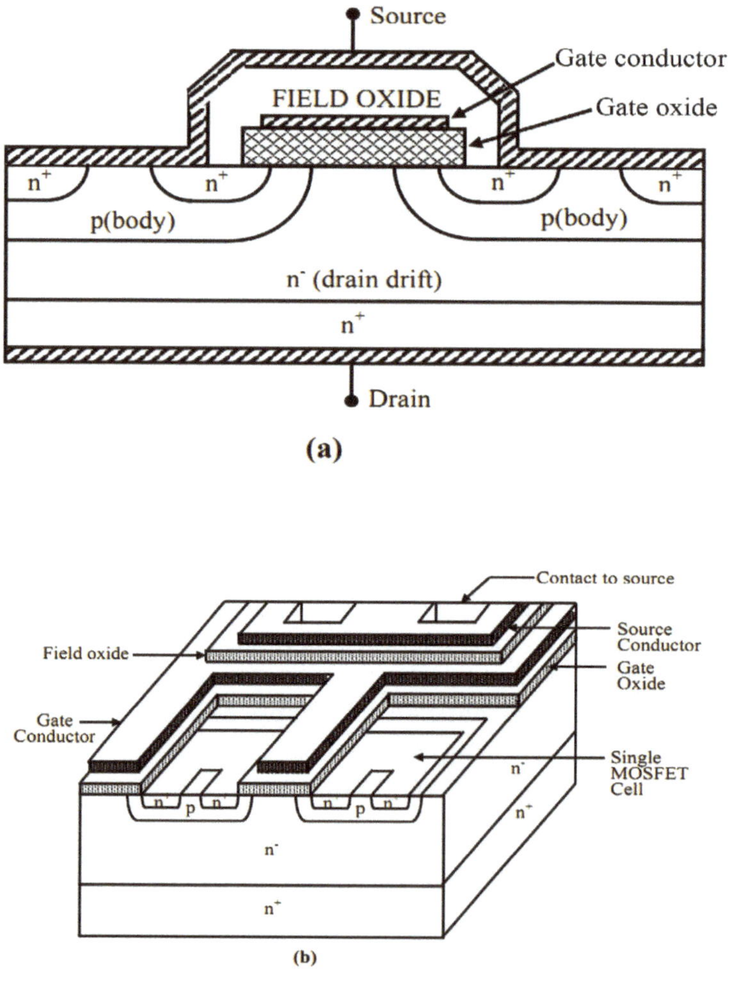

Figure 4.32. *Schematic construction of a power MOSFET (a) Construction of a single cell. (b) Arrangement of cells in a device.*

Source: https://nptel.ac.in/courses/Webcourse-contents/IIT%20Kharagpur/Power%20Electronics/PDF/L-6(DK)(PE)%20((EE)NPTEL).pdf.

A thin layer of silicon dioxide, also known as the gate oxide, separates the gate terminal from the semiconductor body and is positioned over the n- and p-type regions of the cell structure. The source and drain terminals of the entire device are formed by connecting the source and drain regions of every cell on a wafer to the same metallic contacts. Similarly, every gate terminal is linked to every other one. The gate regions enclose the numerous thousands tiny polygon-shaped areas that make up the source. The MOSFET's ON state resistance is also somewhat influenced by the geometric shape of the source regions.

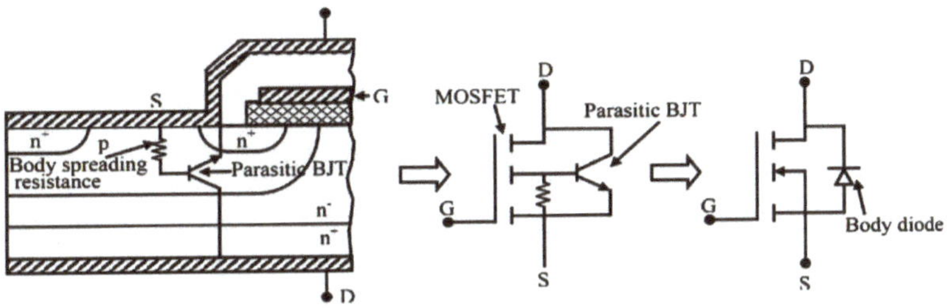

Figure 4.33. *Parasitic BJT in a MOSFET cell.*

Source: https://nptel.ac.in/courses/Webcourse-contents/IIT%20Kharagpur/Power%20Electronics/PDF/L-6(DK)(PE)%20((EE)NPTEL).pdf.

A noteworthy characteristic of MOSFET cells is that, as illustrated in Figure 4.33, each MOSFET cell contains an embedded parasitic BJT, whose base and emitter are shorted by the source metallization, due to the alternating n^+ n^- p n^+ structure. The body spreading resistance of the p type substrate is what causes the nonzero resistance between the base and the emitter of the parasitic npn BJT. Particular attention is given to minimizing this resistance and suppressing the parasitic BJT's switching operation during the MOSFET cell design process. The BJT always remains in cut off when there is an effective short circuit between the body and the source. In the circuit symbol of a power MOSFET, the collector-base junction of the BJT is represented as an anti-parallel diode, also known as the body diode.

4.4.2. Operating Principle of a MOSFET

When at least one of the pn junctions is reverse biased for either polarity of the applied voltage between the source and the drain, it would initially seem that there is no path for any current to flow between the source and the drain terminals. Given how effective the gate oxide is as an insulator, there is also no chance of current injection from the gate terminal. Forms an excellent capacitor is created by the MOSFET's gate region, which is made up of the p-body silicon, the gate oxide layer, and the gate metallization. The SiO_2 and silicon interface forms a depletion region at the interface when a small voltage is applied to this capacitor structure with the gate terminal positive with respect to the source (note that the body and source are shorted), as shown in Figure 4.34 (a).

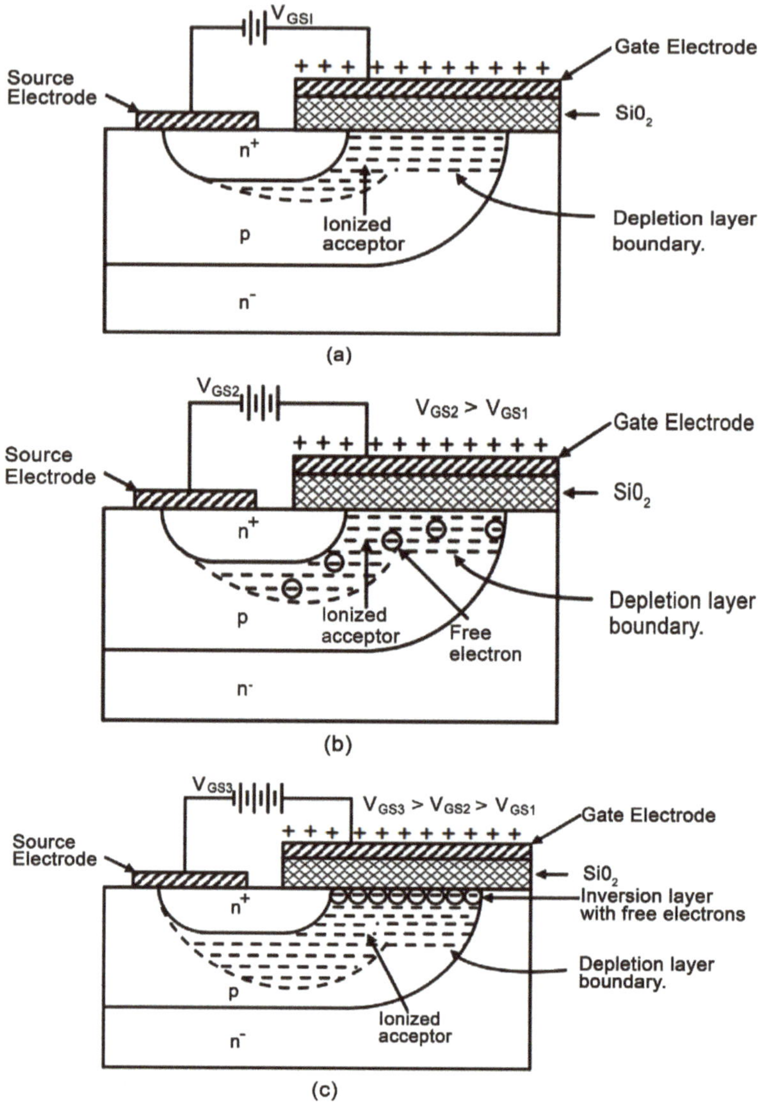

Figure 4.34. *Gate control of MOSFET conduction. (a) Depletion layer formation; (b) Free electron accumulation; (c) Formation of inversion layer.*

Source: https://nptel.ac.in/courses/Webcourse-contents/IIT%20Kharagpur/Power%20Electronics/PDF/L-6(DK)(PE)%20((EE)NPTEL).pdf.

Most hole carriers are repelled from the interface region between the gate oxide and the p-type body by the positive charge induced on the gate metallization. As a result, a depletion region is created and the negatively charged acceptors are exposed. The thickness of the depletion layer increases with further V_{GS} increases. Simultaneously, as Figure 4.34 (b) illustrates, the electric field at the oxide-silicon interface grows and starts to draw free electrons. The direct source of electrons is the thermal ionization process, which produces electron-hole pairs. Before the depletion region, the holes are repelled into the semiconductor bulk. Electrons from the source neutralize the excess holes.

The density of free electrons at the interface equals the density of free holes in the bulk of the body region beyond the depletion layer as V_{GS} increases further. Figure 4.34 (c) illustrates the inversion layer, which is the layer of free electrons at the interface. The inversion layer is a conductive path or channel that allows current to flow between the drain and the source. It possesses all the characteristics of a n-type semiconductor. This device is known as an enhancement type n-channel MOSFET because current conduction occurs through an n-type channel that is produced by the electric field caused by the gate-source voltage. The Gate – Source threshold voltage V_{GS} (th) is the value of V_{GS} at which the inversion layer is thought to have formed. Since the density of free electrons increases further with an increase in V_{GS}, the inversion layer becomes somewhat thicker and more conductive as V_{GS} is increased beyond V_{GS}(th). The inversion layer prevents the nearby depletion layer from raising V_{GS}. The thickness of the depletion layer is now constant.

CASE STUDY

EVALUATION OF MOSFET

The Metal-Oxide-Semiconductor Field-Effect Transistor (MOSFET) stands as a cornerstone of modern electronics due to its versatility, efficiency, and scalability. Evaluating the MOSFET involves understanding its fundamental characteristics, operational principles, and diverse applications. A MOSFET consists of three terminals: the gate, drain, and source, with the gate being insulated from the channel by a thin oxide layer. This structure allows the MOSFET to act as a voltage-controlled device, where the voltage applied to the gate controls the current flow between the drain and source. The primary advantage of MOSFETs over Bipolar Junction Transistors (BJTs) is their high input impedance, which means they draw very little gate current, making them highly efficient for amplification and switching applications.

MOSFETs are broadly classified into two types: N-Channel and P-Channel, each with distinct characteristics and applications. N-Channel MOSFETs, which allow electrons as charge carriers, generally offer better performance in terms of speed and on-resistance compared to P-Channel MOSFETs, which use holes as charge carriers. This makes N-Channel MOSFETs more prevalent in high-speed and high-power applications. The operation of a MOSFET can be divided into three regions: cut-off, triode, and saturation. In the cut-off region, the MOSFET is off, and no current flows through the drain-source channel. In the triode region, the MOSFET operates as a variable resistor, allowing current to flow between the drain and source, controlled by the gate voltage. In the saturation region, the MOSFET acts as a current source, with the current primarily determined by the gate-source voltage (V_{GS}).

MOSFETs excel in switching applications due to their ability to switch on and off rapidly, a critical feature in digital circuits and power electronics. Their fast-switching capabilities are essential in applications such as DC-DC converters, motor drivers, and power supplies, where efficient energy management is crucial. Additionally, MOSFETs are fundamental components in digital logic circuits, forming the building blocks of Complementary Metal-Oxide-Semiconductor (CMOS) technology. CMOS technology, which uses pairs of N-Channel and P-Channel MOSFETs, underpins virtually all modern integrated circuits, from microprocessors to memory devices, due to its low power consumption and high noise immunity.

The evaluation of MOSFET performance also encompasses parameters such as threshold voltage (V_{th}), on-resistance ($R_{DS}(on)$), and transconductance (g_m). The threshold voltage is the minimum gate-to-source voltage required to create a conducting path between the drain and source. On-resistance is a critical parameter in power applications, determining the efficiency of the MOSFET when it is in the on-state. Low R_{DS}(on) values are desirable for minimizing power losses and heat generation. Transconductance reflects the MOSFET's ability to amplify the input signal, a vital attribute in analog applications.

Furthermore, the reliability and thermal management of MOSFETs are crucial in high-power applications. Power MOSFETs can dissipate significant heat during operation, necessitating effective thermal management strategies such as heat sinks and proper circuit layout to ensure stable performance and longevity. Advances in MOSFET technology, such as trench and super-junction structures, have further enhanced their performance, enabling higher current densities and lower on-resistance.

In summary, the MOSFET's high efficiency, fast switching capabilities, and integration into CMOS technology make it indispensable in both analog and digital electronics. Its evaluation involves a comprehensive understanding of its electrical characteristics, performance parameters, and thermal management requirements, highlighting its pivotal role in advancing electronic device functionality and efficiency.

CLASS ACTIVITY

Some types of MOSFETs have a source-drain channel already formed with no applied gate voltage:

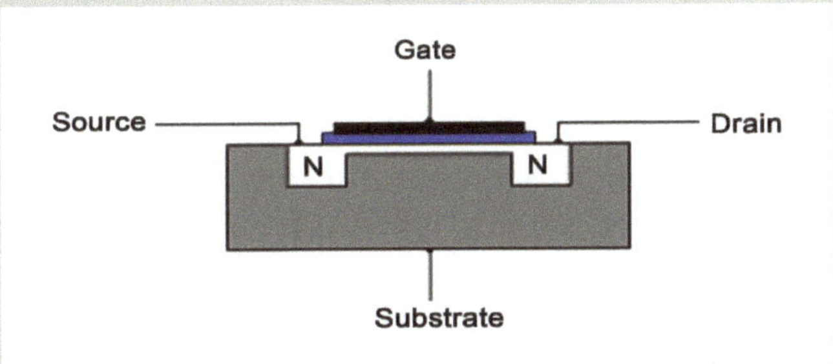

Explain what happens to source-drain conductivity with each of the following applied gate-to-substrate voltages. Modify the illustrations if necessary:

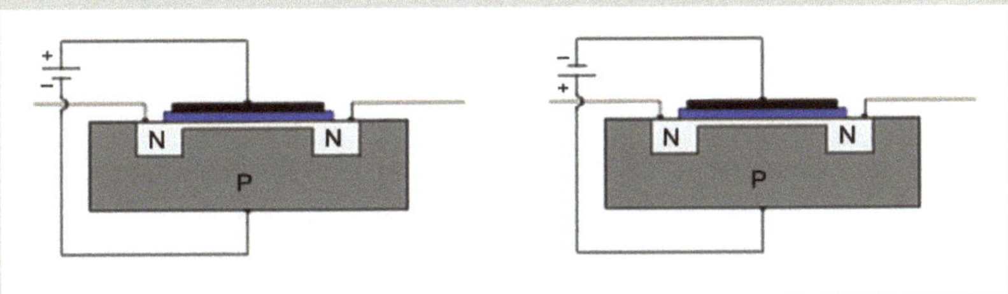

SUMMARY

- A field-effect transistor (FET) is a type of transistor commonly used for weak-signal amplification (for example, for amplifying wireless signals). The device can amplify analog or digital signals. It can also switch DC or function as an oscillator.
- Field-effect transistors are fabricated onto silicon integrated circuit (IC) chips. A single IC can contain many thousands of FETs, along with other components such as resistors, capacitors, and diodes.
- The Junction FET transistor is a type of field effect transistor that can be used as an electrically controlled switch. The electric energy flows through an active channel between sources to drain terminals.
- N channel JFET consists of an n-type bar at the sides of which two p-type layers are doped. The channel of electrons constitutes the N channel for the device.
- MOSFET transistor as its name suggests is a p-type (n-type) semiconductor bar (with two heavily doped n-type regions diffused into it) with a metal oxide layer deposited on its surface and holes taken out of the layer to form source and drain terminals.
- A metal–oxide–semiconductor field-effect transistor (MOSFET, MOS-FET, or MOS FET) is a field-effect transistor (FET with an insulated gate) where the voltage determines the conductivity of the device. It is used for switching or amplifying signals.
- The working of a MOSFET depends upon the MOS capacitor. The MOS capacitor is the main part of MOSFET. The semiconductor surface at the below oxide layer which is located between source and drain terminals. It can be inverted from p-type to n-type by applying positive or negative gate voltages.
- The MOSFET (Metal Oxide Semiconductor Field Effect Transistor) transistor is a semiconductor device which is widely used for switching and amplifying electronic signals in the electronic devices.

REVIEW QUESTIONS

1. What is meant by 'field effect' transistor?
2. Explain the fabrication process of MOSFET.
3. Why is n-channel FET preferred over p-channel FET?
4. Explain the working principle of a JFET.
5. How does the gate-source voltage (V_{GS}) control the current in a JFET?
6. Why are MOSFETs preferred over BJTs for switching applications?
7. Describe the structure and function of a two-terminal MOS capacitor.
8. Explain the operation of an N-Channel MOSFET in the cut-off, triode, and saturation regions.

9. How is a MOSFET used as a switch in digital circuits? Provide an example.
10. What are the advantages of using MOSFETs in digital logic gates compared to BJTs?

REFERENCES

1. Bhalla, A. (2021). Origins of SiC FETs and their evolution toward the perfect switch. *Power Electronics News*.
2. Edgar, J. (1926). Method and apparatus for controlling electric currents (U.S. Patent No. 1745175A).
3. Galup-Montoro, C., & Schneider, M. C. (2007). *MOSFET Modeling for Circuit Analysis and Design*. London/Singapore: World Scientific.
4. Gray, P. R., Hurst, P. J., Lewis, S. H., & Meyer, R. G. (2001). *Analysis and Design of Analog Integrated Circuits* (4th ed.). New York: Wiley.
5. Grundmann, M. (2010). *The Physics of Semiconductors*. Springer-Verlag. ISBN 978-3-642-13884-3.
6. Haugerud, B., Bosworth, L., & Belford, R. (2003). Mechanically induced strain enhancement of metal-oxide-semiconductor field-effect transistors. *Journal of Applied Physics, 94*, 4102–4107.
7. Mistry, K., Armstrong, M., Auth, C., Cea, S., Coan, T., Ghani, T., & Bohr, M. (2004). Delaying forever: Uniaxial strained silicon transistors in a 90nm CMOS technology. In *IEEE Symposium on VLSI Technology Digest of Technical Papers*, June 2004.
8. Mottershead, A. (2004). *Electronic Devices and Circuits*. New Delhi: Prentice-Hall of India. ISBN 978-81-203-0124-5.
9. Ross, B. (2002). *To the Digital Age: Research Labs, Start-Up Companies, and The Rise of MOS Technology*. JHU Press.
10. Shichman, H., & Hodges, D. A. (1968). Modeling and simulation of insulated-gate field-effect transistor switching circuits. *IEEE Journal of Solid-State Circuits, SC-3*(3), 285–289.
11. Steegen, A., Stucchi, M., Lauwers, A., & Maex, K. (1999). Silicide induced pattern density and orientation dependent transconductance in MOS transistors in IEEE Proc. Intl. Electron Devices Meeting, pp. 497-500, Dec 1999.
12. Takahashi, K., Manabe, K., Morioka, A., Ikarashi, T., Yoshihara, T., Watanabe, H., & Tatsumi, T. (2005). High-mobility dual metal gate MOS transistors with high-k gate dielectrics. *Japanese Journal of Applied Physics, 44*, 2210.

CHAPTER

5

Special Semiconductor Devices

LEARNING OBJECTIVES

After studying this chapter, you will be able to:

- Understand the field effect transistors (FETS) types and working principles;
- Discuss on basics, working principle and applications of MOSFET;
- Explain the MOSFET dc circuit analysis;
- Describe the metal oxide semiconductor field effect transistor (MOSFET).

KEY TERMS FROM THIS CHAPTER

Avalanche breakdown	Bipolar devices
Breakdown voltage	Electrode
Gate turn-off thyristors (GTOs)	Insulated-gate bipolar transistor (IGBT)
Latching current	Metal oxide semiconductor (MOS)
NPN transistor	PNP transistor
Semiconductor conductivity	Silicon-controlled switch (SCS)
Snubber network	Thyristor
Unijunction transistor (UJT)	

INTRODUCTION

Essentially, a semiconductor device is an electronic component whose operation depends on the electronic characteristics of semiconductor materials, mainly silicon, germanium, gallium arsenide, and organic semiconductors. Between conductors and insulators is where its conductivity falls. In most applications, vacuum tubes have been replaced by semiconductor devices. Unlike free electrons traveling across a vacuum (usually released by thermionic emission) or free electrons and ions traveling through an ionized gas, they conduct electric current in the solid state.

Semiconductor chips, which are made up of two or more discrete semiconductor devices connected by an interconnecting circuit, can be produced in quantities ranging from hundreds to billions on a single semiconductor wafer, also known as a substrate.

Because it is easy to control their behavior through the intentional introduction of impurities, a process known as doping, semiconductor materials are valuable. Semiconductors can be excellent sensors because they have the ability to control their conductivity through mechanical deformation of a doped monocrystalline silicon grid, exposure to light or heat, or the introduction of an electric or magnetic field. Charge carriers, which are made up of free or mobile electrons and electron holes, are what cause current to flow through a semiconductor. A semiconductor's number of free electrons or holes significantly increased when it is doped with a tiny amount of an atomic impurity, like phosphorus or boron. A doped semiconductor is referred to as a p-type semiconductor (p stands for positive electric charge) when it has excess holes, and an n-type semiconductor (n for negative electric charge) when it has excess free electrons. Negative charges are present with most mobile charge carriers. p- and n-type dopant concentration and location are carefully regulated during the semiconductor manufacturing process. p-n junctions are created when n-type and p-type semiconductors are connected.

Such semiconductor devices that allow for precise control over the flow of big power blocks within a system have been developed recently. These devices function as controlled switches and are capable of controlling power regulation, inversion, and controlled rectification in a load.

The special semiconductor devices are:

- Silicon controlled rectifier (SCR)
- Triode for alternating current (TRIAC)
- Diode For alternating current (DIAC)
- Unijunction transistor (UJT)
- Insulated-gate bipolar transistor (IGBT)

5.1. SILICON CONTROLLED RECTIFIER

A silicon-controlled rectifier (SCR) or semiconductor-controlled rectifier is a solid-state device with four layers used to control current. The term silicon-controlled rectifier is the trade name used by General Electric for this type of thyristor. The principle of four-layer p–n–p–n switching was developed by Moll, Tanenbaum, Goldey, and Holonyak of Bell Laboratories in 1956 The practical demonstration of silicon-controlled switching and detailed theoretical behavior of a device, in agreement with the experimental results were presented by Dr Ian M. Mackintosh of Bell Laboratories in January 1958. The SCR was developed by a team of power engineers led by Gordon Hall and commercialized by Frank W. "Bill" Gutzwiller in 1957.

There is debate among sources about whether silicon-controlled rectifiers and thyristors are the same thing. Some say they are synonymous, while others argue that silicon-controlled rectifiers are a subset of thyristors, which are devices with at least four layers of alternating n- and p-type material. Bill Gutzwiller notes that "SCR" and "controlled rectifier" were used before "thyristor" became popular internationally (Betak & Musil, 2007).

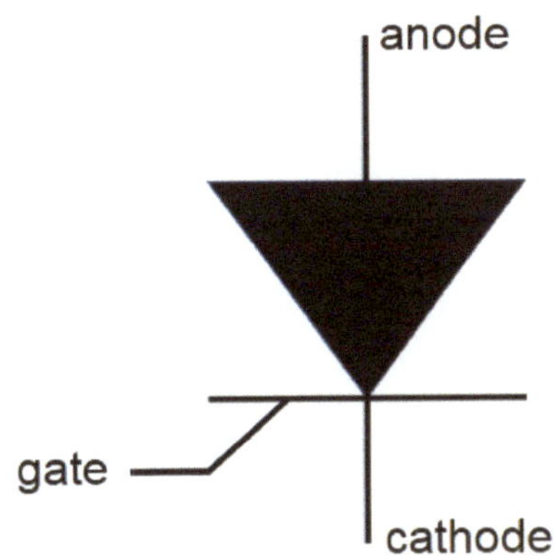

Figure 5.1. *The schematic symbol for a silicon-controlled rectifier (SCR).*

Source: https://blog.mbedded.ninja/electronics/components/transistors/silicon-controlled-rectifiers-scrs/.

TRIACs are bidirectional devices, meaning that charge carriers can flow through them in either direction, whereas SCRs are unidirectional, meaning they can conduct current only in one direction. Unlike TRIACs, which can be triggered by either a positive or negative current applied to their gate electrode, SCRs can only be normally triggered by a positive current flowing into the gate.

5.1.1. Modes of Operation

There are three modes of operation for an SCR depending upon the biasing given to it:

- Forward blocking mode (off state)
- Forward conduction mode (on state)
- Reverse blocking mode (off state)
- **Forward Blocking Mode:** In this operational mode, the anode is given a positive voltage, and the cathode is given a negative voltage, while the gate is disconnected. Junctions J1 and J3 are forward-biased, and J2 is reverse-biased, allowing for only a small leakage current. When the applied voltage reaches a certain value for J2, it undergoes avalanche breakdown and starts conducting. Below this voltage, J2 offers high resistance, and the SCR is considered to be in the off state.

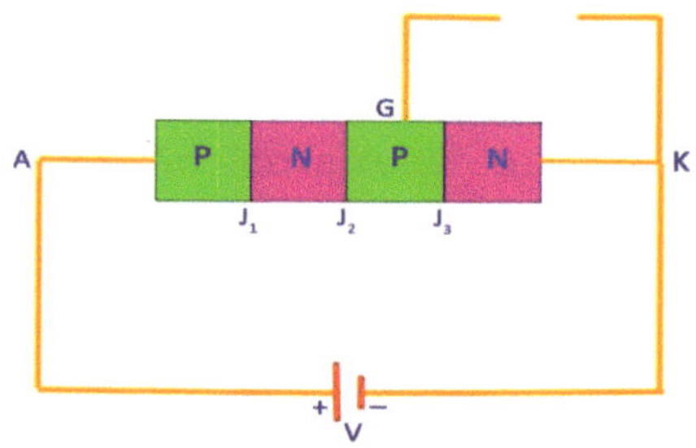

Figure 5.2. *Forward blocking mode of SCR.*

Source: https://how2electronics.com/silicon-controlled-rectifier-scr-construction-working-applications/#google_vignette.

- **Forward Conduction Mode:** There are two ways to transition an SCR from blocking mode to conduction mode: either by surpassing the breakover voltage with the voltage between the anode and cathode, or by delivering a positive pulse to the gate. Once the SCR begins conducting, it will remain in the ON state without the need for additional gate voltage. The minimum current required to keep the SCR in the ON state after removing the gate voltage is known as the latching current.

There are two ways to turn it off:

- o Reduce the current through it below a minimum value called the holding current, or

- o With the gate turned off, short-circuit the anode and cathode momentarily with a push-button switch or transistor across the junction.

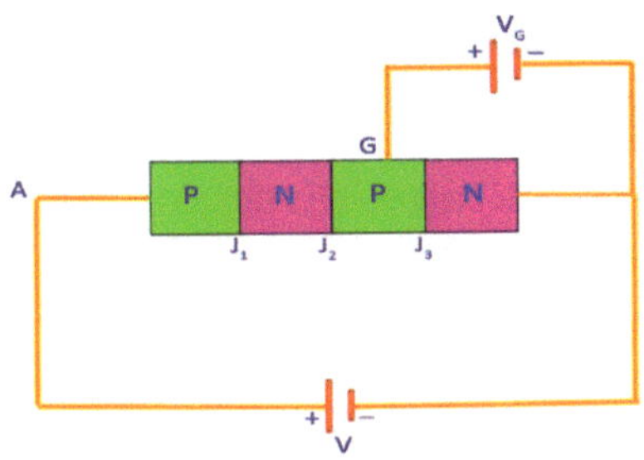

Figure 5.3. *Forward conduction mode of SCR.*

Source: https://how2electronics.com/silicon-controlled-rectifier-scr-construction-working-applications/#google_vignette.

- **Reverse Blocking Mode:** The SCR is in reverse blocking mode when a positive voltage is applied to the cathode and a negative voltage is applied to the anode. This results in J2 being forward biased and J1 and J3 being reverse biased. The apparatus functions as two diodes linked in series. There is a tiny leakage current. This blocking mode is reversed. An avalanche at J1 and J3 causes the reverse current to increase quickly at the critical breakdown level, also known as the reverse breakdown voltage (V_{BR}), if the reverse voltage is raised. Because a long, low-doped P1 region is required, reverse blocking capability in SCRs increases the forward voltage drop. The forward blocking voltage rating and the reverse blocking voltage rating are typically the same. Reverse blocking SCRs are commonly used in current-source inverters (Du et al., 2020).

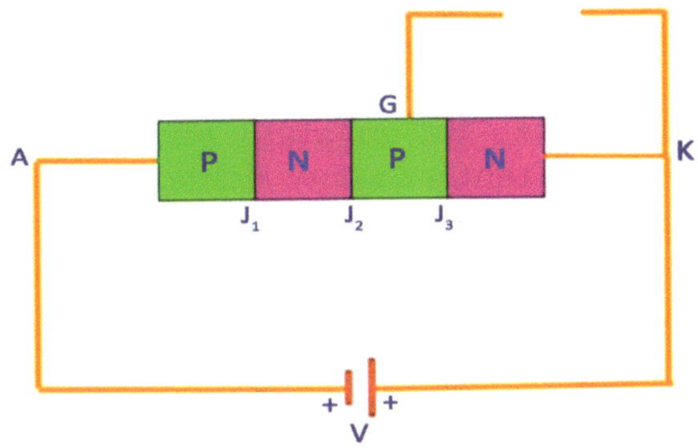

Figure 5.4. *Reverse blocking mode of SCR.*

Source: https://how2electronics.com/silicon-controlled-rectifier-scr-construction-working-applications/#google_vignette.

An asymmetric SCR, or ASCR for short, is a SCR that cannot block reverse voltage. Its reverse breakdown rating is usually in the tens of volts. Where reverse voltage would never occur (such as in switching power supplies or DC traction choppers) or where a reverse conducting diode is applied in parallel (such as in voltage-source inverters), ASCRs are employed.

It is possible to fabricate asymmetric SCRs in the same package as a reverse conducting diode. Reverse conducting thyristors, or RCTs, are what these are called.

5.1.2. Thyristor Turn-on Methods

There are mainly five different methods that turn on SCR.

1. Forward-voltage triggering
2. Gate triggering
3. dv/dt triggering
4. Thermal triggering
5. Light triggering

When the gate circuit is opened and the anode-cathode forward voltage rises, forward-voltage triggering takes place. This is the breakdown of junction J2, also referred to as an avalanche. The thyristor transitions to its on state with a large forward current and low voltage drop at suitable voltages. J1 and J3 in this instance are already forward biased.

If the thyristor is in the forward blocking state, where the applied voltage is less than the breakdown voltage, gate triggering will take place; if not, forward-voltage triggering might happen. Next, a single brief pulse of positive voltage can be applied between the cathode and the gate. This activates the thyristor by providing a single gate current pulse that puts it in the on state. This is the most typical way to activate a thyristor in real life.

When the depletion region's width shrinks with rising temperatures, temperature triggering takes place. A very slight temperature increase near the VPO causes junction J2 to be removed, activating the device.

5.1.3. Simple SCR Circuit

An example of a basic SCR circuit involves connecting an AC voltage source to an SCR with a resistive load. When no current pulse is sent to the gate of the SCR, it remains in a forward blocking state. This allows for controlled initiation of conduction in the SCR. The delay angle α, which is the timing of the gate current pulse in relation to natural conduction ($\omega_t = 0$), determines when conduction begins. Once the SCR conducts, the SCR does not turn off until the current through the SCR, i_s, becomes negative. i_s stays zero until another gate current pulse is applied and SCR once again begins conducting.

5.1.4. Applications

SCRs are primarily utilized in devices that require high power control, sometimes in conjunction with high voltage. They can be used in medium- to high-voltage AC power control applications, such as motor control, power regulators, and lamp dimming, because of the way they operate.

In high-voltage direct current power transmission, high-power AC is rectified using SCRs and related devices. Additionally, they are employed in the management of welding equipment, primarily for gas tungsten arc welding and related procedures. SCRs functions as an electrical switch in a variety of gadgets. The term "solid-state" refers to the electronic controls used by early solid-state pinball machines to operate solenoids, lights, and other components rather than mechanical ones.

Additional uses include timing circuits, computer logic circuits, power switching circuits, controlled rectifiers, SCR crowbars, DC shunt motor speed control, and inverters.

5.1.5. Comparison with SCS

With a few exceptions, a silicon-controlled switch's (SCS) behavior is quite similar to that of a silicon-controlled relay (SCR). When a positive voltage or input current is applied to the gate lead of an SCS, unlike an SCR, it turns off. An SCS, as opposed to an SCR, can become active when the same lead receives a negative voltage or output current.

Almost any circuit that requires a switch that toggles on and off using two separate control pulses can benefit from the use of SCSs. This comprises counters, lamp drivers, logic circuits, and circuits for power switching.

5.2. TRIODE FOR ALTERNATING CURRENT (TRIAC)

An electronic component with three terminals that conducts current in either direction when triggered is known as a TRIAC, or triode for alternating current: also known as a bidirectional triode thyristor or bilateral triode thyristor. TRIAC is a trademark that has been genericized.

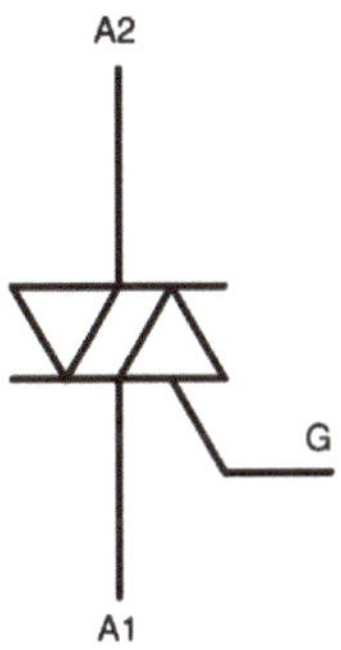

Figure 5.5. *TRIAC schematic symbol.*

Source: https://intradefairs.com/triacs.

TRIACs are a subclass of thyristors, related to silicon-controlled rectifiers (SCRs), in that a small voltage and current can control a much larger voltage and current, similar to a relay. An SCR can only conduct current in one direction; in contrast, TRIACs permit current to flow in both directions. A positive or negative voltage can be applied to the gate of most TRIACs to cause them to activate (a positive voltage is necessary for an SCR to function). After they are activated, SCRs and TRIACs keep conducting even when the gate current stops until the main current falls below a predetermined value, which is known as the holding current. (Junming Zhang, Hulong Zeng, and Ting Jiang, 2012)

Like TRIACs, gate turn-off thyristors (GTOs) offer additional control by shutting off when the gate signal is interrupted. Because TRIACs are bidirectional, they are useful switches for alternating current (AC). Furthermore, phase control—which involves applying a trigger at a specific angle to the AC in the main circuit—allows one to regulate the average current that flows into a load. This is frequently used to dim lights, regulate electric heaters, and adjust the speed of a universal motor. Bipolar devices are TRIACs.

5.2.1. Operation

Think about the triggering in each of the four possible combinations of gate and MT2 voltages with respect to MT1 to comprehend how TRIACs operate. The Figure 5.6 depicts the four distinct cases, or quadrants.

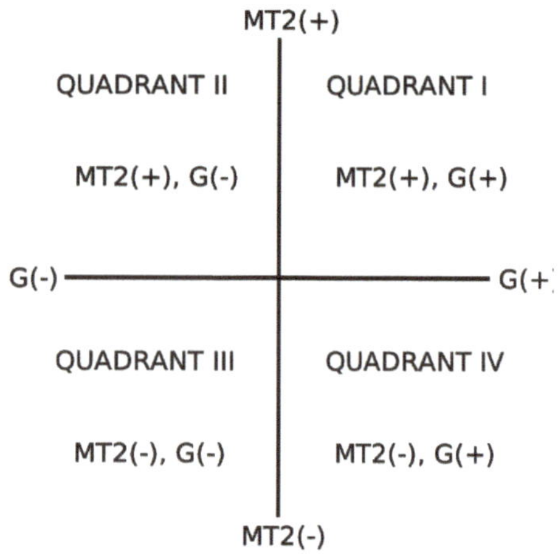

Figure 5.6. *Triggering modes. Quadrants, 1 (top right), 2 (top left), 3 (bottom left), 4 (bottom right); voltages relative to MT1.*

Source: https://intradefairs.com/triacs.

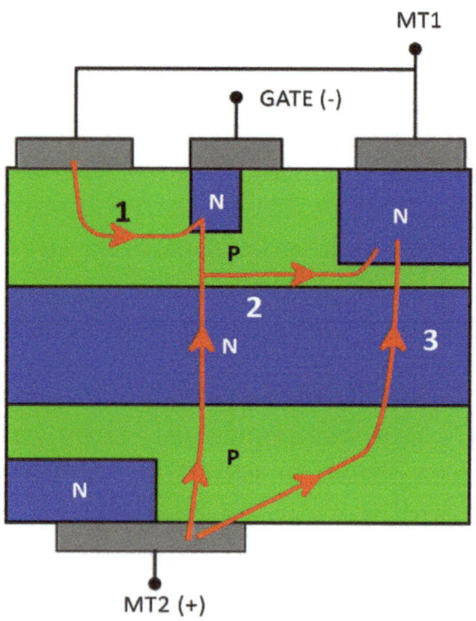

Figure 5.7. *TRIAC semiconductor construction.*

Source: https://intradefairs.com/triacs.

Anode 1 (A1) and Anode 2 (A2) are other names for Main Terminals 1 (MT1) and Main Terminal (MT2), respectively. Quadrant I is the most sensitive (requiring the least amount of gate current), and quadrant 4 is the least sensitive (requiring the most gate current). The relative sensitivity varies depending on the physical structure of a specific TRIAC.

MT2 is positive in quadrants 1 and 2, and current passes through the P, N, P, and N layers from MT2 to MT1. The N region that is connected to MT2 has a negligible impact. MT2 is negative in quadrants 3 and 4, and current passes through P, N, P, and N layers in addition to flowing from MT1 to MT2. While the N region connected to MT1 only takes part in the initial triggering and not the bulk current flow, the N region connected to MT2 is active.

Quadrants 1 and 3 are the only operating modes (both gate and MT2 positive or negative against MT1) because in most applications, MT2 provides the gate current. In other applications, the gate is pulled down to 0V (ground) and MT1 is typically connected to a positive voltage (e.g., +5V) in quadrants 2 and 3. These applications use single polarity triggering from an IC or digital drive circuit.

- **Quadrant 1:** When both the gate and MT2 are positive in relation to MT1, a quadrant 1 operation takes place.

Figure 5.8 provides an illustration of the mechanism. An equivalent NPN transistor is turned on by the gate current, which causes an equivalent PNP transistor to turn on as well by drawing current from its base. A portion of the gate current (dotted line), which would have otherwise flowed through the NPN transistor base and into MT1, is lost via the ohmic path across the p-silicon. In this instance, the stacked n, p, and n layers beneath MT1 behave like an NPN transistor, which turns on when a current flows through its base, thanks to the injection of holes in the p-silicon. The p, n, and p layers over MT2 then behave as a PNP transistor as a result, and when this happens, the transistor turns on because its n-type base becomes forward-biased in relation to its emitter (MT2). Thus, an SCR and the triggering scheme are the same. The corresponding circuit is shown in Figure 5.9.

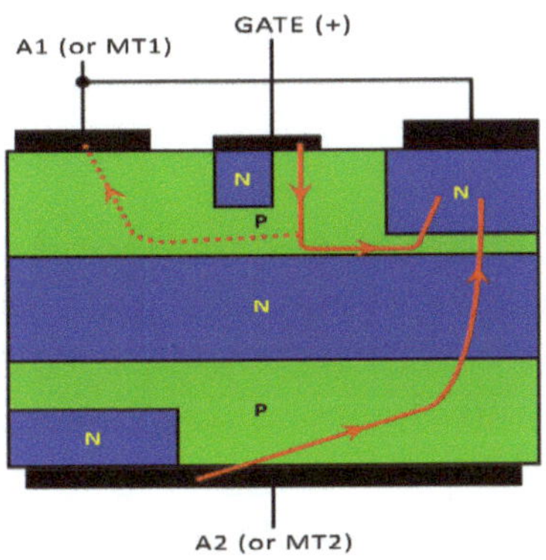

Figure 5.8. *Operation in quadrant 1.*

Source: https://intradefairs.com/triacs.

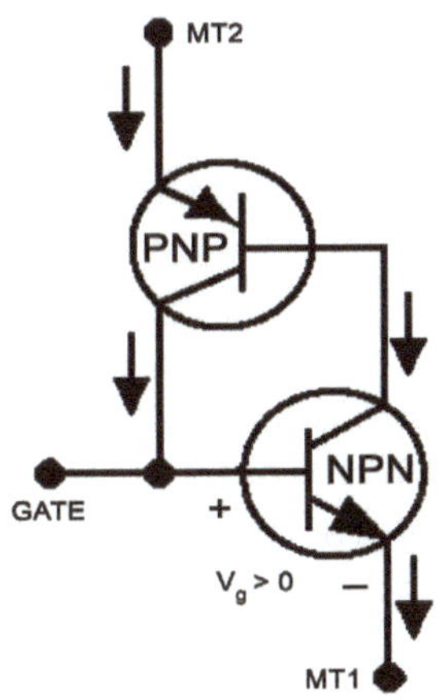

Figure 5.9. *Equivalent electric circuit for a TRIAC operating in quadrant 1.*

Source: https://intradefairs.com/triacs.

But unlike SCRs, the structure is different. Specifically, in TRIAC, a small current always flows straight through the p-silicon from the gate to MT1, bypassing the p-n junction that connects the base and emitter of the corresponding NPN transistor.

A red line in Figure 5.8 represents this current, which explains why a TRIAC requires a higher gate current to turn on than a SCR with a similar rating. This quadrant is typically the most sensitive of the four. This is because it is the only quadrant in which one of the primary device transistors' bases is directly injected with gate current.

- **Quadrant 2:** When MT2 is positive in relation to MT1 and the gate is negative, quadrant 2 operation takes place.

The triggering process is shown in Figure 5.10. The device turns on in three stages, beginning with the p-n junction beneath the gate allowing current from MT1 to enter the gate. This turns on a system made up of a PNP and NPN transistor, with the gate acting as the cathode.

This creates a current between the left and right sides of the p-silicon, which in turn switches on the NPN transistor under the MT1 terminal and consequently also the PNP transistor between MT2 and the right side of the upper p-silicon. As current into the gate increases, the potential of the left side of the p-silicon under the gate rises towards MT1, as the difference in potential between the gate and MT2 tends to lower. Thus, the structure that the majority of the current ultimately crosses is the same as the quadrant-I operation.

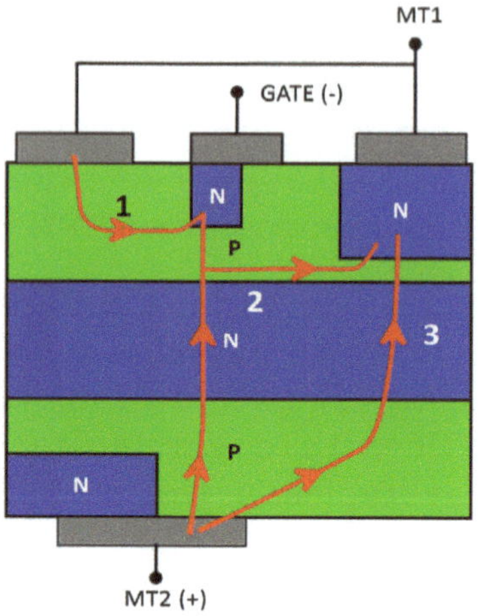

Figure 5.10. *Operation in quadrant 2.*

Source: https://intradefairs.com/triacs.

- **Quadrant 3:** Quadrant 3 operation occurs when the gate and MT2 are negative with respect to MT1.

Figure 5.11 provides an overview of the entire procedure. This process also occurs in distinct steps. The pn junction between the gate and the MT1 terminal becomes forward-biased during the first phase. Electrons are injected in the p-layer beneath the gate because forward-biasing implies the injection of minority carriers in the two layers joining the junction. A portion of these electrons escape to the underlying n-region without recombining. As a result, the n-region's potential is lowered, serving as the base potential for a PNP transistor that turns on (activating a transistor without first lowering its base potential is known as remote gate control). In actuality, the lower p-layer serves as the base of an NPN transistor composed of the final three layers directly above the MT2 terminal, which is activated. As such, the lower p-layer acts as the collector of this PNP transistor and has its voltage increased. As a result, the red arrow in Figure 5.11 that is marked with a 3 represents the final path of current conduction.

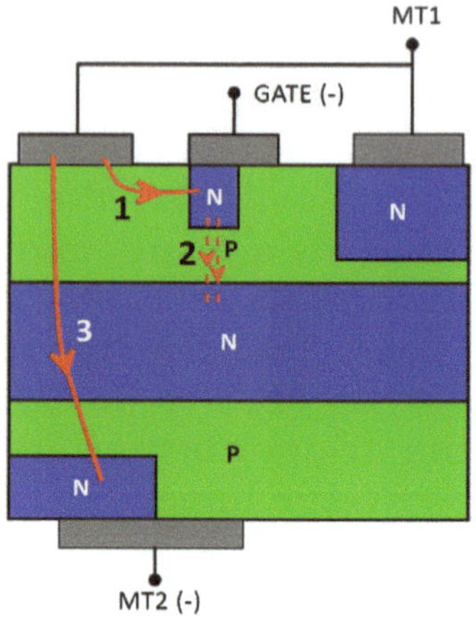

Figure 5.11. *Operation in quadrant 3.*

Source: https://intradefairs.com/triacs.

- **Quadrant 4:** Quadrant 4 operation occurs when the gate is positive and MT2 is negative with respect to MT1.

This quadrant's triggering is comparable to quadrant III's triggering. The procedure, which is shown in Figure 5.12, makes use of a remote gate control. Minority carriers, or free electrons, are injected into the p-region when current moves from the p-layer beneath the gate into the n-layer under MT1. Some of these electrons are gathered by the underlying n-p junction and enter the adjacent n-region without recombining. This lowers the n-layer's potential and activates the PNP transistor made up of the n-layer and the two p-layers adjacent to it, just like in the case of a triggering in quadrant III. In actuality, the lower p-layer serves as the base of an NPN transistor composed of the final three layers directly above the MT2 terminal, which is activated. As such, the lower p-layer acts as the collector of this PNP transistor and has its voltage increased. As a result, the red arrow in Figure 5.12 that is marked with a 3 represents the current's final conduction path.

This is the least sensitive quadrant out of the four. Additionally, only the other three quadrants can trigger certain TRIAC models (logic level and snubberless types).

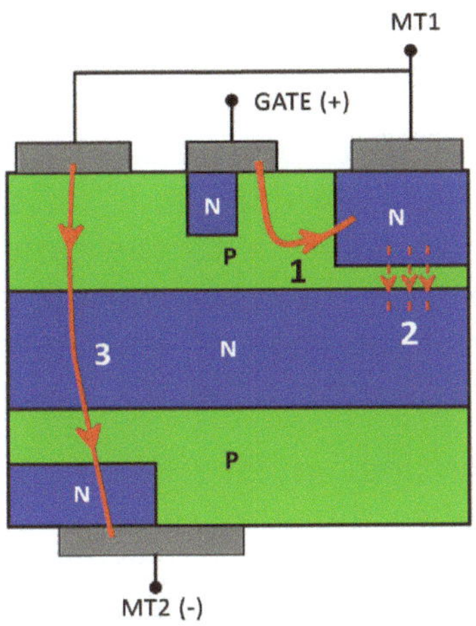

Figure 5.12. *Operation in quadrant 4.*

Source: https://intradefairs.com/triacs.

5.2.2. Issues

When utilizing a TRIAC in a circuit, one should be aware of certain limitations.

- *Gate Threshold Current, Latching Current and Holding Current:* When there is enough current flowing into or out of a TRIAC's gate to activate the pertinent junctions in its operating quadrant, the TRIAC begins to conduct. The term "gate threshold current" refers to the lowest current that can accomplish this, and IGT typically indicates it. In a typical TRIAC, the gate threshold current is generally a few milliamperes, but one has to take into account also that:
  - o I_{GT} depends on the temperature: The higher the temperature, the higher the reverse currents in the blocked junctions. This implies the presence of more free carriers in the gate region, which lowers the gate current needed.
  - o I_{GT} depends on the quadrant of operation, because a different quadrant implies a different way of triggering. As a rule, the first quadrant is the most sensitive (i.e., requires the least current to turn on), whereas the fourth quadrant is the least sensitive.
  - o IGT is dependent on the voltage applied to the two main terminals, MT1 and MT2, when it turns on from an off state. Like high temperature operation, higher voltage between MT1 and MT2 causes greater reverse currents in the blocked junctions, requiring less gate current. I_{GT} is typically provided for a specific voltage between MT1 and MT2 in datasheets.

The device continues to conduct when the gate current is cut off if the current between the two main terminals is greater than what is known as the latching current; if not, the device may turn off. The bare minimum latching current required to maintain the internal structure of the device latched is required to compensate for the absence of gate current. The value of this parameter varies with:

- o Gate current pulse (amplitude, shape and width)
- o Temperature
- o Control circuit (resistors or capacitors between the gate and MT1 increase the latching current because they steal

some current from the gate before it can help the complete turn-on of the device)

- o Quadrant of operation

Specifically, when the gate signal is cut off and the latching current drops to a minimum known as the holding current, the TRIAC has finished triggering if the gate current's pulse width is large enough—typically a few tens of microseconds. Holding current is the minimal current needed to maintain the device's operation after complete commutation of all internal components between the two main terminals.

In datasheets, the latching current is indicated as I_L, while the holding current is indicated as I_H. They are typically in the order of some milliamperes.

- *Static dv/dt:* A high dv/dt between MT2 and MT1 may turn on the TRIAC when it is off. Typical values of critical static dv/dt are in terms of volts per microsecond.

A parasitic capacitive coupling between the gate terminal and the MT2 terminal is what causes the turn-on, allowing currents to enter the gate in response to a significant voltage change at MT2. A viable solution to this constraint is to create an appropriate RC or RCL snubber network. This is often enough to reduce the gate's impedance towards MT1. Between these two terminals, the capacitive current generated during the transient flows out of the device without activating it by connecting a resistor, small capacitor, or both in parallel. To design the right network, one must carefully review the application notes supplied by the manufacturer and conduct testing on the specific device model.

Capacitors and resistors between the gate and MT1 can have typical values of up to 100 nF and 10 Ω to 1 kΩ, respectively. Such a resistor is already included in normal TRIACs, except for low-power models sold as sensitive gates, to protect against spurious dv/dt triggering. This will prevent a multimeter from seeing the gate's ostensible diode-like behavior when testing a TRIAC.

In datasheets, the static dv/dt is usually indicated as $(dv/dt)_s$ and, as mentioned before, is in relation to the tendency of a TRIAC to turn on from the off state after a large voltage rate of rise even without applying any current in the gate.

- *Critical di/dt:* Even though the pulse duration is very short, a high rate of rise in current between MT1 and MT2 (in either direction) during device turn-on can damage or destroy the TRIAC. The cause is that the power dissipation within the device is not distributed evenly during the commutation. When the device is turned on, current flows through it before it has had a chance to spread throughout the junction. Too quick a current rise could result in localized hot spots that could permanently harm the TRIAC. Normally, the device conducts the current imposed by the external circuitry after a few nanoseconds or microseconds, but the full junction switch on takes much longer.

In datasheets, this parameter is usually indicated as di/dt and is typically in the order of the tens of ampere per microsecond.

- *Commutating dv/dt and di/dt:* When a TRIAC has been conducting and tries to turn off with a partially reactive load, like an inductor, the commutating dv/dt rating is applicable. Because the voltage and current are out of phase, the TRIAC tries to turn off when the current falls below the holding value. However, a sudden voltage step occurs between the two main terminals as a result of the phase shift between the current and voltage, turning the device back on.

In datasheets, this parameter is usually indicated as $(dv/dt)_c$ and is generally in the order of up to some volts per microsecond.

Because there is still some excess minority charge in the device's internal layers from the previous conduction, just before it tries to turn off, commutating dv/dt is less than static dv/dt. These charges change the internal potential of the area around the gate and MT1 when the TRIAC begins to turn off, making it simpler for the capacitive current caused by dv/dt to turn on the device once more.

The di/dt of the current flowing from MT1 to MT2 is another crucial component in the commutation from on-state to off-state. The recovery in conventional diodes is comparable in that the reverse current increases with increasing di/dt. Due to the parasitic resistances present in the TRIAC, a high reverse current in the p-n junctions within it may cause a voltage drop between the MT1 region and the gate region, potentially keeping the TRIAC turned on.

In a datasheet, the commutating d*i*/d*t* is usually indicated as $(di/dt)_c$ and is generally in the order of some amperes per microsecond.

When using the TRIAC to drive a load that has a phase shift between current and voltage, like an inductive load, the commutating dv/dt is crucial. Assuming the inductor is to be turned off, the TRIAC tries to turn off when the current drops to zero and the gate is not fed. However, this results in a step in the voltage across it because of the previously mentioned phase shift. If the device's commutating dv/dt rating be surpassed, it will not activate.

5.2.3. Snubber Circuits

Make sure the TRIAC correctly turns off at the end of each half-cycle of the AC in the main circuit when using it to control reactive (inductive or capacitive) loads. Reactive loads can cause a phase shift in current and voltage, which can result in a voltage step that can accidentally turn on the thyristor. This is because TRIACs are susceptible to rapid voltage changes (dv/dt) between MT1 and MT2. Electric motors are generally inductive loads, and off-line power supplies, which are found in many computers and TVs, are capacitive.

A snubber circuit (of the resistor/capacitor or resistor/capacitor/inductor type) placed between MT1 and MT2 can prevent unwanted turn-ons. Additionally, snubber circuits are employed to stop premature triggering, which can be brought on by mains supply voltage spikes, among other things.

To provide MT1 with a low-impedance path and further prevent false triggering, a gate resistor, capacitor, or both in parallel, may be connected between the gate and MT1. This is because turn-ons are caused by internal capacitive currents flowing into the gate because of a high dv/dt, or rapid voltage change. However, because of the capacitor charging, this raises the necessary trigger current or adds latency. In contrast, a resistor placed between the gate and MT1 aids in draining leakage currents from the circuit, enhancing TRIAC performance at elevated temperatures where the maximum permitted dv/dt is lowered. Resistors with values less than 1kΩ and capacitors with values up to 100nF are typically appropriate for this use; however, the specific device model should be adjusted.

One TRIAC may be substituted with two SCRs operating in inverse parallel for more powerful and demanding loads. No matter the nature of the load, the SCRs will all turn off because they will receive a full half-cycle of reverse polarity voltage. But correctly triggering the SCRs is more difficult than triggering a TRIAC because of the separate gates.

If the main circuit current falls below the holding current at trigger time due to a current phase shift, TRIACs may also not activate reliably with reactive loads. To solve the issue, a DC or pulse train can be used to repeatedly trigger the TRIAC until it activates.

5.2.4. Application

Low-power TRIACs are used in a wide range of applications, including light dimmers, electric fan and motor speed controls, and the computerized control circuits found in many contemporary small and large household appliances.

Optoisolators are commonly used when microcontrollers trigger mains voltage TRIACs; for instance, optotriacs can be used to control the gate current. An alternative would be to connect one of the microcontroller's power rails to a main supply when safety permits and electrical isolation of the controller is not required. It is customary in these circumstances to connect the TRIAC's A1 to the positive rail of the microcontroller's power supply, and to connect A2 to the live. The microcontroller can be connected to the TRIAC's gate via a resistor or, in rare cases, a transistor. This allows the microcontroller to sense a drop in voltage and pull sufficient current through the TRIAC's gate to activate it. By doing this, quadrant IV, where TRIACs are normally insensitive, is avoided, and the TRIAC is guaranteed to be triggered in quadrants II and III.

5.3. DIAC (DIODE FOR ALTERNATING CURRENT)

A full-wave or bi-directional semiconductor switch with the ability to operate in both forward and reverse polarity is called a DIAC. The acronym for "diode alternating current" (DIAC) gives it its name. In AC switches, the DIAC is frequently used to help evenly trigger a TRIAC. The primary uses for DIACs are in dimmer applications and fluorescent lamp starter circuits. A DIAC is a three-layer, two-terminal, bi-directional device that can be turned on or off depending on the applied voltage polarity.

5.3.1. Circuit Symbol

The DIAC circuit symbol consists of two triangles positioned between two lines, showcasing the structure of the device as two junctions.

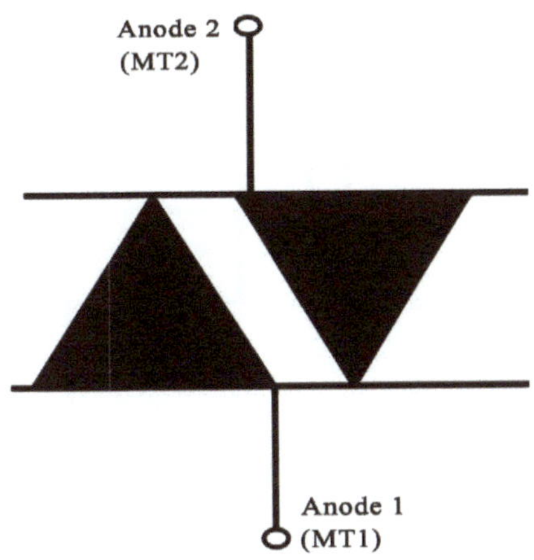

Figure 5.13. *Circuit symbol.*

Source: https://components101.com/articles/diac-symbol-construction-working-and-application-circuits.

The device typically has two terminals known as Anode 1 and Anode 2, or Main Terminals 1 and 2 (MT1 and MT2).

5.3.2. Construction

One of two forms for the DIAC construction is npn or pnp. The two leads are linked to silicon p-regions that are divided from one another by an n-region. The structure of DIAC is similar to that of a transistor difference are:

- There is no terminal attached to the base layer
- The three regions are nearly identical in size. The doping concentrations are identical to give the device symmetrical properties.

It is possible to fabricate the DIAC as a two- or five-layer structure. When reverse breakdown occurs at the reversely biased junction in the three-layer structure, switching takes place. The more popular three-layer version of the device has a break-over voltage of about 30 V. Because of the device's symmetry, operation is nearly symmetrical.

There is also a DIAC structure with five layers. Although it produces an I-V curve that is strikingly similar to the three-layer version, this does not behave quite in the same way. It is analogous to two break-over diodes that are connected back-to-back.

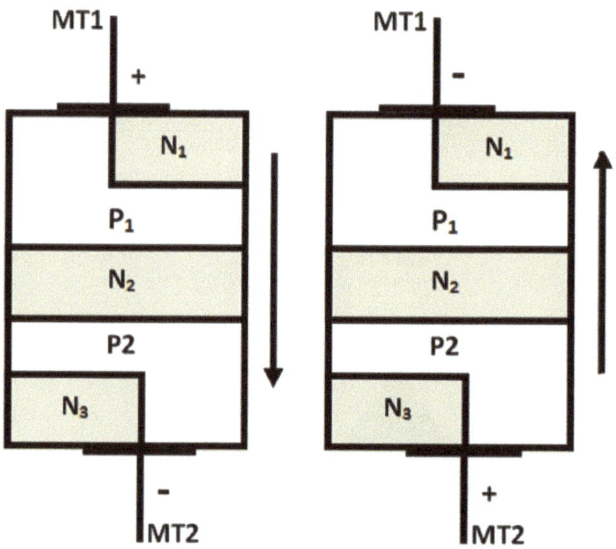

Figure 5.14. *The structure of a DIAC.*

Source: https://components101.com/articles/diac-symbol-construction-working-and-application-circuits.

The DIAC is used in three-layer form for the majority of applications. It offers a noticeable enough improvement in switching properties. The five-layer device might be employed in certain situations.

5.3.3. Operation

When a voltage is applied across the terminals of a DIAC, a small leakage current (I_{bo}) flow. As the voltage increases, the leakage current continues until it reaches a breakover voltage (V_{bo}). At this point, avalanche breakdown of the reverse-biased junction occurs, leading to negative resistance. The current through the device increases as the applied voltage decreases until it reaches the breakback voltage (V_w).

- *V- I characteristics of a DIAC*

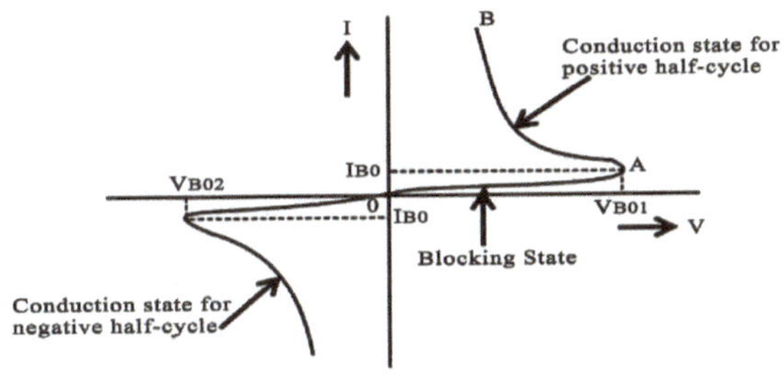

Figure 5.15. *The V- I characteristics of a DIAC.*

Source: https://www.hnhcart.com/blogs/active-components/diac.

The DIAC's V-I characteristic curve is shaped like a Z, with one end in the first quadrant and the other in the third quadrant. During the positive half cycle, current flows from MT1 to MT2, while during the negative half cycle, current flows from MT2 to MT1. The reverse biased junction between the layers initially results in higher resistance in the DIAC, leading to a small leakage current referred to as the blocking condition on the curve.

The breakdown voltage of the majority of DIACs is approximately 30 volts; however, the precise breakdown voltage will vary depending on the type of device. Until the current reaches the designated value known as the holding current—which is the minimal current needed to maintain a device in the ON state—the DIAC will be in the conducting condition.

5.3.4. Applications

Because a TRIAC requires a positive or negative pulse to the gate in order to conduct, the DIAC and TRIAC circuit are frequently combined to provide symmetric firing. There aren't many other uses for DIACs other than activating TRIACs and other kinds of thyristors. Light dimmers, heat controllers, motor speed control, phase control circuits, and numerous other control circuits are among the many applications that employ DIACs as trigger devices. Let's examine some examples of heat control and light dimmer circuits.

- **Heat Control:**

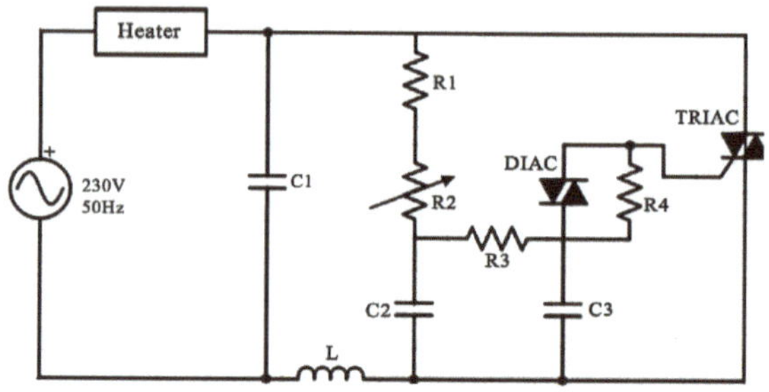

Figure 5.16. Heat Control.

The typical setup of a DIAC-TRIAC combo, which is used to smoothly control AC power to a heater, is shown in the circuit above. Capacitor C1 and choke L together create an LC circuit that lowers the voltage rise across the TRIAC when it turned off. The R4 resistor across the DIAC provides smooth control, and the R2 potentiometer regulates the applied voltage in both half-cycles. The DIAC is used to provide smooth regulation of the heater's heat output because the heat dissipated from the heater increases as the TRIAC conducts longer.

- **Light Dimmer:**

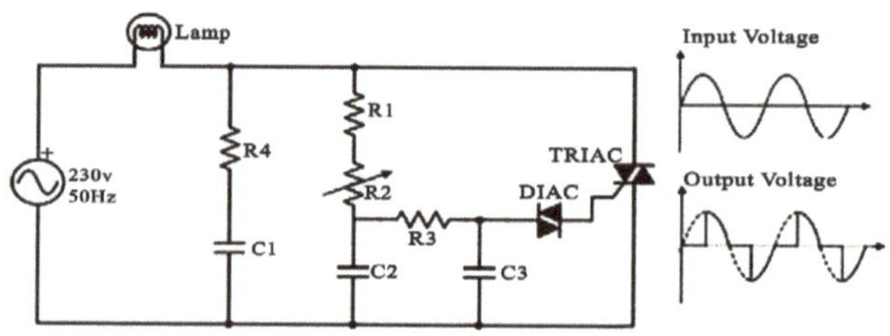

Figure 5.17. Light Dimmer.

The circuit above illustrates the RC phase shift network and DIAC controlled TRIAC for the light dimming application. The voltage applied to the TRIAC's gate terminal is altered by the RC configuration utilizing the resistor R3 and the capacitor C3. When the device is turned off, the rate at which the voltage rises is limited by the series R4-C1 network that is connected across the TRIAC.

As soon as the 230V power source's input voltage is applied, the capacitors C1 and C2 begin to charge at the rate that the variable resistor R2 indicates. The DIAC is triggered and starts to conduct when the voltage across the capacitor C3 exceeds the breakdown voltage of the DIAC. This causes the capacitor C3 to discharge and the TRIAC to receive the gate pulse. The gate pulse activates the TRIAC, causing current to start passing through the lamp. By varying the resistance in R2, which in turn alters the voltage rate at which the TRIAC is triggered in both the positive and negative half-cycles, the rate of charging of the capacitor can also be altered.

5.4. UNIJUNCTION TRANSISTOR

The unijunction transistor (UJT) is a semiconductor device with three leads and a single junction. Its main function is to operate as a switch controlled by electrical signals.

The UJT is no longer utilized as a linear amplifier, but rather finds its applications in free-running oscillators, synchronized or triggered oscillators, and pulse generation circuits at low to moderate frequencies (hundreds of kilohertz). It is commonly used in triggering circuits for silicon-controlled rectifiers. In the 1960s, its low cost per unit and unique characteristics made it popular for a variety of applications such as oscillators, pulse generators, saw-tooth generators, triggering circuits, phase control, timing circuits, and voltage- or current-regulated supplies. While the original unijunction transistor types are now considered outdated, the programmable unijunction transistor, a later multi-layer device, is still widely available.

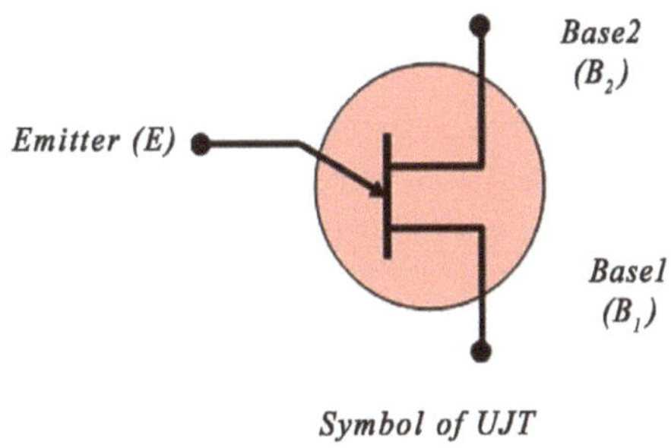

Figure 5.18. *Symbol of UJT.*

Source: https://www.learnelectronicswithme.com/2020/08/ujt-symbol-construction-working-and.html#google_vignette.

5.4.1. Types

1. The original unijunction transistor, also known as UJT, is a basic device consisting of a bar of n-type semiconductor material with diffused p-type material along its length, setting the device parameter ⊠. The 2N2646 model is the most widely utilized version of the UJT.

2. The CUJT is a semiconductor device with a bar of p-type material with diffused n-type material along its length, defining the parameter ⊠. The 2N6114 is a specific model of the CUJT.
3. The Programmable Unijunction Transistor (PUT) is a multi-junction device that, when used with two external resistors, exhibits characteristics similar to the UJT. Similar to the thyristor, the PUT also consists of four p-n layers. It has an anode and a cathode connected to the first and last layers respectively, with a gate connected to one of the inner layers, respectively. Although not directly interchangeable with traditional UJTs, PUTs can perform a similar function. When properly configured in a circuit with two programming resistors to set the parameter ⊠, they behave like a conventional UJT. Examples of PUT models include the 2N6027, 2N6028, and BRY39.

5.4.2. Applications

Unijunction transistor circuits were widely utilized in amateur electronic projects during the 1960s and 1970s due to their ability to construct simple oscillators with only one active device. These circuits were notably used in creating relaxation oscillators for variable-rate strobe lights. With the rise of integrated circuits, oscillators like the 555 timer IC became more prevalent over time.

Apart from its function as the active element in relaxation oscillators, UJTs or PUTs are highly significant when it comes to triggering thyristors, which include silicon-controlled rectifiers (SCR), TRIACs, and so on. A UJT or PUT circuit can be controlled by a DC voltage so that as the DC control voltage is increased, the on-period increases as well. This application is crucial for controlling high AC currents.

Magnetic flux can also be measured with UJTs. The voltage at the PN junction is modulated by the Hall effect. The frequency of UJT relaxation oscillators is impacted by this. However, this behavior is limited to UJT use; PUTs do not exhibit this characteristic.

5.4.3. Construction

The UJT is sometimes referred to as a double-base diode because it has three terminals: an emitter (E) and two bases (B_1 and B_2). A thinly doped silicon n-type bar forms the base. It has two ohmic contacts attached to its ends, B_1 and B_2. The material used in the emitter is heavily doped P-type. The device gets its name from the single PN junction that exists between the emitter and the base. Interbase resistance is the resistance that exists between B_1 and B_2 when the emitter is open-circuit. The device is not symmetrical because the emitter junction is typically located closer to base-2 (B_2) than base-1 (B_1), as symmetrical units do not offer the best electrical characteristics for the majority of applications.

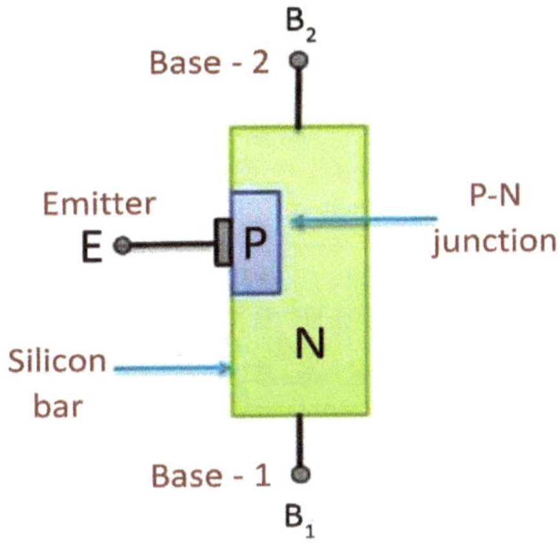

Figure 5.19. *Structure of a p-type UJT.*

Source: https://electronicscoach.com/unijunction-transistor-ujt.html.

5.5. INSULATED-GATE BIPOLAR TRANSISTOR (IGBT)

An electrical switch is a primary application for an insulated-gate bipolar transistor (IGBT), a three-terminal power semiconductor device. It was created with quick switching and high efficiency in mind. The IGBT has four alternating layers (P–N–P–N) that are controlled by a gate structure made of metal, oxide, and semiconductors (MOS).

The IGBT's structure is topologically similar to that of a thyristor with a MOS gate (MOS-gate thyristor), but over the whole operating range of the device, only the transistor action is allowed, and the thyristor action is completely suppressed. In the high-power applications, such as induction stoves, arc-welding machines, lamp ballasts, uninterruptible power supply systems (UPS), and variable-frequency drives (VFDs) for motor control in electric cars, trains, variable-speed refrigerators, and air conditioners, it is utilized for switching power supplies.

The IGBT is also used in switching amplifiers in sound systems and industrial control systems because of its quick on/off capability, which enabling the creation of complex waveforms with low-pass filters and pulse-width modulation. When used as an analog audio amplifier, modern devices handle audio frequencies that are at least ten times higher than their pulse repetition rates in switching applications, extending well into the ultrasonic range. After power MOSFETs, the IGBT was the second most popular power transistor as of 2010.

5.5.1. Basic Structure

This is a basic schematic of a typical DMOS-based N-channel IGBT. This is just one of the various configurations that this device can have. Except for the P+ injecting layer, it is clear that the silicon cross-section of an IGBT and a vertical Power MOSFET are nearly identical. Its P wells and MOS gate structure are comparable to those of N+ source regions. The source or emitter is the N+ layer at the top, and the drain or collector is the P+ layer at the bottom. P-channel IGBTs, for which the doping profile in each layer will be reversed, are also technically possible. A parasitic thyristor made up of a four-layer NPNP structure is present in IGBT. It is not desirable for this thyristor to turn on.

Non-punch through (NPT) IGBTs are those that are made without the N+ buffer layer, whereas punch-through (PT) IGBTs are made with this layer. If the doping level and layer thickness are chosen properly, the buffer layer's presence can greatly enhance the device's performance. Although they have similar physical characteristics, an IGBT operates more like a power BJT than a power MOSFET. The minority carrier injection into the N-drift region and the ensuing conductivity modulation are caused by the P+ drain layer, also known as the injecting layer.

Given its structure, an IGBT's basic equivalent circuit model—which includes MOSFET, JFET, NPN, and PNP transistors—can be created. The JFET connects the collector of the PNP to the base of the NPN, and the collector of the NPN to the base of the PNP. A regenerative feedback loop is formed by a parasitic thyristor, represented by the NPN and PNP transistors. To prevent the thyristor from latching up, which could cause the IGBT latch-up, the resistor RB represents the shorting of the NPN transistor's base-emitter.

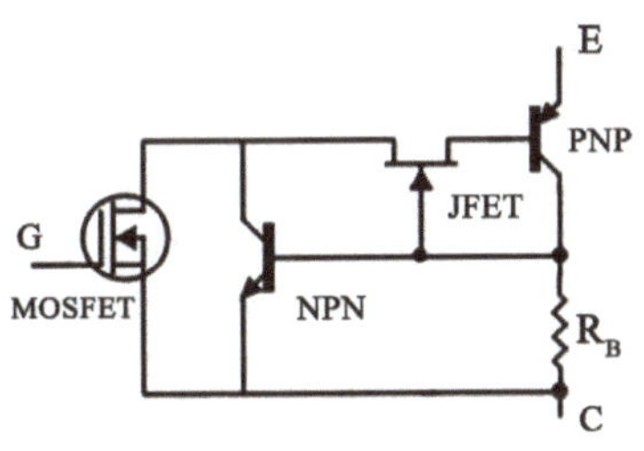

Figure 5.20. *Basic structure of insulated-gate bipolar transistor.*

Source: http://electricaltheorems.blogspot.com/2015/01/insulated-gate-bipolar-transistor-igbt.html.

Any two adjacent IGBT cells' current constriction is represented by a JFET. It provides most voltage support, enabling the MOSFET to be low voltage and, as a result, have a low RDS(on) value. The IGBT circuit symbol consists of three terminals: the Collector (C), the Gate (G), and the Emitter (E).

5.5.2. IGBT Characteristics

The IGBT, being voltage-controlled, only needs a small voltage on the gate to allow conduction, unlike BJTs, which require a continuous supply of base current in sufficient quantity to maintain saturation.

In addition, unlike MOSFETs, which can switch current in both directions (controlled in the forward direction and uncontrolled in the reverse direction), IGBTs are only able to switch current in the "forward direction," that is, from Collector to Emitter.

The insulated gate bipolar transistor's operating principle and gate drive circuits are extremely similar to those of the N-channel power MOSFET. The primary distinction is that, in the IGBT, the resistance provided by the main conducting channel during the device's "ON" state is substantially smaller. Consequently, the current ratings exhibit a significant increase compared to a MOSFET of equivalent power.

The IGBT is a better option for moderate speed, high voltage applications such as pulse-width modulation (PWM), variable speed control, switch-mode power supplies, solar-powered DC-AC inverters, and frequency converters operating in the hundreds of kilohertz range. Its low ON-resistance, ease of drive, and relatively fast switching speeds are its main advantages over other types of transistor devices.

Table 5.1. *IGBT Comparison Table*

Device Characteristic	Power Bipolar	Power MOSFET	IGBT
Voltage Rating	High <1kV	High <1kV	Very High >1kV
Current Rating	High <500A	Low <200A	High >500A
Input Drive	Current, h_{FE} 20-200	Voltage, V_{GS} 3-10V	Voltage, V_{GE} 4-8V
Input Impedance	Low	High	High
Output Impedance	Low	Medium	Low
Switching Speed	Slow (uS)	Fast (nS)	Medium
Cost	Low	Medium	High

Source: https://www.electronics-tutorials.ws/power/insulated-gate-bipolar-transistor.html.

5.5.3. Operation Modes

- **Forward-Blocking and Conduction Modes:** Junctions J1 and J3 are forward-biased, and junction J2 is reverse-biased, when a positive voltage is applied across the collector-to-emitter terminal with the gate shorted to the emitter. This causes the device to go into forward blocking state. On both sides of junction J2, a depletion layer reaches partially into the N-drift and P-base regions. It is possible to convert an IGBT from the forward-blocking state to the forward conducting state by inverting the Si below gate in the P base region and removing the gate-emitter shorting. This creates a conducting channel that connects the N-drift region and the N+ emitter. Electrons are moved from the N+ emitter to the N-drift via this channel. The P+ collector/N-drift becomes forward-biased as a result of the electron flow into the N-drift decreasing the region's potential. A high density of minority carrier holes is injected into the N-drift from the P+ collector under these forward-biased conditions. A condition known as the development of a holes plasma in the N-drift region occurs when the injected carrier concentration is significantly greater than the background concentration. To preserve local charge neutrality, electrons from the emitter contact are drawn to this plasma of holes. In this way, the N-drift region gathers roughly equal excess concentrations of electrons and holes. The conductivity of the N-drift region is significantly increased by these excess electron and hole concentrations. The conductivity modulation of the N-drift region is the term used to describe this mechanism of conductivity rise.

- **Reverse-Blocking Mode:** Junction J1 gets reverse-biased and its depletion layer spreads into the N-drift region when a negative voltage is applied across the collector-to-emitter terminal. The P+ collector/N-drift/P-base regions form an open-based BJT that determines the break-down voltage during reverse-blocking. If there is very little doping in the N-drift region, the device is vulnerable to punch-through. The resistivity and thickness of the N-drift region can be optimized to achieve the required reverse voltage capability.

CASE STUDY

IMPLEMENT AN AC-SWITCH TRIGGERING CIRCUIT USING AN OPTICAL TRIAC

Several different technologies and designs are available among solid-state switches today. The major families are the standard TRIAC (triode for alternating current), the snubberless TRIAC, and the AC-switch (ACS) devices introduced in the early 1990s. These devices are all triggered thanks to a gate current, but this current could be sunk from or sourced to the gate, depending on the technology or design. The triggering circuit must therefore consider the AC-switch type to correctly trigger it. For ACS devices, due to silicon structure, the gate current can only be sunk from the gate.

Sometimes, the control circuit has also to be insulated from the mains voltage. This occurs, for example, when the MCU reference is not the same than the AC switch reference. This is the case with new appliances using an inverter for 3-phase motor control, where the MCU is connected on the DC rail, and the ACS switch is referenced to line.

But it is also required if the appliance designer wants to insulate all the low-voltage circuit from the line. This solution is often expensive, as it would be easier to use an "insulated-enough" user-interface and keep all electronics referenced to the line. This is particularly the case when only a few push-buttons are accessible for the end-user.

The standard insulated solution used to trigger a TRIAC is an opto- TRIAC connected in series with the Triac A2 and G terminals. A resistor is also connected in series to reduce the gate current applied by the opto- TRIAC. Such a solution is good for all TRIAC s. The device is then triggered by a positive gate current when the voltage across the TRIAC, just before turn-on, is positive, and by a negative current in the reverse situation. The TRIAC is then triggered in what are called Q1 and Q3 quadrants.

For ACS devices, as previously said, these devices can only be triggered by a negative current. If an opto- TRIAC is used to drive them, the ACS will only be turned on for a negative bias voltage (where the gate current will also be negative). This will lead to a half-cycle conduction of the ACS.

Certainly, for most applications, such an operating mode is not convenient, but there are new applications where the load only has to conduct in half-cycle mode. For example, this is the case for some pumps, such as used in coffee machines, which feature an internal diode, and for some electromagnets used for door-lock function in washing machines, for which single half-cycle conduction is enough to trigger them.

If the line voltage is applied across the gate and COM terminals, the internal P-N junction may be damaged as its breakdown level is around 10 V. This case could occur during devices switching transients or if the opto- TRIAC fails in short-circuit mode. The solution is then to add a low-voltage or high-voltage diode, respectively, in parallel to the COM-G junction or in series with the opto-Triac (Figure 1).

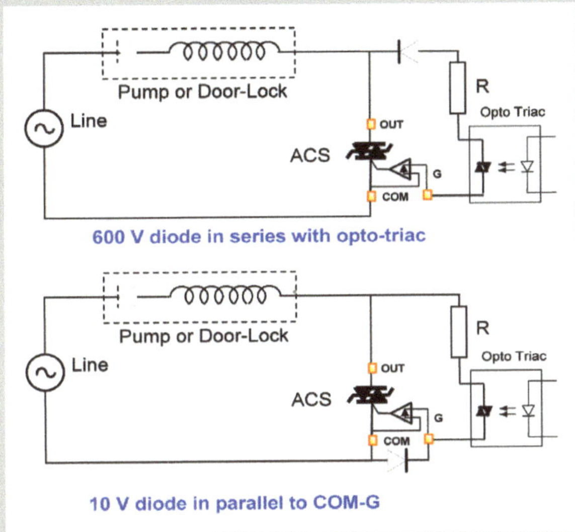

Figure 1. *Solution for half-cycle ACS control with opto-Triac.*

Source: https://www.eetimes.com/implement-an-ac-switch-triggering-circuit-using-an-optical-triac/

Note that in this second case (diode in series with opto), the opto- TRIAC and the diode could be replaced by a reverse-blocking opto-thyristor, with anode terminal connected to ACS gate.

For home appliances, most loads are controlled in full-cycle mode. The previous schematic has then to be adapted to ensure ACS triggering for each cycle. The solution is to add a low-voltage capacitor to store some charge, to apply a gate current at the beginning of the positive conduction. This solution is presented in Figure 2, which also uses two low-voltage diodes.

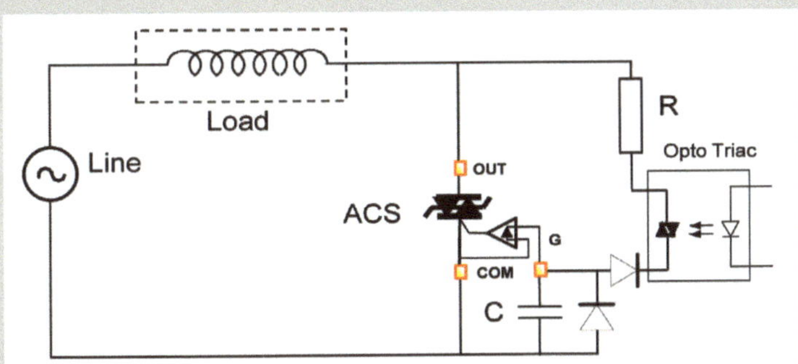

Figure 2. *Solution for full-cycle ACS control with opto- TRIAC.*

Source: https://www.eetimes.com/implement-an-ac-switch-triggering-circuit-using-an-optical-triac/

The operation principle is presented on Figure 3:

1: Opto- TRIAC is turn on and capacitor C is charged up to reach V_{GT} parameter (~0.7 V). ACS is then triggered in Q3 quadrant, for which the I_{GT} current is lower than in Q2.

2: ACS is ON up to next zero current crossing point. G-COM voltage is kept down to -0.7 V, and capacitor C is kept charged.

3: ACS current increases, $V_{G\text{-}COM}$ increases and so capacitor C is discharged through G and COM, applying a peak current up to 10 mA which triggers the ACS.

To apply enough gate current, a high-value capacitor–in the range of 330 µF–has to be used.

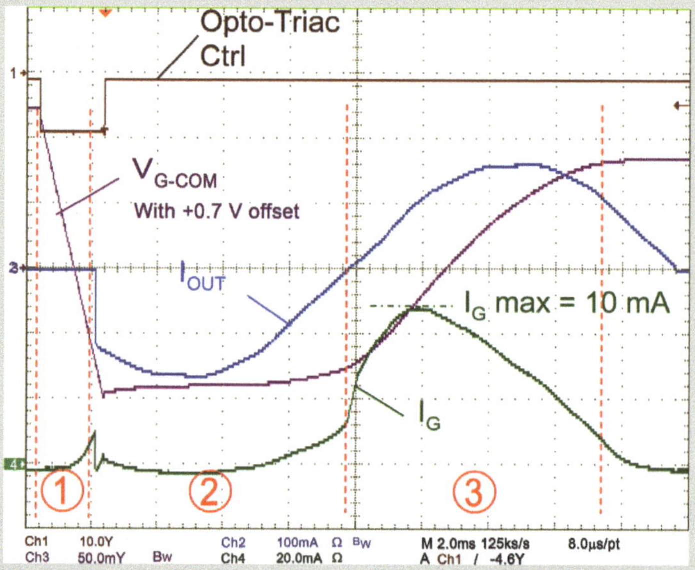

Figure 3. *Operation of circuit of Figure 2.*

Source: https://www.eetimes.com/implement-an-ac-switch-triggering-circuit-using-an-optical-triac/

Note that the ACS is off in every cycle, for the time required to charge back the C capacitor. This leads to ACS turn-on when the voltage across its terminals equals approximately 10 V. This behavior doesn't result in too high EMI disturbance, as the line current is not cut. Indeed, thanks to capacitor C charge, the line current is still approximately sinusoidal.

These three schematics can also be adapted by adding, as is well known, an R-C snubber circuit between resistor R and the opto- TRIAC to increase this last device immunity and enlarge the gate pulse width, to better trigger the AC switch.

CLASS ACTIVITY

SILICON CONTROLLED RECTIFIERS (SCR) – ADALM2000

- **Objective**

The objective of this activity is to examine the structure and operation of the Silicon Controlled Rectifier or SCR. SCRs are mainly used in devices where the control of high power, possibly at high voltage, is needed. The ability to switch large currents on and off makes the SCR suitable for use in medium to high-voltage AC power control applications, such as lamp dimming, regulators and motor control. In addition, unintentional SCRs can form in integrated circuits and when these SCRs get triggered circuit malfunction or even reliability problems and damage can result.

- **Materials**

ADALM2000 Active Learning Module

Solder-less Breadboard

2 – 1 KΩ Resistors

2 – 100 KΩ Resistors

1 – 0.1 uF capacitor (marked 104)

1 – small signal NPN transistor (2N3904)

1 – small signal PNP transistor (2N3906)

- **Directions**

Build the model of the equivalent circuit of an SCR as shown in Figure on your solder-less breadboard.

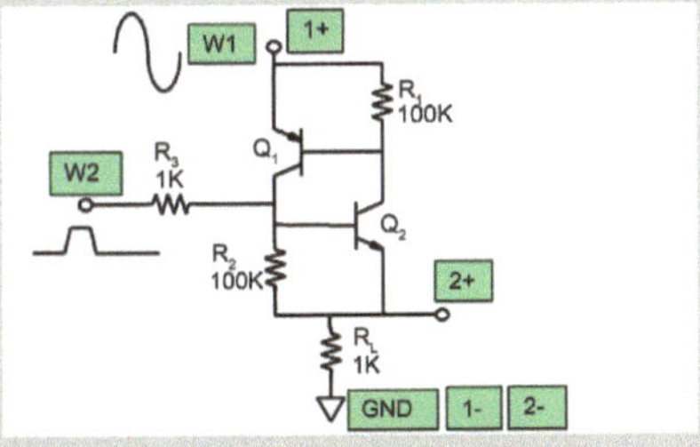

SUMMARY

- A semiconductor device is an electronic component that relies on the electronic properties of a semiconductor material (primarily silicon, germanium, and gallium arsenide, as well as organic semiconductors) for its function. Its conductivity lies between conductors and insulators.
- A silicon-controlled rectifier or semiconductor-controlled rectifier is a four-layer solid-state current-controlling device.
- SCRs are unidirectional devices (i.e., can conduct current only in one direction) as opposed to TRIACs, which are bidirectional (i.e., charge carriers can flow through them in either direction).
- When a negative voltage is applied to the anode and a positive voltage to the cathode, the SCR is in reverse blocking mode, making J1 and J3 reverse biased and J2 forward biased.
- An SCR incapable of blocking reverse voltage is known as an asymmetrical SCR, abbreviated ASCR. It typically has a reverse breakdown rating in the tens of volts.
- Forward-voltage triggering occurs when the anode–cathode forward voltage is increased with the gate circuit opened. This is known as avalanche breakdown, during which junction J2 will break down.
- SCRs are mainly used in devices where the control of high power, possibly coupled with high voltage, is demanded.
- A TRIAC (triode for alternating current; also, bidirectional triode thyristor or bilateral triode thyristor) is a three-terminal electronic component that conducts current in either direction when triggered.
- A TRIAC starts conducting when a current flowing into or out of its gate is sufficient to turn on the relevant junctions in the quadrant of operation.
- Low-power TRIACs are used in many applications such as light dimmers, speed controls for electric fans and other electric motors, and in the modern computerized control circuits of many households small and major appliances.
- The DIAC is a full-wave or bi-directional semiconductor switch that can be turned on in both forward and reverse polarities. The DIAC gains its name from the contraction of the words Diode Alternating Current.
- The DIAC circuit symbol is generated from the two triangles held between two lines as shown below. In some way this demonstrates the structure of the device which can be considered also as two junctions.
- A unijunction transistor (UJT) is a three-lead electronic semiconductor device with only one junction. It acts exclusively as an electrically controlled switch.
- Unijunction transistor circuits were popular in hobbyist electronics circuits in the 1960s and 1970s because they allowed simple oscillators to be built using just one active device.

- Because the IGBT is a voltage-controlled device, it only requires a small voltage on the Gate to maintain conduction through the device unlike BJT's which require that the Base current is continuously supplied in a sufficient enough quantity to maintain saturation.

REVIEW QUESTIONS

1. Discuss the modes of operation of silicon-controlled rectifier.
2. Describe different methods that turn on methods of SCR.
3. Explain the concept of quadrants in the operation of a TRIAC.
4. What are the key parameters to consider when selecting a TRIAC for a specific application?
5. What is a DIAC? Describe its basic structure and operation.
6. How does a DIAC differ from a standard diode?
7. What is a unijunction transistor (UJT)? Describe its basic structure and operation.
8. What are the key parameters of a UJT, such as interbase resistance and peak point voltage?
9. What is an Insulated-Gate Bipolar Transistor (IGBT)? Describe its basic structure and operation.
10. Explain how the IGBT combines the characteristics of both MOSFETs and BJTs.

REFERENCES

1. Betak, P., & Musil, V. (2007). Snap-back characteristics tuning of SCR-based semiconductor structures. *WSEAS Transactions on Electronics, 4*(9), 175–180. ISSN 1109-9445.
2. Cardesin, J., et al. (2009). LED permanent emergency lighting system based on a single magnetic component. *IEEE Transactions on Power Electronics, 24*(5), 1409–1416.
3. Cheng, J.-h., et al. (2000). A small-signal model utilizing amplitude modulation for the class-D converter at fixed frequency. *IEEE Transactions on Power Electronics, 15*(6), 1204–1211.
4. Dong, A., Xiong, J., Mitra, S., Liang, W., Gauthier, R., & Loiseau, A. (2018). Comprehensive study of ESD design window scaling down to 7nm technology node. In *Proceedings of the 40th Electrical Overstress/Electrostatic Discharge Symposium (EOS/ESD)*, Reno, NV, USA, pp. 1–8.
5. Du, F., Hou, F., Song, W., Chen, L., Nie, Y., Qing, Y., Xu, Y., Liu, J., Liu, Z., & Liou, J. J. (2020). An improved silicon-controlled rectifier (SCR) for low-voltage ESD application. *IEEE Transactions on Electron Devices, 67*(2), 576–581.
6. Du, F., Qing, Y., Hou, F., Zou, K., Song, W., Chen, R., Liu, J., Chena, L., Liou, J. J., & Liu, Z. (2021). Compact and low leakage devices for bidirectional low-voltage ESD protection applications. *IEEE Electron Device Letters, 42*(7), 391–394.

7. He, L., Salcedo, J. A., Parthasarathy, S., Hajjar, J.-J., & Sundaram, K. (2020). A new low-capacitance high-voltage-tolerant protection clamp for high-speed applications. *IEEE Transactions on Electron Devices, 67*(7), 3030–3034.

8. Ker, M.-D., & Lo, W.-Y. (2000). Design on the low-leakage diode string for using in the power-rail ESD clamp circuits in a 0.35-µm silicide CMOS process. *IEEE Journal of Solid-State Circuits, 35*(5), 601–611.

9. Krames, M. R., Shchekin, O. B., Mueller-Mach, R., Mueller, G. O., Zhou, L., Harbers, G., & Craford, M. G. (2007). Status and future of high-power light-emitting-diodes for solid-state lighting. *Journal of Display Technology, 3*(2), 160–175.

10. Liu, W., Liou, J. J., Yeh, H.-C., Wang, H., Li, Y., & Yeo, K. S. (2012). Bidirectional diode-triggered silicon-controlled rectifiers for low-voltage ESD protection. *IEEE Electron Device Letters, 33*(10), 1360–1362.

11. Mao, C., Lauritzel, P., Lin, P., & Budihardjo, I. (1994). A systematic approach to modeling of power semiconductor devices based on charge control principles. In *Proceedings of the 25th IEEE Power Electronics Specialists Conference (PESC)*, vol. 1, pp. 31–37.

12. Mergens, M., Russ, C., Verhaege, K., Armer, J., Jozwiak, P., Mohn, R., Keppens, B., & Trinh, C. (2003). Diode-triggered SCR (DTSCR) for RF-ESD protection of BiCMOS SiGe HBTs and CMOS ultra-thin gate oxides. In *Proceedings of the IEEE International Electron Devices Meeting*, Washington, DC, USA, pp. 21.3.1–21.3.4.

13. Patil, N., Celaya, J., Das, D., Goebel, K., & Pecht, M. (2009). Precursor parameter identification of insulated gate bipolar transistor (IGBT) prognostics. *IEEE Transactions on Reliability, 58*(2), 211–276.

14. Raderecht, P. S. (1971). A review of the shorted emitter principle. *International Journal of Electronics, 31*, 541–564.

15. Rim, C. T., & Cho, G. H. (1990). Phasor transformation and its application to the DC/AC analyses of frequency/phase-controlled series resonant converters (SRC). *IEEE Transactions on Power Electronics, 5*(2), 201–211.

16. Rogers, D. J., Green, T. C., & Silversides, R. W. (2014). A low-wear on load tap changer diverter switch for frequent voltage control on distribution networks. *IEEE Transactions on Power Delivery, 29*(2), 860–869.

17. Shammas, N. Y. A., Chamund, D. J., & Taylor, P. (2004). Forward and reverse recovery behavior of diodes in power conversion applications. In *Proceedings of the 24th International Conference in Microelectronics*, Serbia, pp. 3–10. IEEE.

18. Temple, V. A. K. (1986). MOS controlled thyristors–A new class of power devices. *IEEE Transactions on Electron Devices, ED-33*, 1609–1618.

19. Xie, X., Wang, J., Zhao, C., Lu, Q., & Liu, S. (2012). A novel output current estimation and regulation circuit for primary side controlled high power factor single-stage flyback LED driver. *IEEE Transactions on Power Electronics, 27*(11), 4602–4612.

20. Zhang, J., Zeng, H., & Jiang, T. (2012). A primary-side control scheme for high-power-factor LED driver with TRIAC dimming capability. *IEEE Transactions on Power Electronics, 27*(11), 4619–4629

CHAPTER 6

Feedback Amplifier

LEARNING OBJECTIVES

After studying this chapter, you will be able to:

- Understand conceptual development through block diagrams;
- Focus on negative feedback;
- Perform calculations of open-loop gain, closed-loop gain and feedback factors;
- Identify topologies of the feedback amplifier;
- Explain the effect of feedback on input and output impedance;
- Discuss sensitivity;
- Know the effect of positive feedback.

KEY TERMS FROM THIS CHAPTER

Amplifier
Desensitivity
Load resistor
Null detector
Positive saturation voltage
Self-stabilizing system
Voltage buffers
Closed-loop gain
Feedback fraction
Negative feedback
Output signal
Potentiometer
Signal voltage
Voltage follower
Comparator
Input resistance
Positive feedback
Rail voltage
Topology
Impedance

INTRODUCTION

The basic idea behind amplifier operation is to increase an input signal's voltage, current, and power. They come in a types of varieties based on the electrical circuit's requirements. The two most crucial factors are power output and efficiency. A feedback amplifier is one type of amplifiers, in which a portion of the output is fed back into the input.

An amplifier type known as a feedback amplifier has feedback between the input and output signals. It is referred to as a feedback amplifier because the idea of feeding the output signal back into the input circuit is known as feedback. With effective control, it depends on the input and output. Positive and negative feedback are the two categories into which feedback amplifiers fall.

6.1. CONCEPTUAL DEVELOPMENT THROUGH BLOCK DIAGRAMS

A feedback amplifier is an electronic circuit that uses feedback from its output to its input to regulate and improve performance. To operate, a sample of the output signal is taken, compared to the input signal, and any necessary adjustments are made to the amplification process. The feedback amplifier's block diagram and operation will be shown next.

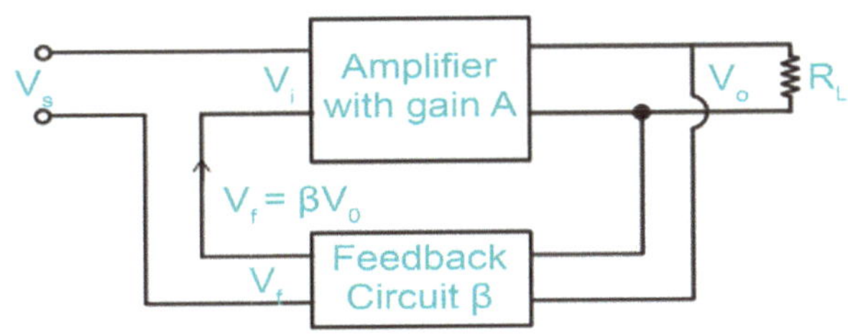

Figure 6.1. *A block diagram for a feedback amplifier.*

Source: https://testbook.com/physics/feedback-amplifier.

An amplifier and a feedback circuit make up its two components. The feedback fraction, or gain, of the amplifier is denoted by the letter "A" and is calculated as the product of the input voltage V_i and the output voltage V_o, and the voltage feedback is given by, $V_f=\beta V_o$ from the output of the amplifier.

For negative feedback:

$V_i=V_s-V_f=V_s-\beta V_o$

For positive feedback:

$V_i=V_s+V_f=V_s+\beta V_o$

Let's consider the case of negative feedback. The output voltage V_o should be equal to $V_s-\beta V_o$ times gain of the amplifier "A"

$AV_s-A\beta V_o=V_o$

$AV_s = V_o(1+A\beta)$

$V_o/V_s = A/1+A\beta$

where V_s is applied signal voltage. The overall gain with feedback from the amplifier is,

$A_f = V_o V_s$

where V_o is the output voltage and V_s is the input signal voltage. The gain feedback for a positive feedback amplifier,

$$A_f = A/1-A\beta \qquad (6.1)$$

and for a negative feedback amplifier,

$$A_f = A/1+A\beta \qquad (6.2)$$

Equations (6.1) and (6.2) are the standard equations for determining the gain feedback of amplifiers, based on the type of amplifier.

6.2. NEGATIVE FEEDBACK

Negative feedback occurs when the feedback energy is out of phase with the input signal. Negative feedback lowers the amplifier's gain. For this reason, it is sometimes referred to as inverse or degenerative feedback. But there are benefits to receiving unfavorable feedback.

- Higher input impedance
- Lower output impedance
- Better stabilized gain
- Reduction of noise
- Improved frequency response

When an op-amp's output is connected to its inverting input and a voltage signal is applied to its noninverting input, we observe that the op-amp's output voltage closely tracks the input voltage:

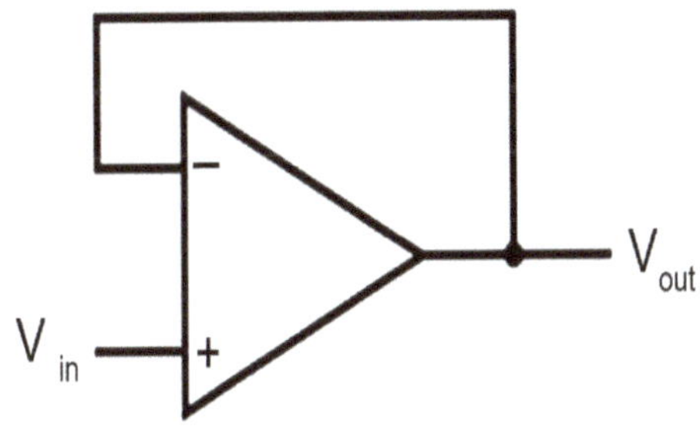

Figure 6.2. *Output voltage of the op-amp closely follows that input voltage.*

Source: https://www.allaboutcircuits.com/textbook/semiconductors/chpt-8/negative-feedback/.

V_{out} will rise in proportion to the differential gain as V_{in} rises. But when V_{out} rises, the output voltage is fed back to the input that is being inverted, which reduces the voltage differential between the inputs and lowers the output. When a voltage input is applied, the op-amp will produce an output voltage that is almost exactly equal to V_{in}, but only low enough to allow for the remaining voltage difference between V_{in} and the (-) input to be amplified and produce the output voltage.

The circuit will soon come to a stable point—referred to as equilibrium in physics—where the output voltage is precisely the appropriate value to maintain the proper differential. Negative feedback, which is the process of taking the output voltage of the op-amp and coupling it to the inverting input, is essential to creating a self-stabilizing system (this holds true for dynamic systems in general, not just op-amps).

Because of its stability, the op-amp can now operate in its linear (active) mode rather than being completely saturated and either "on" or "off," as it did when it was used as a comparator and had no feedback at all.

Because the op-amp's gain is so high, the voltage on the inverting input can be maintained almost equal to V_{in}. We can write an equation relating the output voltage to the input voltage and the gain, *G*:

$$V_{out}=G\cdot(V_{in}-V_{out})$$

Then, solving for the output voltage, we get the following:

$$V_{out} = \frac{V_{in}}{1+(\frac{1}{G})}$$

Let's say that our op-amp has a differential voltage gain of 200,000 and Vin equals 6 V, we can calculate the output voltage using our equation:

$$V_{out} = \frac{6}{1+(\frac{1}{20{,}000})} = 5.999700015V$$

This produces precisely the right differential voltage (6 V – 5.99997000015 V = 29.99985 µV) at the output terminal to cause 5.99997000015 volts to appear, and the system maintains balance there. As you can see, 29.99985 µV is not a significant differential, so we can assume for computational purposes that negative feedback is holding the differential voltage between the two input wires exactly at 0 volts.

The effects of negative feedback

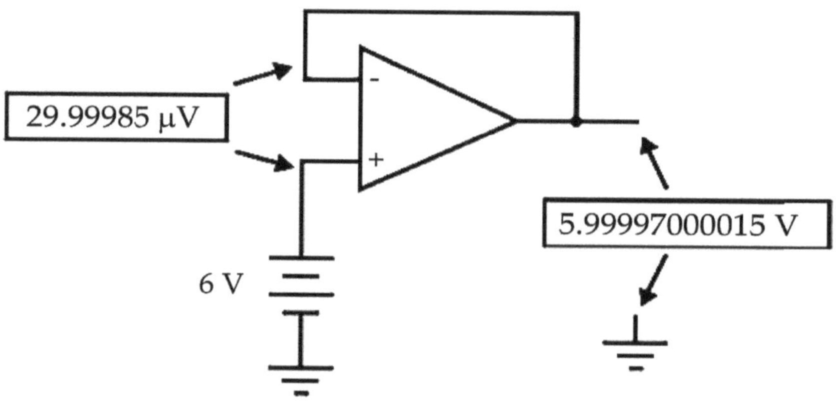

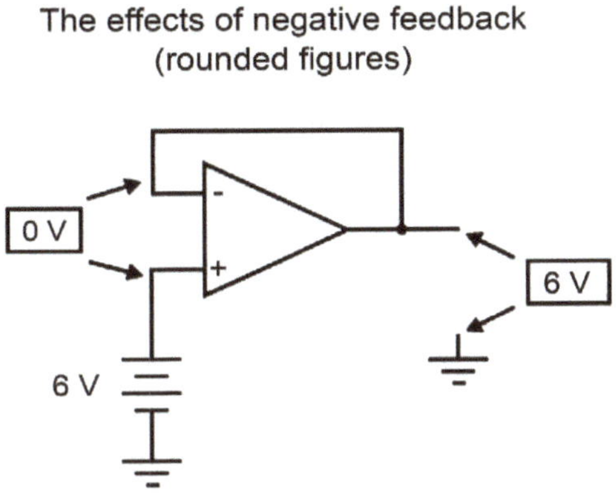

Figure 6.3. *The effects of negative feedback.*

Source: https://www.allaboutcircuits.com/textbook/semiconductors/chpt-8/negative-feedback/.

6.2.1. Advantage of Negative Feedback in Op-Amps

Employing an op-amp with negative feedback has the major benefit of not caring how big the op-amp's actual voltage gain is. The output voltage would only hold slightly closer to Vin if the op-amp's differential gain were 250,000 rather than 200,000 (requiring less differential voltage between inputs to achieve the necessary output). For all intents and purposes, the output voltage in the previously shown circuit would still be equal to the non-inverting input voltage. Therefore, the circuit designer can create an amplifier circuit with precise gain without having to rely on the factory to set op-amp gains precisely. With negative feedback, the system self-corrects. With a steady gain of 1, the entire circuit mentioned above will just follow the input voltage.

6.2.2. How Does the Circuit in the Op-Amp Works?

Using our differential amplifier model once more, we can imagine the operational amplifier as a variable voltage source that is controlled by a very sensitive null detector. This type of detector could be a meter movement or another sensitive measurement device that is used in bridge circuits to detect when there is zero volts, or an imbalance. In order to "balance" the inverting and non-inverting input voltages and provide zero voltage across the "null detector," the "potentiometer" inside the op-amp producing the variable voltage will move to whatever position it must:

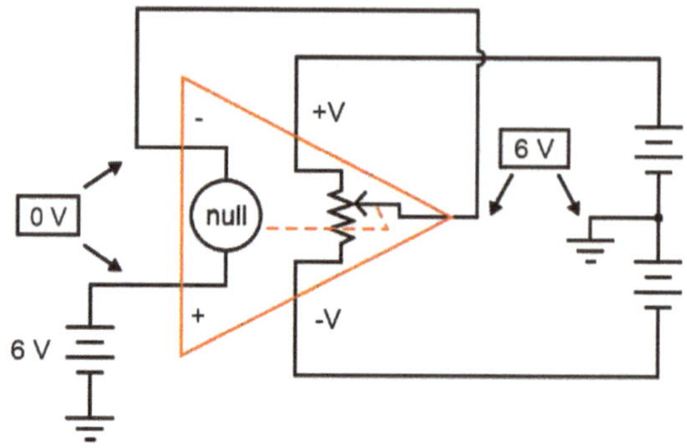

Figure 6.4. *Circuit in the op-amp.*

Source: https://www.allaboutcircuits.com/textbook/semiconductors/chpt-8/negative-feedback/.

The output voltage equals the input voltage, in this case 6 volts, as the "potentiometer" adjusts to provide the output voltage required to satisfy the "null detector" at an "indication" of zero volts. The "potentiometer" inside the op-amp will shift positions to maintain the "null detector" in balance (indicating zero volts) if the input voltage varies in any way. This will always produce an output voltage that is roughly equal to the input voltage.

This is accurate within the voltage range that the op-amp is capable of producing. It will precisely follow the input voltage between the limits of +15 volts and -15 volts if it has a power supply that is capable of providing +15V/-15V and an optimal amplifier that can swing its output voltage that far. Consequently, the circuit mentioned above is referred to as a voltage follower as a result. It shares characteristics with its one-transistor equivalent, the common-collector (also known as "emitter-follower") amplifier, including a high input impedance, low output impedance, high current gain, and a voltage gain of 1. Voltage followers, sometimes referred to as voltage buffers, are devices that increase a voltage signal's capacity to source current when it is too weak (due to a high source impedance) to drive a load directly. The op-amp model in the previous figure illustrates how the output voltage is effectively isolated from the input voltage, meaning that the power supply that powers the op-amp provides the current on the output pin instead of the input voltage source.

It should be noted that a lot of op-amps are unable to precisely adjust their output voltages to match the rail voltages of the power supply (+V/-V). One of the models that cannot is the 741; when saturated, its output voltage peaks between one and two volts of the +V and -V power supply voltages, respectively. As a result, a 741 op-amp's output may therefore reach +14 volts or as low as -13 volts (approximately) with a split power supply of +15/-15 volts, but no higher. Its bipolar transistor design is to blame for this. The terms "positive saturation voltage" and "negative saturation voltage," respectively, refer to these two voltage limits. Other op-amps can swing their output voltages within millivolts of either power supply rail voltage. One such op-mp is the model 3130, which has field-effect transistors in the final output stage. As a result, the supply voltages and their positive and negative saturation voltages are nearly equal.

6.3. CALCULATIONS OF OPEN-LOOP GAIN, CLOSED-LOOP GAIN AND FEEDBACK FACTORS

Op amps are devices with a very high gain. Op-amps with negative feedback are used to achieve stable operation.

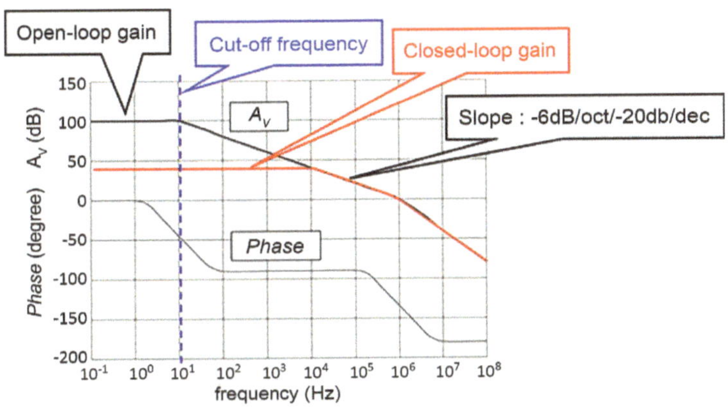

Figure 6.5. *Bode plot of an op-amp.*

Source: https://toshiba.semicon-storage.com/ap-en/semiconductor/knowledge/faq/linear_opamp/what-are-open-loop-and-closed-loop-gains-of-an-op-amp.html#:~:text=Above%20the%20cutoff%20frequency%20where,%2DR2%2FR1.

When a device is configured for a negative feedback circuit, the gain is referred to as closed loop gain; otherwise, it is known as open loop gain. Closed-loop gain is typically set by the feedback network and is not device-specific.

When frequency doubles above the cutoff frequency, the open-loop gain halved, with the gain falling by -3dB below the passband. Its slope, on a logarithmic scale, is -20 dB/dec or -6 dB/oct. Fig. illustrates the closed-loop gain (Gv) of the negative feedback circuit. The formula for 6.6 is -R2/R1.

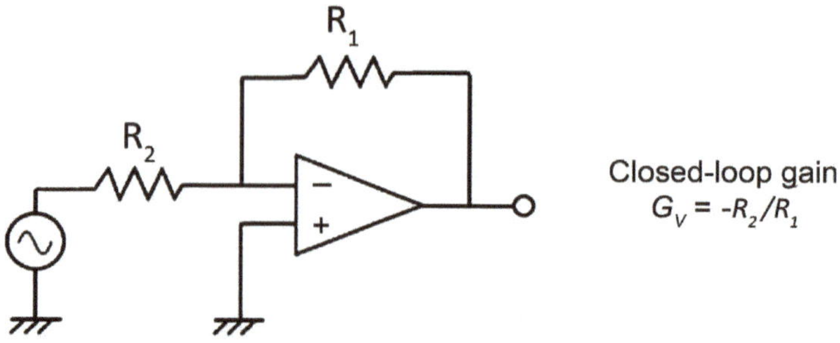

Figure 6.6. *Example of a negative feedback circuit (inverting amplifier).*

Source: https://toshiba.semicon-storage.com/ap-en/semiconductor/knowledge/faq/linear_opamp/what-are-open-loop-and-closed-loop-gains-of-an-op-amp.html#:~:text=Above%20the%20cutoff%20frequency%20where,%2DR2%2FR1.

6.4. TOPOLOGIES OF THE FEEDBACK AMPLIFIER

Four fundamental amplifier topologies are available for connecting the feedback signal. It is possible for the voltage and current to be fed back into the input in parallel or series, respectively.

- Voltage Series Feedback Amplifier
- Voltage Shunt Feedback Amplifier
- Current Series Feedback Amplifier
- Current Shunt Feedback Amplifier

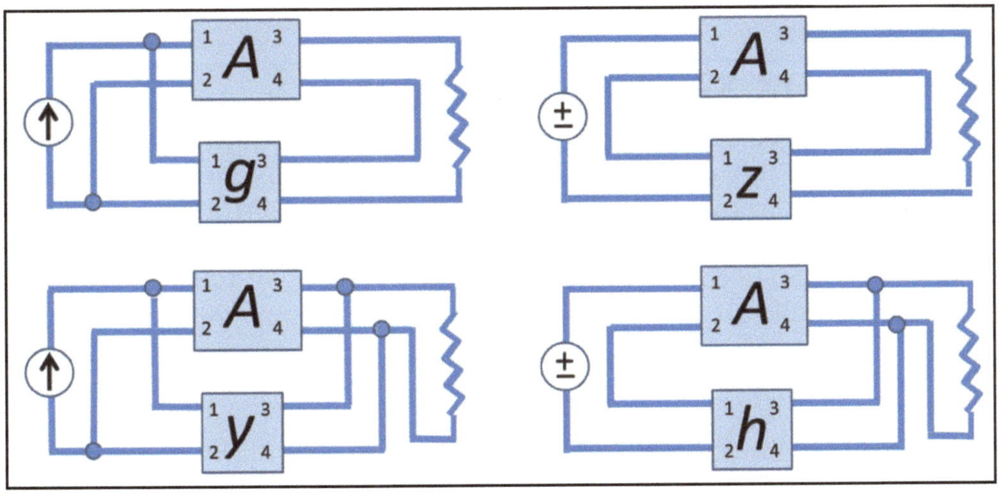

Figure 6.7. *Feedback amplifier topologies.*

Source: https://www.elprocus.com/what-is-a-feedback-amplifier-types-characteristics-and-applications/.

6.4.1. Voltage Series Feedback Amplifier

With this kind of circuit, a portion of the output voltage can be applied in series with the input voltage via the feedback circuit. The voltage-series feedback-amplifier block diagram is displayed above. It is clear from this diagram that the feedback circuit is connected in series with the input but in parallel with the output. Because of the series connection with the input, the output impedance will be decreased, and the input impedance will be increased when the feedback circuit is connected in parallel through the output.

6.4.2. Voltage Shunt Feedback Amplifier

This kind of circuit allows for the parallel application of a portion of the output voltage to the input voltage via the feedback circuit. The voltage shunt feedback-amplifier block diagram is displayed above, and it makes clear that the feedback circuit is situated in a shunt using both the input and the output. The output and input impedances will both decrease when the feedback circuit is connected in a shunt through the input and the output.

6.4.3. Current Series Feedback Amplifier

In this kind of circuit, the feedback circuit applies a portion of the output voltage in series with the input voltage. It is clear from the block diagram of the current series feedback-amplifier above that the feedback circuit is connected in series with both the input and the output. Both the output and the input impedance will rise when the feedback circuit is connected in series through the input and the output.

6.4.4. Current Shunt Feedback Amplifier

A portion of the output voltage is applied to the input voltage in a shunt through the feedback circuit in this kind of circuit. It is clear from the block diagram of the current shunt feedback-amplifier above that the feedback circuit is situated in a shunt by means of both the input and the output. The output impedance will rise when the feedback circuit is connected in series through the output but in parallel with the input; conversely, the input impedance will fall due to the parallel connection with the output.

6.5. EFFECT OF FEEDBACK ON INPUT AND OUTPUT IMPEDANCE

An amplifier's input and output impedances are significantly impacted by feedback as well; the kind of modification depends on the architecture of the amplifier-feedback network combination.

Feedback can be set up to return data about the voltage applied to the load or the current flowing through it, as shown in Figure 6.8. Physical arguments make it obvious that these two output topologies need to change the impedance in different ways when it comes to the load. Feedback tends to lessen changes in output voltage brought on by disturbances (changes in load current), which suggests that the amplifier in Figure 6.8(a) has a lower output impedance. This is especially true if the information fed back to the input relates to the output voltage. On the other hand, if load current information is fed back, output current changes brought on by disturbances (changes in load voltage) are lessened, indicating that this kind of feedback increases output impedance.

Figure 6.9 displays two potential input topologies. The input signal is applied in series with the amplifier's differential input in Figure 6.9(a). We are guaranteed that any necessary output signal level can be attained with a small amplifier input current if the amplifier characteristics are satisfactory. As a result, low current consumption from the input signal source indicates high input impedance. The topology depicted in Figure 6.9(b)lowers input impedance because there is little voltage across the amplifier-input and parallel input-signal connections.

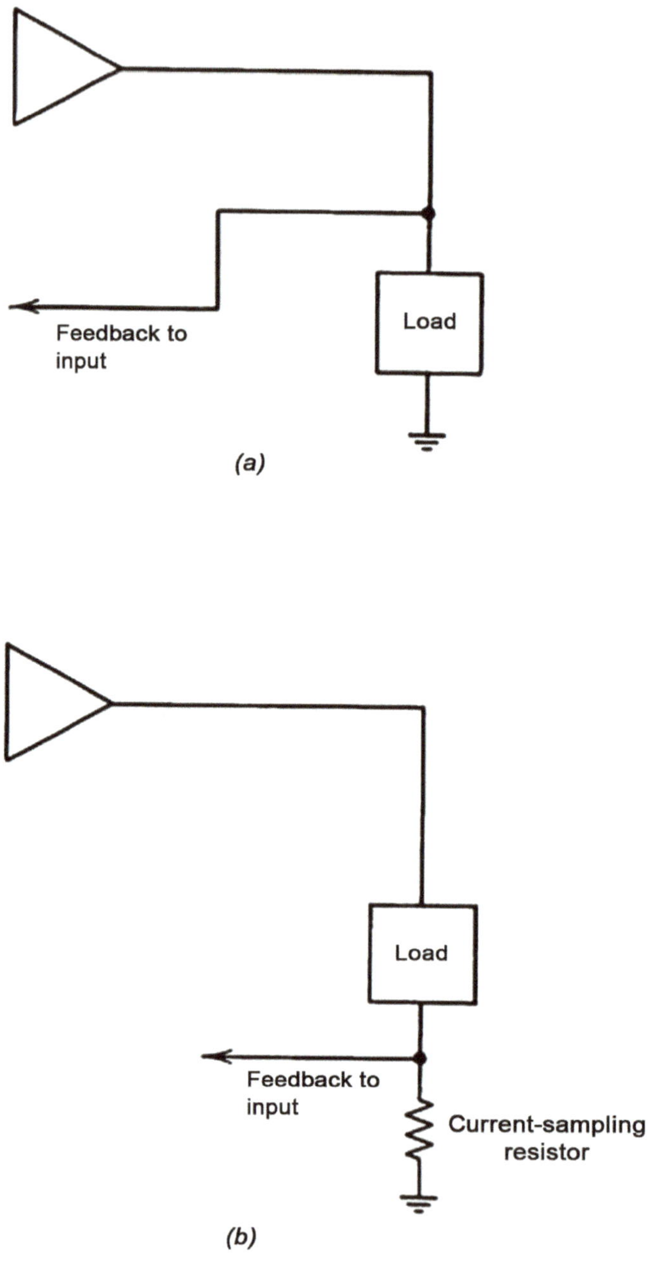

Figure 6.8. *Two possible output topologies. (a) Feedback of load-voltage information. (b) Feedback of load-current information.*

Source: https://eng.libretexts.org/Bookshelves/Electrical_Engineering/Electronics/Operational_Amplifiers%3A_Theory_and_Practice_(Roberge)/02%3A_Properties_and_Modeling_of_Feedback_Systems/2.05%3A_EFFECTS_OF_FEEDBACK_ON_INPUT_AND_OUTPUT_IMPEDANCE.

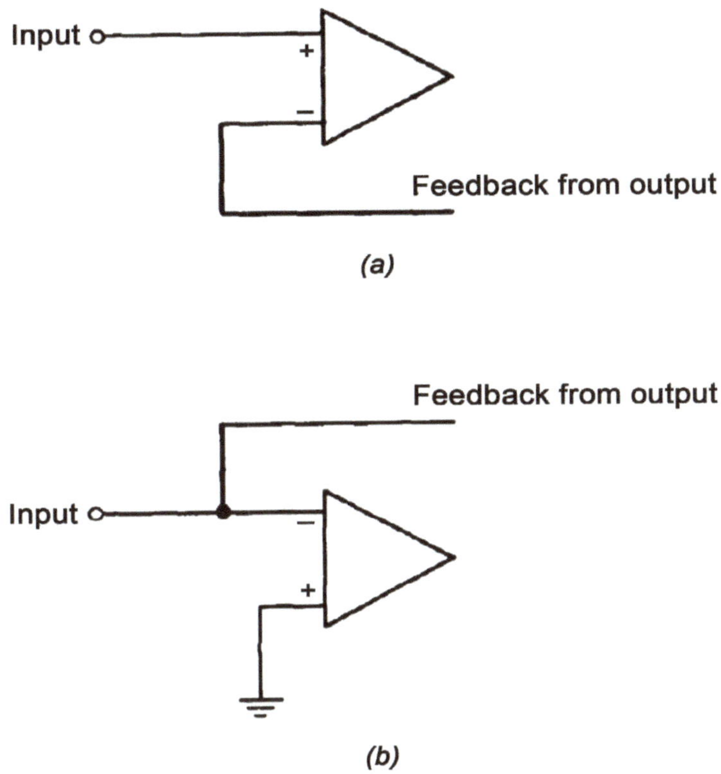

Figure 6.9. *Two possible input topologies. (a) Input signal applied in series with amplifier input. (b) Input signal applied in parallel with amplifier input.*

Source: https://eng.libretexts.org/Bookshelves/Electrical_Engineering/Electronics/Operational_Amplifiers%3A_Theory_and_Practice_(Roberge)/02%3A_Properties_and_Modeling_of_Feedback_Systems/2.05%3A_EFFECTS_OF_FEEDBACK_ON_INPUT_AND_OUTPUT_IMPEDANCE.

The following example demonstrates how the loop transmission is directly related to the amount by which feedback scales input and output impedances. Figure 6.10 shows an operational amplifier connected for high input and high output resistances. This topology's input resistance is just the ratio of B/C. The process of determining output resistance involves connecting a voltage source in series with the load resistor and computing V_D/V_i, which is the ratio of the change in V_D to the resulting change in load current. The following equations hold if it is assumed that the components of I_l and the current through the sampling resistor *RS* are due to I_i and that $Ri \gg Rs$, indicating that the amplifier provides system gain rather than a passive network.

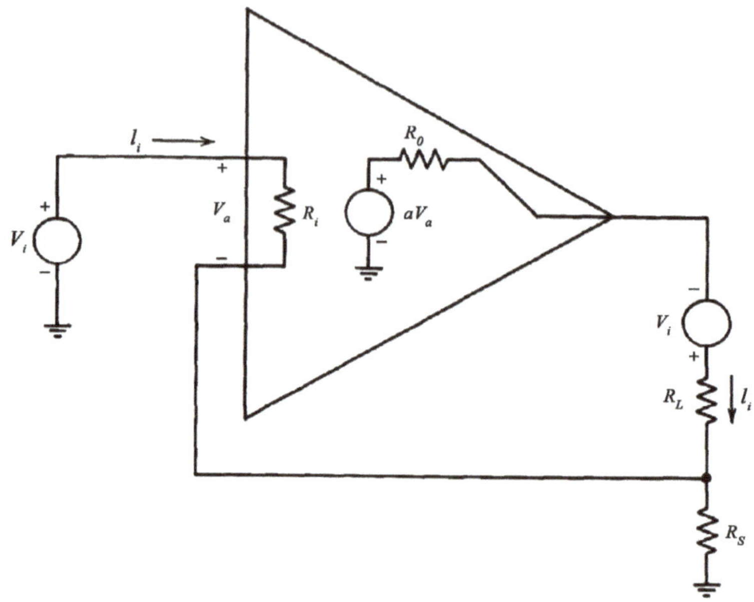

Figure 6.10. *Amplifier with high input and output resistances.*

Source: https://eng.libretexts.org/Bookshelves/Electrical_Engineering/Electronics/Operational_Amplifiers%3A_Theory_and_Practice_(Roberge)/02%3A_Properties_and_Modeling_of_Feedback_Systems/2.05%3A_EFFECTS_OF_FEEDBACK_ON_INPUT_AND_OUTPUT_IMPEDANCE

$$V_a = V_i - R_S I_l$$

$$I_l = \frac{aV_a + V_l}{R_o + R_L + R_S}$$

$$I_i = \frac{V_a}{R_i}$$

Figure 6.11 shows these equations in block diagram format. This block diagram verifies the anticipated result that, since the input voltage is compared with the output current sampled via resistor *RS*, the ideal trans-conductance (ratio of I_l to V_i) is simply equal to *GS*. By observing that, the input resistance is calculated:

$$\frac{I_i}{V_i} = \frac{1}{R_{in}} = \frac{1}{R_i\{1 + [aR_S/(R_o + R_L + R_S)]\}} \qquad (6.3)$$

or

$$R_{in} = R_i\left(1 + \frac{aR_S}{R_o + R_L + R_S}\right)$$

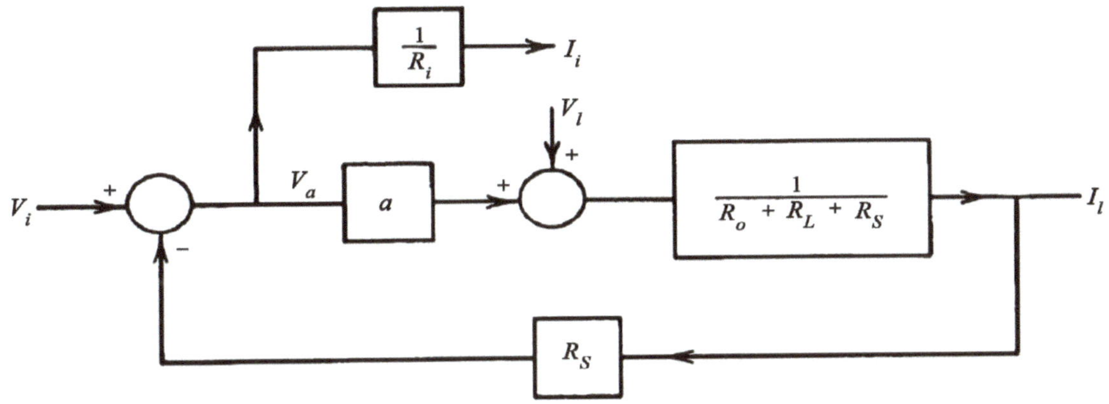

Figure 6.11. *Block diagram for amplifier of Figure 6.10.*

Source: https://eng.libretexts.org/Bookshelves/Electrical_Engineering/Electronics/Operational_Amplifiers%3A_Theory_and_Practice_(Roberge)/02%3A_Properties_and_Modeling_of_Feedback_Systems/2.05%3A_EFFECTS_OF_FEEDBACK_ON_INPUT_AND_OUTPUT_IMPEDANCE

It should be noted that in this example, the output resistance is computed by connecting a voltage source in series with the load resistor. This method is used to highlight how *RL* affects loop transmission, which in turn affects output resistance. An alternate development could swap out the test generator for the resistor to assess the resistance to the load.

$$\frac{I_l}{V_l} = \frac{1}{R_{out}} = \frac{1}{(R_o + R_L + R_S)\{1 + [aR_S/(R_o + R_L + R_S)]\}}$$

Yielding:

$$R_{out} = (R_o + R_L + R_S)\left(1 + \frac{aR_S}{R_o + R_L + R_S}\right) \qquad (6.4)$$

The essential features of Equations 6.3 and 6.4 are the following. If the system has no feedback (e.g., if a=0), the input and output resistances become:

$$R'_{in} = R_i$$

and,

$$R'_{out} = R_o + R_L + R_S$$

Feedback increases both of these quantities by a factor of $1+[aR_S/(Ro+R_L+R_S)]$, where $-aR_S/(Ro+R_L+R_S)$ is recognized as the loop transmission. As a result, we can observe that the resistances in this example are raised by the same factor as the sensitivity increase caused by feedback—that is, one less the loop transmission. The general outcome means that, for any of the topologies depicted in the above figures, input or output impedances can always be computed by determining the impedance of interest without feedback and scaling it (up or down, depending on the topology) by a factor of one minus the loop transmission.

Although feedback provides an easy way to adjust the input or output impedance of an amplifier, it is theoretically possible to achieve results that are at least as good, if not better, without feedback. Take Figure 6.12, for instance, which illustrates three different configurations for an operational amplifier that provide unity voltage gain and high input impedance.

The follower connection of Figure 6.12 (a)provides a voltage gain:

$$\frac{V_o}{V_i} = \frac{a}{1+a}$$

or approximately unity for large values of a. The relationship between input impedance and loop transmission shows that the input impedance for this connection is:

$$\frac{V_i}{I_i'} = Z_i(1+a)$$

The connection shown in Figure 6.12 (b) precedes the amplifier with an impedance that, in conjunction with the input impedance of the amplifier, attenuates the input signal by a factor of 1/(1+a). This attenuation creates a composite voltage gain that is the same as the follower connection when combined with the amplifier's inherent voltage gain. Similarly, the series impedance of the attenuator input element adds to the input impedance of the amplifier itself so that the input impedance of the combination is identical to that of the follower.

When utilizing an ideal transformer as an impedance-modifying element instead of the feedback approach, the input impedance can be enhanced. The input impedance of the transformer-amplifier combination is and the overall voltage gain is the same as that of the follower connection with a transformer turns ratio of (a+1):1.

$$\frac{V_i}{I_i''} = Z_i(1+a)^2$$

This value greatly exceeds the value obtained with the follower for large amplifier voltage gain.

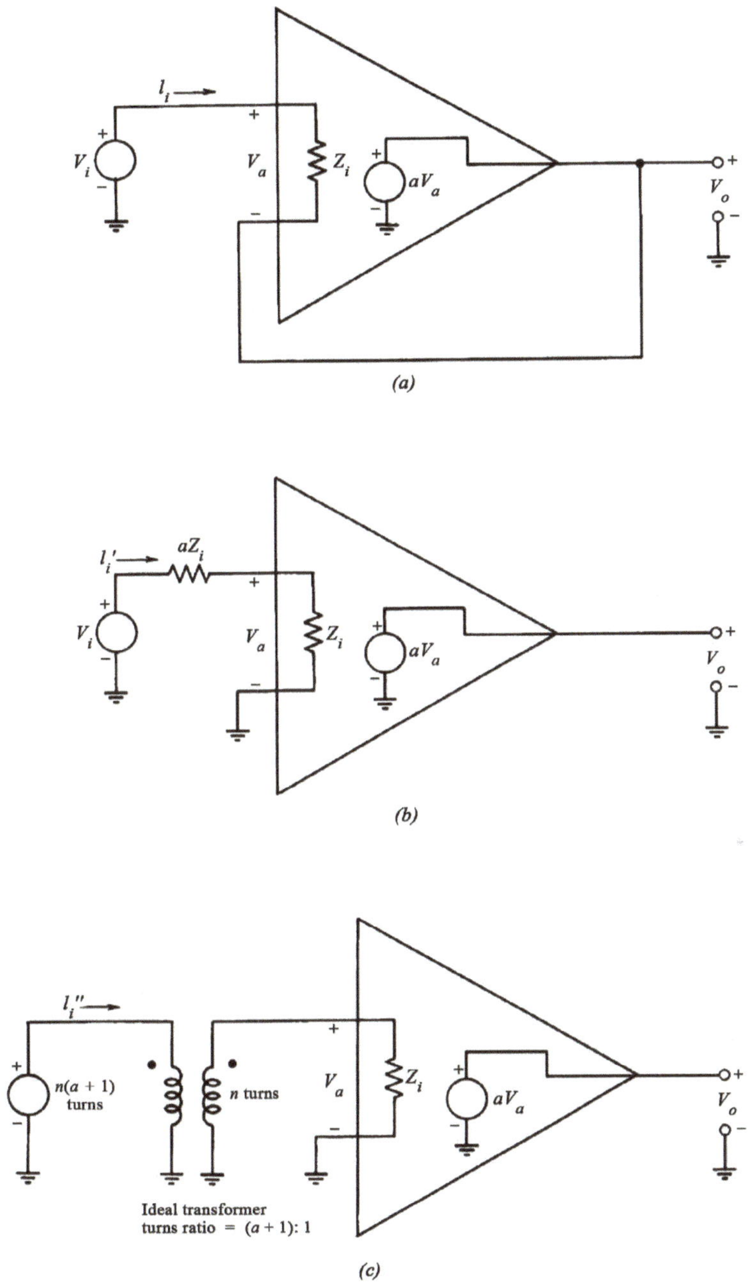

Figure 6.12. *Unity-gain amplifiers. (a) Follower connection. (b) Amplifier with input attenuator. (c) Amplifier with input transformer.*

Source: https://eng.libretexts.org/Bookshelves/Electrical_Engineering/Electronics/Operational_Amplifiers%3A_Theory_and_Practice_(Roberge)/02%3A_Properties_and_Modeling_of_Feedback_Systems/2.05%3A_EFFECTS_OF_FEEDBACK_ON_INPUT_AND_OUTPUT_IMPEDANCE.

The aforementioned example is by no means meant to suggest that attenuators or transformers are preferable to feedback when adjusting impedance levels. The two earlier methods' practical drawbacks—such as the noise accentuation that results from significant input-signal attenuation and transformers' limited frequency response—often make them unusable. Nonetheless, the example does help to show that the impedance scaling is caused by the amplifier's power gain rather than the application of feedback. The fact that the input impedance of the amplifier connection can be increased indefinitely by connecting a step-up transformer after it and raising the voltage attenuation of the network or the transformer that comes before the amplifier to achieve an overall gain of one will serve to further underscore this point. This observation reflects the amplifier's zero output impedance, which allows it to provide infinite power gain on its own.

Accepting this reality of impedance scaling can be done in a rather philosophical way by realizing that feedback is most often employed due to its basic benefit of lessening the sensitivity of a system to variations in the gain of its forward-path element. By selecting the right topology, one can achieve both sensitivity and the benefits of impedance scaling.

6.6. SENSITIVITY

The closed loop gain of a negative feedback amplifier is expressed as:

$$A' = \frac{A}{1+\beta A} \tag{6.5}$$

The result of taking the logarithms of the two sides of the above equation is:

$$log_e A' = log_e A - log_e (1+\beta A) \tag{6.6}$$

Differentiating the above equation, gives:

$$\frac{dA'}{A'} = \frac{dA}{A} - \frac{\beta dA}{(1+\beta A)}$$

$$= \left[\frac{1}{A} - \frac{\beta}{1+\beta A}\right] dA$$

$$= \left[\frac{1+\beta A-\beta A}{A(1+\beta A)}\right] dA \quad \text{or}$$

$$\frac{dA'}{A'} = \frac{1}{(1+\beta A)}\left(\frac{dA}{A}\right). \tag{6.7}$$

The above Eqn. (6.7) relates the fractional change in the closed-loop gain of the negative feedback amplifier namely (dA'/A') with the fractional change in its open-loop gain, namely (dA/A).

6.6.1. Sensitivity of the Negative Feedback Amplifier (S)

The ratio of the percentage change in voltage gain with feedback to the percentage change in voltage gain without feedback determines the sensitivity of the negative feedback amplifier; this sensitivity is expressed as:

$$S = \frac{dA'/A'}{dA/A} \tag{6.8}$$

But inverting the above Eqn. (6.7), we have:

$$\frac{dA'/A'}{dA/A} = \left(\frac{1}{1+\beta A}\right). \tag{6.9}$$

Substituting Eqn. (6.9) in Eqn. (6.8) gives the expression for the sensitivity as:

$$S = \left(\frac{1}{1+\beta A}\right). \tag{6.10}$$

Note: A negative amplifier's gain stability will be large when its sensitivity is small, and vice versa.

- *Example Problem:* An amplifier has a feedback ratio of 0.05 and an open loop gain of 800. Determine the percentage change in the closed loop gain if the temperature causes a 20% change in the closed loop gain.
- *Solution:* We are aware that gives the percentage change in the closed loop gain.

$$\frac{dA'}{A'} = \frac{1}{(1+\beta A)}\left(\frac{dA}{A}\right).$$

From the problem given, A = 800, β = 0.05 and dA/A = 20% or dA/A = 20/100. Substituting these in the above equation gives:

$$\frac{dA'}{A'} = \frac{1}{(1+0.05\times 800)}\left(\frac{20}{100}\right),$$
$$\simeq \frac{1}{200}, \quad \text{or}$$
$$\frac{dA'}{A'} \simeq 0.5\%$$

As a result, the closed loop gain of the negative feedback amplifier changed by 0.5%.

Note: We can conclude that negative feedback increases the gain stability of an amplifier because the percentage change in the closed loop gain of the amplifier is smaller than the percentage change in the open loop gain.

6.7. EFFECT OF POSITIVE FEEDBACK

Positive feedback strengthens and encourages a system's behavior, whereas negative feedback regulates and corrects it. It is also referred to as regenerative feedback because of this. When the fed-back signal is in phase with the input signal and increases the magnitude of the input signal, this is known as positive feedback, i.e., it enhances the system's overall gain.

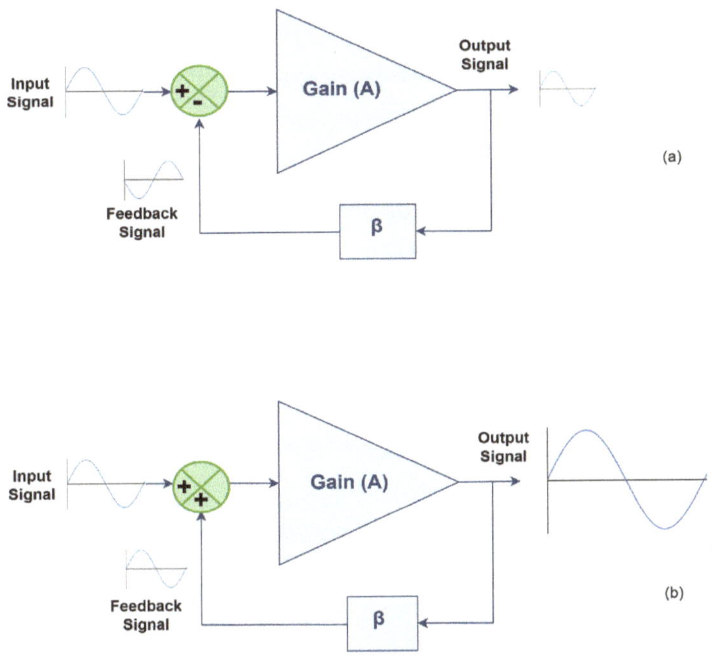

Figure 6.13. *Comparison of 2 types of feedback configurations containing negative feedback (a) and positive feedback (b).*

Source: https://www.electronics-lab.com/article/the-positive-feedback-in-electronics/.

The mechanisms used to feed an output signal to the input terminal for an in-phase (positive feedback) and 180 phase-shifted (negative feedback) feeding signal are depicted differently in Figure 6.13. Similar to an amplifier, the main 2-port linear network has an open-loop gain of 'A,' while the feedback network has a transfer function with a magnitude of 'β.' Naturally, there is a brief delay before the feedback signal reaches the input terminal and combines with the input signal, which is represented as a sinusoidal waveform in this sketch.

One important characteristic of positive feedback is the escalation of minor disruptions. Positive feedback propagates changes in a system in the same direction until the output reaches its maximum value. Positive feedback amplifies the input signal under specific gain conditions to the point where the device's output oscillates between its maximum and lowest states (upper and lower rails). As a result, the maximum amount of positive feedback that is permitted is usually strictly limited to alter the input signal very slightly.

While positive feedback is rarely used in amplifiers because it typically increases distortion and instability, negative feedback is frequently used in practical domains to create controlled amplifiers and filters. In oscillator circuits, however, the positive feedback mechanism is used. It should be noted that under some circumstances, positive feedback can be detrimental to semiconductor junctions and cause thermal runaway.

6.7.1. Positive Feedback in OP-AMPs

When an op-amp amplifier has positive feedback, the feedback voltage applied to the noninverting input will push the op-amp more strongly in the direction of the signal that is generated at the output, i.e., toward saturation. The simplest schematic circuit of an op-amp positive feedback configuration is depicted in Figure 6.14, where the output terminal (Vout) is connected directly to the noninverting input terminal (V+). The external input voltage source (Vs) in this configuration can be connected to the inverting terminal (V–). The op-amp is biased by two positive and negative DC power supplies.

The magnitude and polarity of the voltage at the non-inverting input (V+) will be the only factors that control the output voltage if the inverting input is grounded (keeping the V–terminal at zero volts by removing the input voltage source of Vs). The op-amp will drive its output positive if this voltage happens to be positive. The positive output saturation that results from feeding that positive voltage back to the noninverting input is equal to the positive power supply voltage of Vcc, i.e., the top rail

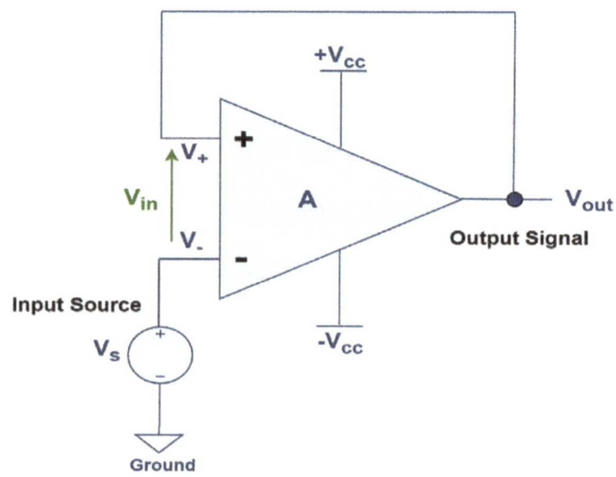

Figure 6.14. *An op-amp circuit with simple positive feedback.*

Source: https://www.electronics-lab.com/article/the-positive-feedback-in-electronics/.

Thus:

- **Case 1:** if V– = 0 and V+ > 0 then Vout = +Vcc

In contrast, the op-amp's output will drive in a negative direction if the voltage on the noninverting input happens to go negative. This will feed back to the noninverting input and result in full negative saturation (-Vcc or the lower rail). Thus:

- **Case 2:** if V− = 0 and V+ < 0 then Vout = -Vcc

You can finish this straightforward analysis by applying input external excitation (Vs), varying the voltage at the inverting terminal (V−), and getting the same outcomes. As a result, the circuit we have here has a bistable output, which denotes stability in either of the two states (saturated positive or saturated negative). It has a tendency to latch on to that state or stay unchanged once it reaches one of those saturated states. The current state will change to the next by applying a voltage to the inverting (V−) input that has the same polarity but a little higher magnitude.

By putting two resistors between the input and output terminals in the feedback path, the performance of this feedback system can be more precisely controlled. By adding Rf and Rin to the op-amp and placing an input voltage source (Vs) at the noninverting terminal (V+) while the inverting terminal (V−) is grounded, Figure 6.15 illustrates such a configuration.

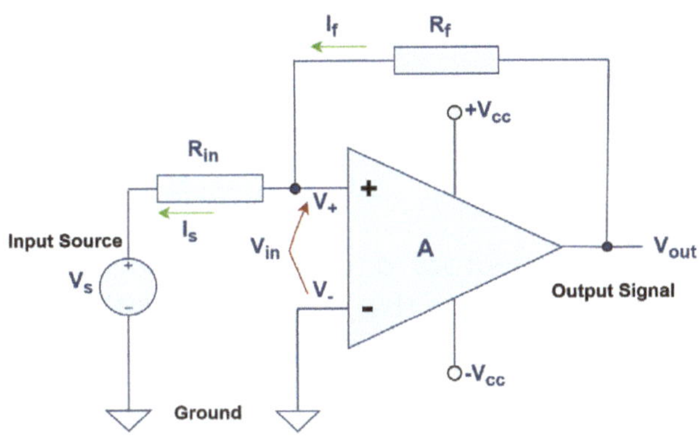

Figure 6.15. *Configuration of a positive feedback op-amp circuit as a comparator.*

Source: https://www.electronics-lab.com/article/the-positive-feedback-in-electronics/.

A digital comparator is a device that uses this circuit to transform continuous analog signals into discrete, two-state signals. Hysteresis is another effect that can be produced in a comparator by changing the size of the feedback resistor. Hysteresis essentially provides the comparator with two thresholds. The comparator circuit is more resistant to noise voltages that could cause unintended swings in the output by obtaining these thresholds. Unless there is a significant change in the magnitude of the input voltage, the hysteresis keeps the output in its current state.

The output signal (Vout) and the input signal (Vin) amplitude waveforms are shown versus time in Figure 6.16. The difference between noninverting and inverting voltages, or Vin, is defined as follows:

Vin = V+ − V−

We can first assume that the output of the op-amp is at positive saturation (+Vcc) in order to analyze the system's operation. V− equals 0 V since it is grounded. The op-amp's high impedance input characteristic means that the main current only passes through the resistors of Rf and Rin, leaving the op-amp empty. Thus, $I_f = I_s$, the two currents are equal.

Using Ohm's law:

If = (Vout − Vs)/(Rf + Rin)

By using the superposition principle, we first assume that Vs = 0, i.e., Rin is grounded. So:

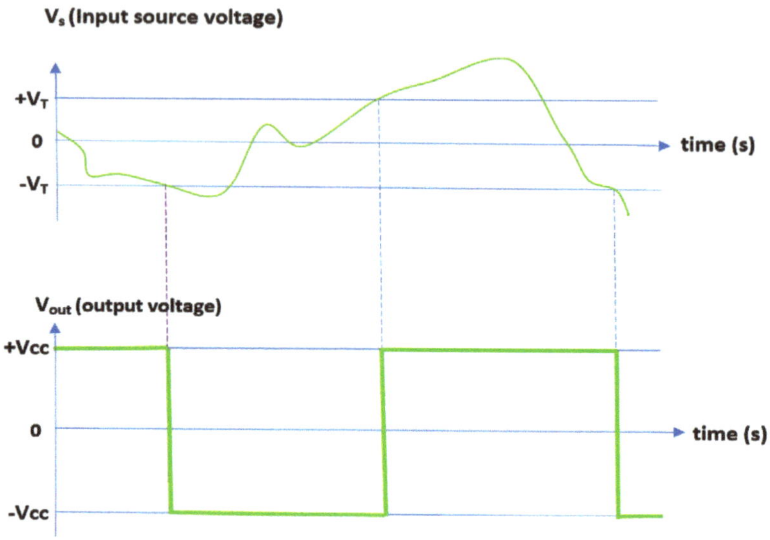

Figure 6.16. *Waveforms of the input signal (Vs) and the output signal (Vout) versus time.*

Source: https://www.electronics-lab.com/article/the-positive-feedback-in-electronics/.

Vin = Rin.

It stays at positive saturation (+Vcc) in the output. We now presume that Vs is not equal to zero. When Vs is progressively lowered, Vin will eventually reach 0 V, at which point the output's state will change to -Vcc. This voltage at the input is known as the negative threshold voltage (Vs = -VT). The following formulas can be used to find the negative threshold voltage:

If = -VT/Rin = Vout/Rf = -Vcc/Rf

and then,

VT = Vcc (Rin/Rf).

Consequently, the ratio of the input to feedback resistors can be used to control the width of the hysteresis region. Now, if 0 V is applied to the input (Vs = 0) while the output is at negative saturation (-Vcc), Vin turns negative and the output stays at (-Vcc). Nonetheless, there comes a point at which Vin drops to zero and the output switches states if the input voltage (Vs) is non-zero and increases. This voltage, which is equal to +Vcc, is known as the positive threshold voltage (Vs = +VT). (Rin/Rf). For most uses, Rf is significantly bigger than Rin.

The relationship between the input margin of threshold voltages and the output saturation voltages is depicted in Figure 6.17. Arrows indicate potential routes for altering the output voltage as a result of modifications to the voltage input source.

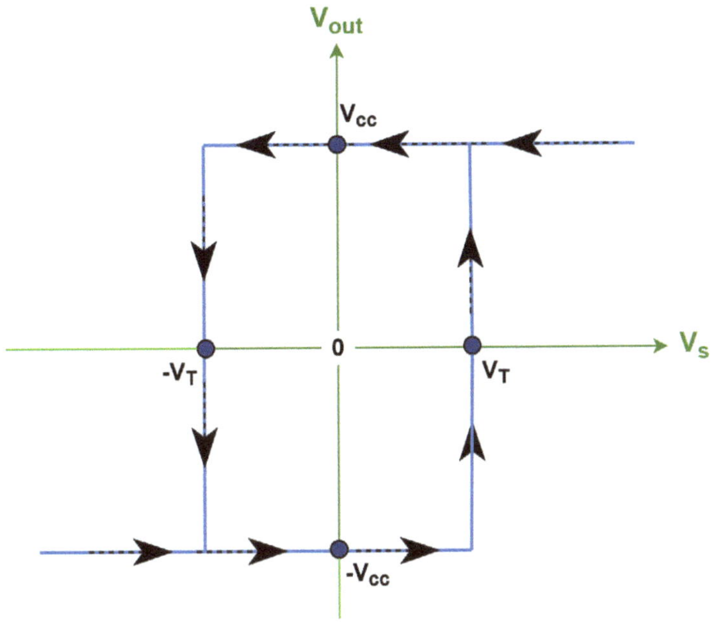

Figure 6.17. *Hysteresis and threshold voltages in positive-feedback comparator.*

Source: https://www.electronics-lab.com/article/the-positive-feedback-in-electronics/.

As Figure 6.16 makes abundantly evident, even with considerable distortion and undesired swings in the input signal, extracting a clean square wave in the output is possible by adding threshold margins and a small amount of positive feedback to the comparator.

6.7.2. The Oscillation Mechanism of Feedback Networks

Yet, there are drawbacks to feedback that are equal to its benefits. Given that each network's gain varies in both magnitude and phase with frequency, it is likely possible under some circumstances to introduce enough phase-shift to generate positive feedback. In other words, the amplifier may oscillate unexpectedly and become unstable, producing an output signal that is independent of the input or absent altogether.

Oscillation is a drawback that is mostly detrimental to amplifier circuits. However, oscillators are useful and frequently used in timing and modulation circuits in applications such as communication systems. For better understanding, it is important to notice the denominator in the basic negative feedback equation, $Af = A/(1 + \beta A)$. When $\beta A = -1$ or a magnitude of 1 at a phase angle of 180°, the denominator becomes 0, and the gain with feedback 'Af' mathematically becomes infinite or very high physically. Thus, a very small signal (like a noise voltage) can provide a measurable output voltage, and the circuit acts as an oscillator even without an input excitation.

Block diagrams of a general feedback system are displayed in Figure 6.18. If there is a false voltage at the amplifier input, "Vin," the amplifier stage will produce an output voltage of Vout = AVin, and the feedback stage will produce a voltage of Vf = β(AVin). where the loop gain is denoted by βA.

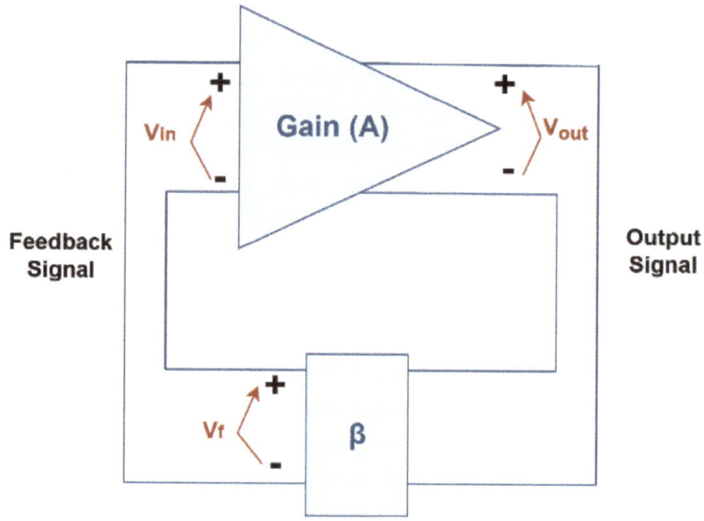

Figure 6.18. *A general feedback system without external excitation.*

Source: https://www.electronics-lab.com/article/the-positive-feedback-in-electronics/.

Vf can be produced equal to Vin if the feedback network and base amplifier circuits supply βA with the proper magnitude and phase. The circuit will then continue to function even if the imaginary voltage Vin is eliminated because the feedback voltage is enough to power the feedback circuits and amplifier. Therefore, if the condition βA = 1 is satisfied, the output waveform will continue to exist even in the absence of any input excitation. This is referred to as the oscillation Barkhausen criterion.

In actuality, the oscillator can be started without an input signal. The system begins to oscillate by amplifying the constant noise voltage when only βA is made equal to or greater than 1. These oscillations are occasionally referred to as parasitic oscillations.

Figure 6.19 illustrates how a steady-state oscillation condition builds up because of the noise signal. There are three distinct sections to this waveform: (a) During this time, specific circumstances are met for an initial noise voltage to be amplified and manifest at the output. (b) The positive feedback process will cause this tiny non-sinusoidal signal—which results from βA not being exactly 1—to gradually increase in magnitude over the course of the following time. (c) As a result of circuit saturation, the output waveform enters a steady-state oscillation with a constrained envelope.

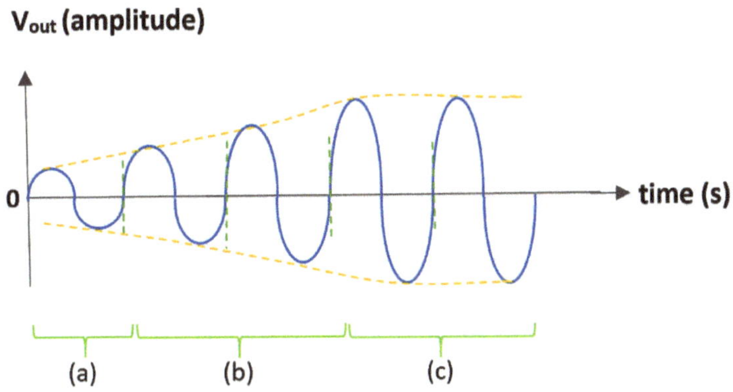

Figure 6.19. *Generation of the periodic waveform due to oscillation in the circuit.*

Source: https://www.electronics-lab.com/article/the-positive-feedback-in-electronics/.

Positive feedback is helpful when building oscillator circuits, but undesirable oscillations make an amplifier ineffective. Furthermore, undesired but inevitable feedback paths are a common feature of electronic circuits. Performance can be negatively impacted by the signals that are fed back via these paths. Sometimes, analogous effects are seen as a result of stray or parasitic capacitance between the components of an electronic part or circuit.

CASE STUDY

DESIGN OF HIGH-PERFORMANCE LOW NOISE AMPLIFIERS MMICS

- RFIC Solutions Inc.

With its headquarters located in San Jose, California, and a design center in India, RFIC Solutions Inc. is a fabless RF design house that specializes in wireless solutions. We use state-of-the-art GaAs, InGaP/GaAs, InP, CMOS, and SiGe Semiconductor processes to design highly integrated system on chip (SOC) and system on a package (SOP); custom integrated circuits (ICs); and IP cores using MESFET, pHEMT, and HBT devices. LNA, PA, Switch, full transceivers/RFICs, and RF Modules for any kind of wireless syst

m, including PCS, WLAN, WiMax, and cellular applications, are among the areas of expertise. RFIC Systems, Inc. is a complete RF and microwave solution provider. Our business plan includes the provision of full RF chip supply as well as design intellectual property, design services, PCB design services, layout services, and foundry services.

- **The Client**

The client was a South Korean company that specializes in RF technology and products for the wireless industry and develops creative solutions. They manufacture cutting-edge RF components, such as gain blocks, local drivers, IF ICs, power amplifiers, discrete devices, and RF integrated circuits (RFICs), and they supply next-generation products.

- **Design Challenges**

By concentrating on business and revenue growth through Low Noise, high linearity RFIC amplifier designs, the client aimed to take the lead in the market. The customer wanted an amplifier design with minimal noise and high linearity. Regarding their WCDMA band LNA design, PCS, and cellular designs, they desired high performance design solutions. Prioritizing a prompt and economical solution was paramount. The client also requested a single chip solution, or a high-performing broadband LNA chip, that could

e adjusted for performance in the aforementioned bands in terms of NF, Gain, and OIP3.

In less than six months, including IC fabrication and testing, RFIC Solutions was selected to design all four LNAs on the GaAs process with extremely stable performance for low noise and linearity over temperature, process, and supply voltage variations. The frequency range covered by the designs was 40MHz–4GHz.

- **Solution**

The client selected RFIC Solutions due to our creative design service approach, which makes use of some of our current Design IPs and project management procedures. Our project plan was effective in terms of both time and engineering resources. The creative and economical pricing model offered by RFIC Solutions satisfied our client's specified design requirements. The RFIC solutions India design center, where a group of committed engineers offers our clients high-caliber RFIC design solutions, made the affrdable design service possible.

RFIC Solutions, Inc. finished four LNA designs in the allotted time and satisfied all requirements regarding process, temperature, and supply voltage variations in the design that was ready for tape out. To offer consistent performance across changes in supply voltage and temperature, RFIC Solutions employed tried-and-true design methodologies. The designs reduced the cost of board assembly by requiring only bypass and coupling capacitors as off-chip components.

During the GDSII tape out of the designs, RFIC Solutions gave the client-side engineering team complete support. High-performance layouts with the capacity to tune the circuit with little changes to the mask or via the idea of "mask sharing" were supplied by our specialized RFIC layout design team. The client was able to submit all of the designs on time for fabrication. As part of our comprehensive design service flow, RFIC Solutions also assisted the client in testing designs using test plans, component, and assembly diagrams.

Up until the GDSII tape out, a team of engineers from our design center kept the client informed about any changes to the specifications. Our customer was extremely satisfied with our interactive approach because it allowed them to include any design changes they wanted. By utilizing creative and reliable design techniques along with effective project management techniques, we ensured that the design modifications were monitored and completed with great efficiency. Following fabrication, the designs were put together and tested, and every one of them demonstrated compliance with the simulated specifications, with noise figures as low as 0.65dB and OIP3 as high as 38dBm. Over the 40MHz to 4GHz band, the broadband LNA required some tuning to get close to desired performance, as it deviated from the specifications by 3.5dB.

- **Benefits for Client**

RFIC Solutions, Inc. was able to offer our client a cost-effective service within the design window, enabling them to launch new, high-performance LNA chips into the market seven months after the project's inception. The client achieved:

- 0.5 to 1dB of better Noise Figure performance over the other competition.
- High linearity performance over the other competition.
- Cost reduction of 40% due to RFIC solutions cost-effective design service.
- Efficient & focused usage of client's own engineering resources.
- Product launch within scheduled time.

CLASS ACTIVITY

A SERIES-SHUNT FEEDBACK AMPLIFIER USING OP-AMP

- **Purpose:** This lab's goals are to acquaint the student with fundamental feedback topology and technical vocabulary and investigate how negative feedback affects an Op-Amp system's gain, bandwidth, and terminal resistances.
- **Required:**
  - Oscilloscope
  - Function Generator
  - Digital Multimeter
  - LM741 Operational Amplifier
  - Various ¼ watt Resistors
- **Theory:** Positive and negative feedback are the two categories into which feedback can be divided. In all stable systems, the first category is the most utilized. Positive feedback is mostly used by other systems that run under unstable operating conditions. For instance, Oscillator employs positive feedback in specific scenarios. The circuit's or the system to be controlled's output terminals are where the feedback process begins. A tiny amount of the output—voltage or current—is taken, its sign reversed, and then added to the input signal. The negative feedback system's general block diagram is displayed in Figure 1.

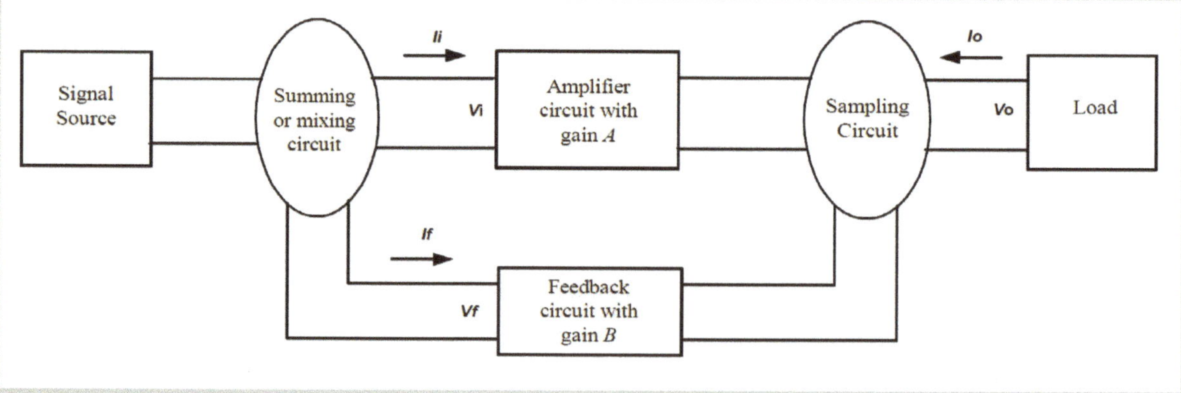

Figure 1: *Basic structure of amplifier with feedback network.*

Source: http://www.engr.newpaltz.edu/~zunoubm/S20/electronics2lab/Lab7_S13.pdf.

When the amplifier circuit is designed with the general feedback concept in mind, the output signal sample will be either voltage or current with a 180-degree phase shift from the input signal. It has been discovered that by choosing the right feedback topology, negative feedback can increase the amplifier's bandwidth, stability, noise immunity, and control over its input and output resistance. The interface connecting the input, feedback, and output circuits is commonly referred to as the feedback topology. For instance, a series connection is used for the input-feedback circuit interface and a shunt connection is used for the output-feedback circuit interface in a series-shunt topology. The way that the signals are sampled at the output circuits or mixed (summed) at the input can also be referred to as the feedback topology. Typically, the input circuit and series connection are used to mix the voltage at the input. Similar to this, a shunt connection between the input circuit and the current is used to mi the current at the input circuit.

Through series connections with the output circuits and shunt connections with the output side, respectively, the voltages and currents are sampled. In actuality, these feedback topologies are possible:

- Voltage sampled – series mixed (voltage) at the input Series-Shunt feedback topology.
- Current sampled – series mixed (voltage) at the input Series-Series feedback topology.
- Current sampled – shunt mixed (current) at the input Shunt-Series feedback topology.
- Voltage sampled – shunt mixed (current) at the input Shunt-Shunt feedback topology.

The fundamental feedback topologies with an Op-Amp are displayed in Figure 2 below. Take note of how the sampling point at the output and the mixing point at the input are connected differently.

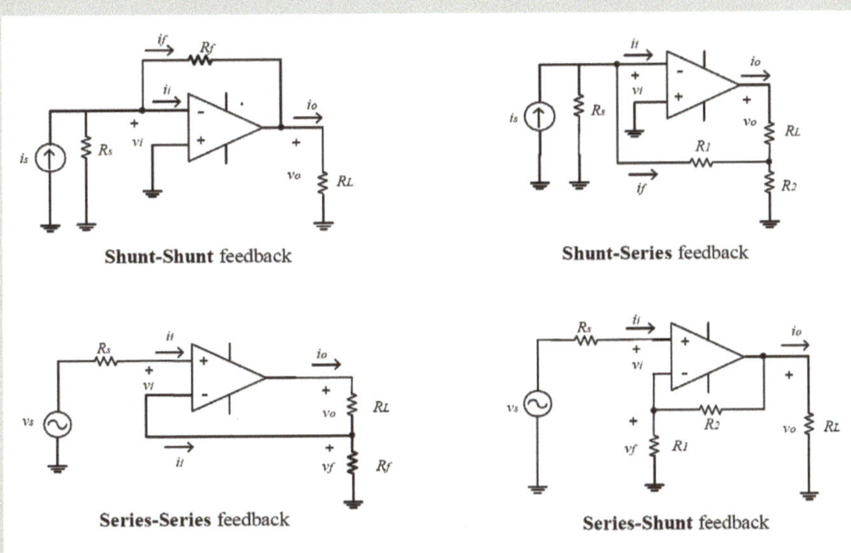

Figure 2: *Basic structure of amplifier with feedback network.*

Source: http://www.engr.newpaltz.edu/~zunoubm/S20/electronics2lab/Lab7_S13.pdf.

PRE-LAB – Series-Shunt Feedback Amplifier. One benefit of using feedback is to control the characteristics of the amplifier under test. To find A, Af, β, Rif, and Rof through analysis, the following steps are carried out:

1. Draw the basic amplifier without the feedback circuit, and then replace the active device by the proper equivalent circuit.
2. Determine the loading element values and construct the loaded amplifier circuits.
3. Determine the amplifier gain A, Ri and Ro of the loaded amplifier as shown in series-shunt example in the textbook.
4. Determine the feedback parameter β.
5. Determine the expressions for Rif and Rof.

- **Experiment:** The basic amplifier block with limited open loop gain A = 100 will be built in this lab exercise using two operational amplifiers, as illustrated in Figure 3. One amplifier stage A1 with gain 100 and a unity gain stage A2 (buffer) make up the basic amplifier block. Because a buffer is being used at the input, the input resistance Ri is effectively infinite. Additionally, there is nearly no output resistance.

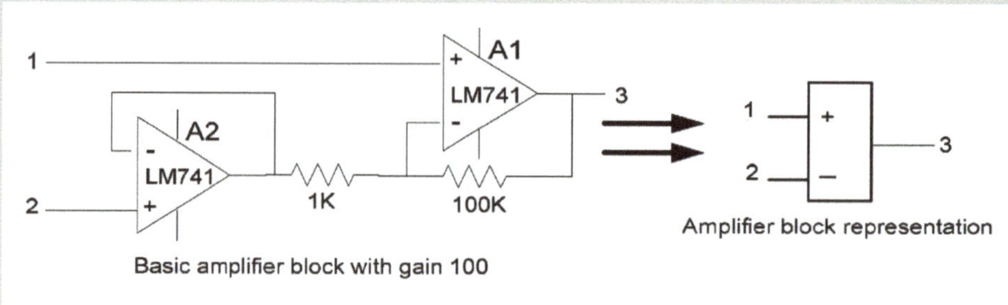

Figure 3. *Basic amplifier block with open loop gain 100.*

Source: http://www.engr.newpaltz.edu/~zunoubm/S20/electronics2lab/Lab7_S13.pdf.

Using a power supply ± 15 V, assemble the basic amplifier block depicted in Figure 3. Verify that the power supply is connected to both LM741 operational amplifiers. As indicated in Figure 4(a), attach a voltage divider to the basic amplifier's input terminal. Next, use an oscilloscope to perform the following measurements. Keep in mind that you will be utilizing two distinct load resistance values (RL, ∞ and 10KΩ) in the upcoming measurements.

- **Open Loop Measurements**

1. With RL= ∞ (open circuit at C), apply a sinusoidal waveform input signal with frequency 100Hz and suitable amplitude to get undistorted output waveform. (Troubleshooting- If you see no output at point C, then look for the signal at point A. Usually there should be a signal there unless the breadboard has a bad contact. Subsequently, measure the signal at the non-inverting input of

op-amp A1 (see Figure3) which is the same point as B in Figure 4(a). In case you don't see any signal, A2 of Figure 3 is likely bad. Otherwise measure the signal at inverting input of A1. Here, you should have the same signal as B of Figure 4(a). If you don't see any signal, A1 is likely bad. A1 and A2 are referred to in Figure 3.

2. At nodes A, B, and C, measure the voltage from peak to peak. Note down these values for your reference.
3. Raise the input signal's frequency until node C's voltage falls to 0.707 times its value at 100 Hz. Record the amplifier's maximum 3dB frequency.
4. Step 2 should be repeated after connecting a 10KΩ load (at node C) to the amplifier block's output.

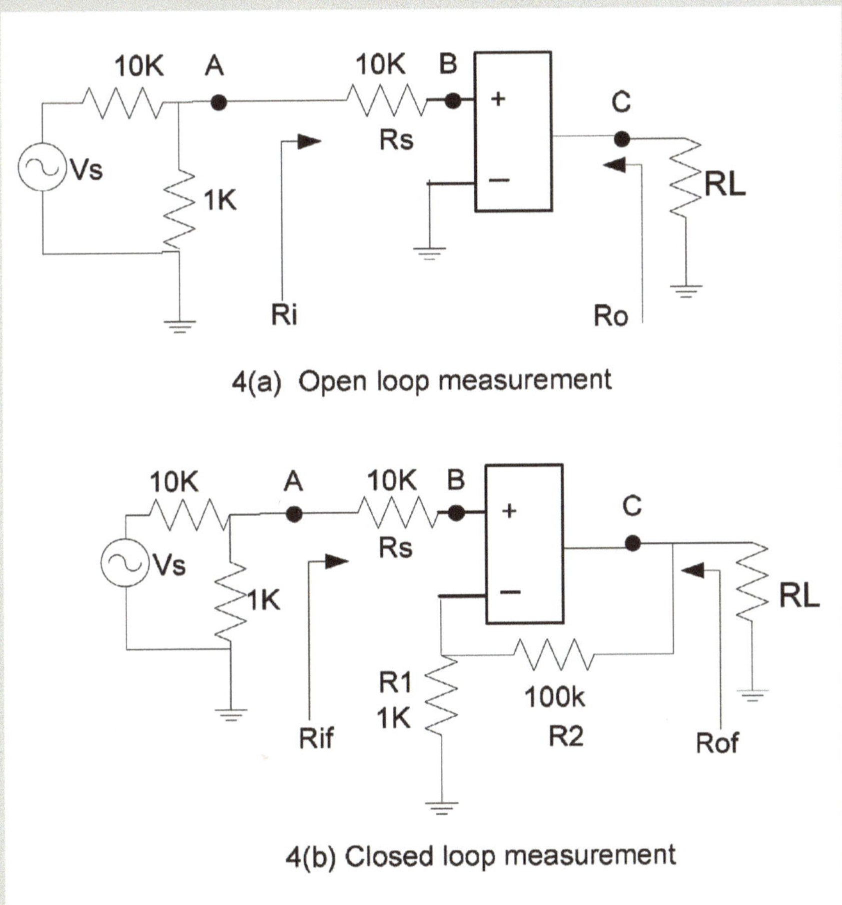

Figure 4: *Basic amplifier and Series-Shunt feedback amplifier.*

Source: http://www.engr.newpaltz.edu/~zunoubm/S20/electronics2lab/Lab7_S13.pdf.

5. Tabulate your data for the load resistance values (∞ and 10KΩ).

6. Use the above record of the measurements to compute the gain A and the output resistance Ro. What is the upper 3dB point in each case? Comment on the results. (Please Note that the input resistance is very hard to measure (about 10MΩ) because of the presence of the buffer amplifier at the input of the basic amplifier block).
7. Perform PSPICE simulations of the above steps (where applicable).

- **Closed Loop Measur**ements (Amplifier with Feedback)

8. Using the previously constructed amplifier block, assemble the Series-Shunt feedback amplifier as depicted in Figure 4(b).
9. Utilizing the load resistance RL initially disconnected (RL=∞), apply the input signal as outlined in step 1.
10. At nodes A, B, and C, measure the voltage from peak to peak. Note down these values for your reference.
11. Raise the input signal's frequency until node C's voltage falls to 0.707 times its value at 100 Hz. Record the amplifier's maximum 3dB frequency.
12. Replicate step 9 after attaching the 10KΩ load resistance.
13. Tabulate your data for the load resistance values (∞and 10KΩ).
14. Determine the gain Af between nodes A and C, the input resistance Rif, and the output resistance Rof. using the above measurement record. What is the amplifier's upper 3dB frequency? Discuss the findings.
15. Steps 8–13 must be repeated after shunting R2 with a 10kΩ resistor.
16. Discuss the impact of feedback on the input and output resistances, the upper 3dB frequency, the amplifier gain, and the results.
17. Run PSPICE simulations of the aforementioned procedures (if relevant).

SUMMARY

- An electronic circuit that sends feedback from its output to its input to control and enhance performance is called a feedback amplifier. It works by taking a sample of the output signal, comparing it to the input signal, and then modifying the amplification process as necessary.
- When the feedback energy is out of phase with the input signal, it is called negative feedback. Negative feedback reduces the gain of the amplifier. That is why, it is sometimes called degenerative or inverse feedback.
- One great advantage in using an op-amp with negative feedback is that the actual voltage gain of the op-amp doesn't matter, so long as its very large. If the op-amp's differential gain were 250,000 instead of 200,000, all it would mean is that the output voltage would hold just a little closer to Vin (less differential voltage needed between inputs to generate the required output).
- Op amps are very high gain devices. To achieve stable operation, op-amps are used with negative feedback. The gain of the device alone is called open loop gain, and the gain when configuring a negative feedback circuit is called closed loop gain. Closed-loop gain is not device-specific and is usually determined by the feedback network.
- Feedback also has important effects on the input and output impedances of an amplifier, with the type of modification dependent on the topology of the amplifier-feedback net.
- While negative feedback controls and corrects the behavior of a system, positive feedback encourages and strengthens the behavior of the system. For this reason, it is also known as regenerative feedback. Positive feedback generally occurs when the fed-back signal is in phase with the input signal and it causes the magnitude of the input signal to be increased, i.e., it improves the overall gain of the system.
- With positive feedback in Op-amp amplifiers, the feedback voltage to the noninverting input will drive the op-amp harder in the direction of the output generated signal, i.e., toward saturation.
- The advantages of feedback, however, are accompanied by corresponding disadvantages. Since the magnitude and phase of the gain of each network vary with frequency, it is probable in certain conditions to introduce sufficient phase-shift to cause positive feedback. Under these circumstances, the amplifier may become unstable and generate an output signal independent of the input (or without input signal); in other words, it oscillates unexpectedly.

REVIEW QUESTIONS

1. What is feedback in amplifiers?
2. What are the four types of feedback amplifiers?
3. What is the difference between positive and negative feedback amplifier?

4. What is feedback voltage?
5. What are the properties of a feedback amplifier?
6. Why is the feedback used in feedback amplifier design?
7. What is positive feedback in an amplifier?
8. Where is positive feedback in amplifiers used?
9. What is the feedback factor?
10. What are the merits and demerits of negative feedback amplifier?
11. How to calculate gain for closed-loop?
12. How does feedback affect input and output impedance?

REFERENCES

1. Black, H. S. (1934). Stabilized feedback amplifiers. *Bell System Technical Journal, 13*(1), 1–18. doi:10.1002/j.1538-7305. 1934.tb00652. x.
2. Jaeger, R. C., & Blalock, T. N. (2006). *Microelectronic Circuit Design* (3rd ed.). McGraw-Hill Professional.
3. Kal, S. (2004). *Basic Electronics: Devices, Circuits, and IT Fundamentals*. Prentice-Hall of India Pvt Ltd.
4. Kuo, B. C., & Golnaraghi, F. (2003). *Automatic Control Systems* (8th ed.). Wiley.
5. Palumbo, G., & Pennisi, S. (2002). *Feedback Amplifiers: Theory and Design*. Springer Science & Business Media.
6. Pederson, D. O., & Mayaram, K. (2007). §5.2 Feedback for a general amplifier. In *Analog Integrated Circuits for Communication: Principles, Simulation and Design*. Springer Science & Business Media.
7. Sedra, A. S., & Smith, K. C. (2004). *Microelectronic Circuits* (5th ed.). Oxford University Press.
8. Self, D. (2013). *Audio Power Amplifier Design* (6th ed.). Focal Press.

CHAPTER 7

Fundamentals of Electronic Circuit Fabrication

LEARNING OBJECTIVES

After studying this chapter, you will be able to:

- Define the electronic circuit design;
- Discuss on ohm's law, Kirchhoff's laws, and equivalent circuits;
- Identify the noise in electronic circuits.

KEY TERMS FROM THIS CHAPTER

Automotive industries
Digital circuits
Electronic circuitry
Magnetism
Printed circuit boards (PCBs)
Simulation software
Transistor
Darlington pair
Electricity
Integrated circuits (ICs)
Miniaturization
Serial communication
Telecommunications

INTRODUCTION

An electronic circuit is a path or network of connected electronic parts that enables the passage of electricity to carry out a particular task. From straightforward devices like calculators and flashlights to sophisticated systems like computers and communication networks, these circuits are essential to contemporary technology. One can control the flow and behavior of electric current by using resistors, capacitors, inductors, diodes, and transistors, which are the main components of an electronic circuit. Diodes permit current to flow only in one direction; inductors resist changes in current; transistors function as switches or amplifiers to control the current flow; capacitors store and release electrical energy; and resistors set a limit on the amount of current that comes through a circuit. In general, there are two types of electronic circuits: digital and analog. For tasks involving light, radio waves, and sound, analog circuits are indispensable because they process continuous signals. Discrete signals are handled by digital circuits, which are the foundation of computers and other digital systems. Binary code is used to represent data in these circuits. In order to carry out calculations, store data, and manage complex instructions, these circuits use logic gates and flip-flops, which modify binary signals.

Typically, the process of designing an electronic circuit entail drawing a schematic diagram that uses standardized symbols to show the components and their connections. These schematics are used by engineers to plan the circuit's layout and make sure it satisfies the necessary requirements and runs consistently. These designs are often implemented physically using printed circuit boards (PCBs), which offer a stable surface on which components can be mounted and electrical connections can be made via conductive pathways etched into the board. With the ability to combine and miniaturize multiple components into a single chip, integrated circuits (ICs) represent a significant advancement in electronic circuitry. The electronics industry has undergone a revolution thanks to this invention, which made it possible to create small, effective, and potent gadgets. Almost every piece of electronic equipment, from complex communication devices and industrial machinery to domestic appliances, contains integrated circuits (ICs).

Kirchhoff's circuit laws, Ohm's Law, and other fundamental theories explain how the laws of electricity and magnetism control the operation of an electronic circuit. Kirchhoff's laws address the conservation of current and voltage in electrical networks, whereas Ohm's Law links voltage, current, and resistance in a circuit. These ideas aid engineers in predicting and analyzing the behavior of circuits to ensure they function as intended under a variety of circumstances. Innovation is still being fueled by developments in electronic circuit design and technology in a number of industries, including computing, telecommunications, healthcare, and automobiles. In order to improve performance, lower power consumption, and increase miniaturization, modern circuits are becoming more complex by integrating cutting-edge materials and techniques. Further significant advancements in circuit design and functionality are anticipated as a result of the continuing development of quantum computing and nanotechnology, which will further expand the capabilities of electronic devices.

7.1. ELECTRONIC CIRCUIT DESIGN

Analyzing and creating electronic circuits is the process of designing electronic circuits. A circuit designer should be able to predict the voltage and current at each node in the circuit when creating an analog or digital circuit. It is possible to manually analyze any linear circuit using mathematical computations, as well as simple non-linear circuits. While complex circuits are analyzed using software. More time, money, and risk can be avoided in the process of creating circuit prototypes by using electronic circuit design simulation software to help developers design circuits more precisely and efficiently.

7.1.1. Circuit Board Simulator

The behavior of an actual electronic circuit is replicated by the electronic circuit simulator using mathematical models. Circuit operation can be modeled with simulation software, which is also a very valuable analytical tool. Most integrated circuit design is done through simulation due to the limitations of the breadboard and the cost of tools like photo masks. The simulator for analog circuits is called SPICE. Digital simulations are primarily performed using Verilog and VHDL. While circuit board simulators simplify the process of developing large circuits, there are certain complexities involved in the simulation. Circuit simulators do not consider process variations that arise during the fabrication of a design. Despite being slight, the variations have a big impact on the final product.

7.1.2. Basics of an Electronics Circuit Design Process

Every basic electronic gadget is constructed as a single unit. All discrete transistors, diodes, resistors, capacitors, and inductors existed prior to the development of digital circuits (ICs). Based on its input, any circuit or system can generate the desired result. Here, we provide some fundamental information about the electronic circuit design process. The distinction between digital and analog circuits.

- **Analog Circuit:** Electronic circuit designs that use current or voltage variations over time to represent data are known as analog circuit designs. An analog circuit's main constituents are wires, transistors, capacitors, resistors, and diodes. Electrical signals in analog circuits have a continuous value. These circuits are shown in schematic diagrams where wires are denoted by lines and individual components by distinct symbols. There are circuits in series, parallel, or both types in every analog circuit.

- **Digital Circuits:** Electrical signals are treated as discrete values in digital electronic circuit design. Zeros and ones are used to represent the data. Transistors are widely used in digital circuits, where they are connected to produce logic gates that enable Boolean logic. Positive feedback is produced by connecting transistors, which is how latches and flip-flops work. Consequently, digital circuits can provide logic and memory, which allows them to carry out calculations. Microprocessors and other general-purpose computing chips like application-specific integrated circuits are made using digital circuits.
- **Schematic Circuit Diagrams:** The representation of parts and connections in a circuit using standard symbols in lieu of the component's actual image is called a schematic circuit diagram. Electrical and electronic equipment are designed, built, and maintained with the use of circuit diagrams. The schematic diagrams are arranged on a page from left to right and top to bottom, though this is not standardized. For example, in a signaling circuit, the speaker is on the right and the antenna is on the left. Likewise, the top of the page has the positive power supply, while the bottom has the ground and negative supply. Standardized techniques are also used in relay logic line diagrams to depict schematic diagrams. The components strung between the two vertical power supply rails—one on the left and one on the right—represent a ladder. For this reason, ladder logic diagram is another name for it.
- **Electronic Switch Circuit:** An electrical switch is a tool used to stop current flow in a circuit. In essence, these are binary devices—they can be fully ON or fully OFF. In addition to controlling operation, ON/OFF switches also activate various circuit features. Mechanical devices with two or more terminals connected to metal contacts are called switches. Switch closed when contacts are in contact with each other. As a result, the switch is ON and current flows. The switch is open and there is no current flow when the contacts are apart.
- **Toggle Switch:** The toggle switch is actuated by a lever angled in one or more positions. The lever flips up or down to close or open the contact. The light switches used in the household are an example of a toggle switch.
- **Push Button Switch:** Push button switch is a two-position device actuated with a button to open and close the contacts. Each time you push the button the contact alternates between the open and close.
- **Joystick:** A joystick switch is triggered by a lever free to move in more than one axis of motion. The circle and dot notation on the switch symbol indicate the direction of joystick lever motion that is required to trigger the contact. Joystick hand switches are used to control crane, robot and in games.
- **Liquid Level Switch:** The switch mechanism is triggered by a floating object once the liquid level

reaches a predetermined level. The floating object closes the circuit when the liquid level reaches a certain point. This closed-circuit conducts, enabling it to carry out the intended function. Other switches found in electronic circuits include the nuclear level switch, pressure switch, proximity switch, lever actuator limit switch, and speed switch.

- **Power Supplies:** For electronic circuit design to be successful, power supplies are essential. Nevertheless, they must fulfill certain requirements and can be costly. Thankfully, there are several tools available to assist designers in selecting power supplies more wisely. Among these are simulation tools that study power supply designs to find significant insights into electronics design. Then, they can assist you in selecting the appropriate power supply and component package based on the power requirements of your circuit. Regulated and unregulated power supplies are the two main categories. Each has unique benefits and drawbacks. For instance, an unregulated power supply has a wide input voltage and a narrow output range, whereas a regulated power supply offers a stable output voltage.
- **Use of Microcontrollers:** Microcontrollers are useful tools for creating compact, multipurpose electronic circuits. Programmable pins on a microcontroller can function as inputs or outputs. For instance, pin 9 of the STM32F042 microcontroller is designated PA3. This pin can be used as an input output pin for a capacitive touch sensor controller, as well as a timer output and a means of receiving serial communication input. Complementary metal-oxide-semiconductor (CMOS) chips are the standard component of microcontrollers, which makes them prone to static charges. They can therefore be used in electronic circuits, despite the fact that static charges can harm these gadgets. Input and output ports on the microcontroller allow it to be connected to external devices. These could be motion, push button, or temperature sensors. The CPU then receives these signals and determines what to do with the data. The CPU can send signals to LED lights or motors based on the inputs, and we can convert inputs to outputs in a variety of ways. One way to regulate the temperature in a room is by connecting a temperature sensor to a motor.

Microcontrollers have memory components, such as data memory, in addition to their input and output ports, which are used to temporarily store data while instructions are being executed. The data memory, however, is ephemeral and is only preserved when the device is powered on. Information is received by input ports, typically in binary format. After that, the output devices receive instructions from the processor and carry out tasks not handled by the microcontroller.

- **Using Decoupling and Coupling Capacitors:** In order to reduce noise and improve circuit performance, capacitors are crucial components of electronic circuits. Capacitors come in two varieties: coupling and decoupling. A small, reactive capacitor that is positioned on the board between an IC and its load is called a decoupling capacitor. It acts as a sound barrier to block incoming AC noise. When designing an electronic circuit, the location of a decoupling capacitor is crucial. They should be placed as close to the chip as possible to effectively decouple it. Nonetheless, the fundamental principles of the circuit dictate where a decoupling capacitor should be placed. Before decoupling capacitors can supply the necessary current, they must have time to charge and discharge. To maintain a steady voltage, they must also be able to withstand abrupt voltage fluctuations. For the intended use and the circuit, the capacitors must be the proper size. Another way to lessen an IC's sensitivity to power ripple and noise is to decouple capacitors. By serving as a charge reservoir and shunting the transient current to ground, decoupling, or separating two circuits, lowers the noise. It also aids in keeping the IC's power supplies steady.

- **Using Pull-Up and Pull-Down Resistors in Electronic Components**: Electrical components known as pull-up and pull-down resistors serve as voltage and current limiters. They prevent the input from falling into a logical low state while permitting a tiny amount of current to escape. They work well in circuits that have pushbuttons as well as other electronic circuits. To define the voltage in the absence of a driving signal is the function of pull-up resistors. Its value can change based on the use, but it usually falls between 4 and 7 kilo ohms. In digital logic circuits, pull-up resistors are crucial components. High, low, and high impedance are the three states of a digital logic circuit. A pin enters the high-impedance state if it is not pulled to its logic level. This problem can be resolved by utilizing pull-up resistors. Similar to pull-up resistors, pull-down resistors pull a low-impedance pin. Pull-down resistors are used in electronic circuits by connecting them between a microcontroller pin and a ground terminal. To transition between the two states, connect a pull-down resistor between a pin and the ground.
- **Using Transistor ARRAYS/PAIRS in Electronic Circuit Design:** The transistor is a fundamental component in circuit design, serving as a building block. A grasp of these basics and how to integrate them is essential for efficient circuit construction. However, assembling these fundamental components can be challenging, requiring practice to master. One common example of a transistor array/pair circuit is the Darlington Pair. This configuration combines multiple transistors on a chip, with each transistor's emitter connected to the base of the next. The Darlington Pair offers high current gain and is particularly useful for driving low-impedance loads.
- **Resistor Wattage in Electronic Circuit Designing***:* In electronic circuits, resistors aid in controlling the flow of electrical current. They can be categorized based on their wattage and come in a range of sizes. It is possible to determine

a resistor's power supply rating using a common formula. Thus, a resistor's wattage can vary from a few milliwatts to several kilowatts. A resistor's wattage is determined by its output voltage and current. Youor depends on its voltage and current output, increasing with higher currents and voltages Generally speaking, you should use at least twice as much as the required power supplies. Therefore, to obtain a higher power rating, you can connect two resistors in parallel. Resistors additionally have tolerance bands. They typically show tolerances of +5 and -10%. A resistor's tolerance band is a ring made of gold or silver. When a gold resistor is within the tolerance band, it is deemed to be in good operating condition. You should replace it if the tolerance is higher than this. Resistors come in a large range of packaging. SIP-9 packages, for instance, come with eight separate 47-ohm resistors. Every resistor has a unique pin for connecting to the outside world. The other ends are attached to a shared pin. For passive SMD components, the SIP-9 package is the industry standard, though some designs employ alternative packages. Modern designs are increasingly going in the direction of very small packages. This enables designers to fit more features into a smaller area.

- **Understanding the Discharge Time of Batteries in Electronic Circuits:** Understanding the battery's discharge time is important when building a battery system. Batteries ought to be charged until their discharge time matches the lowest voltage of the gadget. The open circuit voltage is the name given to this voltage. The open circuit voltage is 0 V when the battery is not under any load. Batteries shouldn't be discharged any further than this. The State of Charge (SOC) and Depth of Discharge (DOD) of a battery are necessary to determine the discharge time. Another name for battery discharge time is discharge current. This figure serves as the foundation for figuring out the battery's capacity. We use amp-hours to measure the discharge rate. Stated differently, there exists a direct correlation between the rate of charge flow and the duration of battery discharge. The rates at which different types of batteries charge vary. Lead acid batteries, for instance, should never run entirely empty before being recharged. Furthermore, the electrodes' effective surface area affects the charge rate. The discharge time of a battery is an essential part of its design. To select the appropriate battery, you must be aware of the battery's discharge time. Although it can vary, a battery's capacity is measured in amp-hours. The discharge time of a battery should be considered when designing one for high-power applications.

7.1.3. Electronic Circuit Development Strategies

In the product development cycle, there are several important factors to take into account when creating an electronic circuit. The integrity of the production process is among the most crucial elements. It is imperative to notify the mechanical and purchasing teams of any modifications made to the schematic. Because electronic circuits are essential and crucial, engineers require software that is user-friendly and capable of recording all modifications. Compact design and dependable component interconnections are prerequisites for circuits. Overlapping cables on smaller devices result in interferences that can lead to device malfunctions. Conductors, a load, and a current source make up a basic circuit. A power source is connected to the source and load. The main function of a circuit is to permit the flow of electricity. A device with two terminals that supplies the potential difference needed for current to flow through a circuit is known as a voltage source. The load, or electrically consuming device, is another essential component. A lightbulb is the most basic load, although numerous different loads can be found in more intricate circuits.

7.1.4. Factors to Consider When Creating a Circuitry Design

A circuitry design must take several things into account. You should first comprehend the fundamental components of the circuit and how they function. Once you are aware of this, you can start circuit design. Combining these fundamental building pieces is not always easy and may require some experience.

- **Integrated Circuits Parameters:** The specifications of different integrated circuit elements are referred to as IC parameters. Through the regulation of the IC's speed, power dissipation, and heat generation, these parameters impact its performance. As a result, these factors need to be considered throughout the entire electronic circuit design process. The following considerations must be made when implementing sophisticated circuitry: Physical design entails building circuit elements on a silicon wafer. A "chip floor plan," which lists the chip's features, inputs, and outputs, is the first step in the process. Next, to prepare a silicon substrate for manufacturing, we insert circuit elements into it. Custom layout methods frequently satisfy design specifications. This is accomplished using a piece of software known as an integrated circuits layout editor.
- **Passive Components:** Passive components play a crucial role in electronic circuits and devices by regulating electricity flow, storing energy, and producing inductance. They can also enhance voltage and current levels. Common examples include incandescent bulbs and loudspeakers, which use transducers to create sound waves and are essential in radio frequency applications. In PCB design, incorporating passive components is necessary, and designers must ensure access to accurate PCB footprints in their libraries to avoid limitations of ECAD software libraries. Adding common PCB footprints to the library is recommended for optimal functionality.
- **Ground Plane:** When creating circuits, it is crucial to consider the ground plane. The ground plane serves as the conductive pathway connecting a component to the rest of the circuit. It is important for the ground plane to be free of conductive loops to prevent electromagnetic interference, which can lead to ground loops

and external magnetic fields if not addressed. The ground plane should run beneath electronic components but should not completely cover the bottom layer. If full coverage is not feasible, adjustments may need to be made to the component and trace layout.

Implementing a ground plane also enhances signal integrity. By utilizing a ground plane, the circuit is less likely to experience crosstalk between signal traces, reducing noise and enhancing signal integrity for high-speed transmission lines.

- *Current Return Path:* Circuit design is a difficult task that necessitates an understanding of the basic components and techniques. You can create an electronic circuit design that functions by comprehending these. Regretfully, assembling these building blocks is not always simple and can require multiple tries to get them right.
- **Avoiding Parallel Tracks:** Crosstalk can occur when parallel tracks are used in circuit design. The distance between and length of the tracks determine how much crosstalk there is. To minimize the amount of crosstalk, it is best to keep traces separated by less than 90 degrees. Additionally, when designing circuits, try to stay away from using parallel tracks.
- **High-speed Circuit Design Rules:** When creating high-speed military circuits, it is important to avoid placing signals near the edges of the PCB to prevent routing errors. However, this precaution can impact trace impedance and signal integrity. It is recommended to separate analog and digital ground planes and place two reference planes, one for each, with components placed underneath for easier routing.

Another key aspect of high-speed electronic circuit design is avoiding stub traces, as they can act as antennas and lead to reflections. Pull-up or pull-down resistors in high-speed signal paths are common sources of stubs. Careful consideration should be given to signal return paths to prevent noise coupling and EMI issues. Additionally, it is wise to avoid using unused pads in the layout.

7.2. OHM'S LAW, KIRCHHOFF'S LAWS, AND EQUIVALENT CIRCUITS

Ohm's law uses the well-known formula V = IR to relate the voltage and current in a circuit element. With this relationship in place, a design engineer can precisely analyze a circuit with both active and passive components by applying Kirchhoff's laws increasingly complex circuits. We can simplify many complex circuit blocks by deriving from these fundamental laws the most widely used equivalent circuits, such as the Thevenin and Norton forms, as long as we make reasonable assumptions when simplifying each case. A voltage source connected in series with a resistor, for instance, can be used to simplify a transducer, like a microphone, and provide the well-known Thevenin or Norton form shown in Figure 7.1. Similar to this, the key features outlined in a regulated direct current (DC) power supply's specification can be understood by comparing it to its Thevenin equivalent. It is important to remember that Thevenin's equivalent cannot be used to determine the dissipation inside a circuit. Additionally, independent sources of voltage or current are required if we are to calculate the equivalent resistance, R_s, by substituting their internal resistances for the original sources. Another reminder is to use caution when calculating R_s in circuits that have circuits with dependent voltages or currents.

7.2.1. Time and Frequency Domains

One can characterize every electrical signal as a function of frequency or time. We refer to signals that we observe as a function of time as time domain measurements. Frequency domain measurements are measurements made when we see the frequencies that are present in signals. The term "spectral" describes any signal's frequency content. Every practical element we employ as a foundation operates within a specific frequency range in accordance with design specifications. As a result, we can specify a bandwidth for every circuit block as well as an overall product bandwidth. Time and frequency are simply related when signals are periodic; that is, one is the inverse of the other. The signal's spectrum can then be found using the Fourier series. The spectrum of nonperiodic signals is obtained by applying a Fourier transform. However, integration over all time, or from $-\infty$ to $+\infty$, is required when executing a Fourier transformation. This is not feasible, so instead we use a Discrete Fourier Transform (DFT) on a sampled version of the signal to approximate the Fourier transform. As the number of samples and sampling rate increase, the computational load for direct DFT—which is typically computed in a processor subsystem—increases quickly. In 1954, Cooley and Tukey developed the fast Fourier transform (FFT) algorithm as a workaround for this issue. Computational loads are greatly decreased with the FFT. Let f(t) be any random time function. An infinite sum of sine and cosine terms can be found when f(t) is also periodic, with a period of T. This expansion can be expressed in the form of the Fourier series of f(t).

$$f(t)=\frac{a_0}{2}+\sum_{n=1}^{\infty}\left[a_0\cos\left(\frac{2\pi nt}{T}\right)+b_n\sin\left(\frac{2\pi nt}{T}\right)\right], \qquad (1)$$

where Fourier coefficients a_n and b_n are real numbers independent of t and which may be obtained from the following expressions:

$$a_n=\frac{2}{T}\int_0^T f(t)\cos\left(\frac{2\pi nt}{T}\right)dt, \text{ where } n=0,1,2,\ldots, \qquad (2)$$

$$b_n=\frac{2}{T}\int_0^T f(t)\sin\left(\frac{2\pi nt}{T}\right)dt, \text{ where } n=0,1,2,\ldots.$$

Further, n = 0 gives us the DC component, n = 1 gives us the fundamental, n = 2 gives us the second harmonic, and so on.

Another way of representing the same time function is:

$$f(t)=\frac{a_0}{2}+\sum_{n=1}^{\infty}d_n\cos\left(\frac{2\pi nt}{T}+\phi_n\right), \qquad (3)$$

where $d_n=\sqrt{a_n^2+b_n^2}$ and $\tan\phi_n = -b_n/a_n$. The parameter d_n is the magnitude and ϕ_n is the phase angle.

Equation (3) can be rewritten as

$$f(t)=\sum_{n=-\infty}^{+\infty}c_n e^{j2\pi nt/T}, \qquad (4)$$

Where,

$$c_n = \frac{1}{T}\int_0^T f(t)e^{-j2\pi nt/T}dt \text{ and } n = 0, \pm1, \pm2.$$

The series expansion of Equation (4) is referred to as the complex exponential Fourier series. The c_n are called the complex Fourier coefficients.

This representation suggests that a periodic signal includes all harmonically related frequencies (both positive and negative) to the fundamental frequency. The inclusion of negative frequencies is due to the mathematical model of the signal provided by Equation (4), which necessitates their use. In addition to negative frequencies, complex exponential functions like $e^{j2\pi nt/T}$ are also required in this representation, despite not having a physical interpretation. The use of these mathematical tools is simply to provide a comprehensive description of a periodic signal, suitable for both theoretical and practical applications.

7.2.2. The Fourier Transform

The Fourier transform and its inverse are the foundation for the transformation from the time domain to the frequency domain and back again. The Fourier transform of an arbitrary function f(t) can be defined as follows when f(t) is not necessarily periodic:

$$F(f) = \int_{-\infty}^{\infty} f(t)e^{-j2\pi ft}dt. \tag{5}$$

The time function f(t) is obtained from F(f) by performing the inverse Fourier transform:

$$f(t) = \int_{-\infty}^{\infty} F(f)e^{j2\pi ft}df. \tag{6}$$

Hence, a pair of Fourier transforms is formed by f(t) and F(f). For functions that meet specific minimum requirements, the Fourier transform can be applied to both periodic and non-periodic functions. These requirements are easily met by all signals found in the real world. The Fourier expandability of a function is determined by these requirements, also referred to as the Dirichlet conditions. A spectrum analyzer represents a signal in the frequency domain, while an oscilloscope shows a signal's amplitude as a function of time. The two kinds of representations for some typical signals that are encountered in practice (forming pairs of Fourier transforms).

7.2.3. Discrete and Digital Signals

Either the amplitude of a discrete signal is discrete and continuous in time, or the signal is discrete in time and continuous in amplitude. Switched capacitor filters and charge-coupled device (CCD) arrays can both produce discrete signals. On the other hand, a digital signal, like those used in digital signal processing (DSP) applications, is discrete in both time and amplitude.

- *Discrete Time Fourier Transform*: The Discrete Time Fourier transform (DTFT) maps a discrete time function h[k] into a complex function $H(e^{j\omega})$, where inverse transforms are

$$H\left(e^{j\omega}\right)=\sum_{k=-\infty}^{\infty} h[k]e^{-jk\omega} \tag{7}$$

$$h[k]=\frac{1}{2\pi}\int_{-\pi}^{\pi} H\left(e^{j\omega}\right)d\omega. \tag{8}$$

Keep in mind that the inverse transform is an integral over the real frequency variable, just like in the Fourier transform. There is no need for a complex inversion integral. Because of the continuous variable ω, the DTFT is only helpful for calculations done by hand. The Fourier transform's function in continuous time system analysis and the DTFT's in discrete time system analysis are extremely similar.

- *Discrete Fourier Transform*: An approximation to the DTFT is the Discrete Fourier transform (DFT). The DFT maps a discrete time sequence of N-point duration into an N-point sequence of frequency domain values. Formally, the defining equations of the DFT are:

$$H\left(e^{j\frac{2\pi n}{N}}\right)=H[n]=\sum_{k=0}^{N-1} h[k]e^{-j\frac{2\pi nk}{N}} \text{ for } k, n = 0, 1, \dots, N-1 \tag{9}$$

$$h[k]=\frac{1}{N}\sum_{n=0}^{N-1} H[n]e^{j\frac{2\pi nk}{N}} \text{ for } k, n = 0, 1, \dots, N-1. \tag{10}$$

As shown in Figure 7.1, these equations represent the DTFT relationships with the frequency discretized to N points separated by 2π/N radians around the unit circle. Furthermore, there is a time limit of N points on the sequence. Two finite series of N points are used in each of the definitions of the DFT. They are the foundation of the N points DFT when combined. Because we are only evaluating the DTFT at N points in the Z domain, the restriction to N points and reduction from integral to summation take place. Usually, limitations in the problems or analysis resources dictate the value of N. In powers of two, N is most frequently used. Nevertheless, this is not necessary for many algorithms. Almost any integer value can represent N.

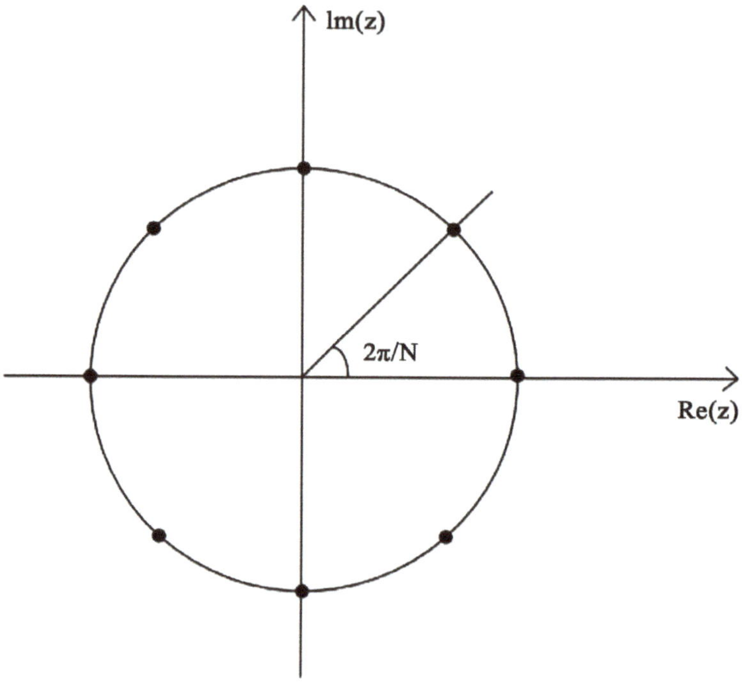

Figure 7.1. *Unit circle with N = 8.*

Source: http://library.navoiy-uni.uz/files/nihal%20kularatna%20-%20electronic%20circuit%20design-%20from%20concept%20to%20implementation%20(2008)(504s).pdf.

- *Fast Fourier Transform:* The FFT is an effective algorithm for computing the DFT, not a transform. The FFT algorithms take advantage of the symmetry and periodic nature of the exponential term, the signal, and the summation kernel to eliminate the needless computations involved in calculating the DFT of a sequence. FFTs are substantially faster than DFTs directly for large values of N. The approximate computation loads for the DFT computed using the FFT decimation-time algorithm and the DFT calculated directly for different sizes are displayed in Table 7.1. Take note that the savings increase significantly as N rises above 16. Specifically, the decimation-in-time algorithm on realistic FFT algorithms. The FFT function is a common feature of contemporary digital storage scopes, and the majority of these methods and algorithms are employed within the scope's processor subsystem.

Table 7.1. *Comparison of Approximate Computational Loads for Direct DFT and Decimations in Time FFT*

Number of Points	Multiplications		Additions	
	DFT	FFT	DFT	FFT
4	16	4	12	8
8	64	12	56	24
16	256	32	240	64
32	1024	80	992	160
64	4096	192	4032	384
128	16,384	448	16,256	896
256	65,536	1024	65,280	2048
512	262,144	2304	261,632	4608
1024	1,048,576	5120	1,047,552	10,240
2048	4,194,304	11,264	4,192,256	22,528
4096	16,777,216	24,576	16,773,120	49,152

Source: http://library.navoiy-uni.uz/files/nihal%20kularatna%20-%20electronic%20circuit%20design-%20from%20concept%20to%20implementation%20(2008)(504s).pdf.

7.2.4. Feedback and Frequency Response

Feedback can be either negative or positive. In amplifier design, negative feedback is commonly used to achieve one or more of the following:

- Desensitize the gain—make the gain less sensitive to variations in the value of circuit components.
- Reduce nonlinear distortion—make the output gain constant and independent of the signal level.
- Reduce the effect of noise—make improvements to the signal-to noise ratio (SNR).
- Control the input/output impedances.
- Extend the bandwidth.

Feedback to an amplifier is illustrated in Figure 7.2, a signal flow diagram where x represents either voltage or current signals. The basic amplifier with an open loop gains of A now receives a modified input signal, $x_i = x_s - x_f$, from the feedback block addition. $x_f = \beta xo$ is the result of the feedback block with a feedback factor of β, where xi is the open-loop amplifier's input, xo is its output, and x_f is its feedback signal.

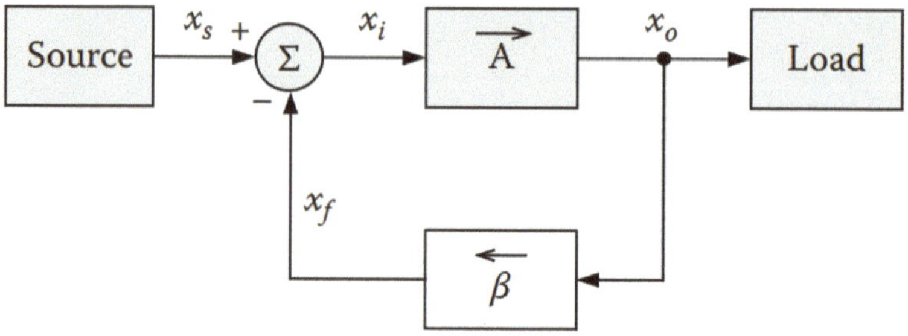

Figure 7.2. *General structure of the feedback amplifier.*

Source: http://library.navoiy-uni.uz/files/nihal%20kularatna%20-%20electronic%20circuit%20design-%20from%20concept%20to%20implementation%20(2008)(504s).pdf.

With the assumptions that the transmission of the signals through the basic amplifier is only in the forward direction and the transmission through the feedback block is only in the backward direction, it can be shown that the overall gain of the feedback system A_f is given by the relationship:

$$A_f = \frac{x_o}{x_s} = \frac{A}{1+A\beta}. \quad (11)$$

Quantity Aβ is called the loop gain and quantity (1 + Aβ) is called the amount of feedback. In a practical design, open loop gain Aβ is large, Aβ >> 1, then from Equation (11), it follows that $A_f \approx 1/\beta$, which is a very practical and useful property in design, because the close loop gain is now entirely dependent on the feedback network.

- **Gain Desensitization:** We can also infer several helpful characteristics of the feedback system in Figure 7.2 from the discussion above. We can demonstrate, for instance, that the percentage change in A_f because of changes in certain circuit parameters is less than the percentage change in A as a result of the quantity of feedback (1 + Aβ). This is provided by the partnership.

$$\frac{dA_f}{A_f} = \frac{1}{(1+A\beta)}\frac{dA}{A}. \quad (12)$$

- **Noise Reduction:** An amplifier's SNR can be raised by using negative feedback to lower interference or noise. Figure 7.3 provides a summary of this situation. For the scenario shown in Figure 7.3a, we can demonstrate that the SNR is given by V_s/V_n if we relate the total noise in the amplifier to its input by a noise source of value V_n. Now, the total output V_o is given by Figure 7.3b: if the system is modified by adding a feedback block with a feedback factor of β and a theoretically noise-free amplifier with gain Ap.

$$V_o = V_s \frac{AA_p}{1+AA_p\beta} + V_n \frac{A}{1+AA_p\beta}. \quad (13)$$

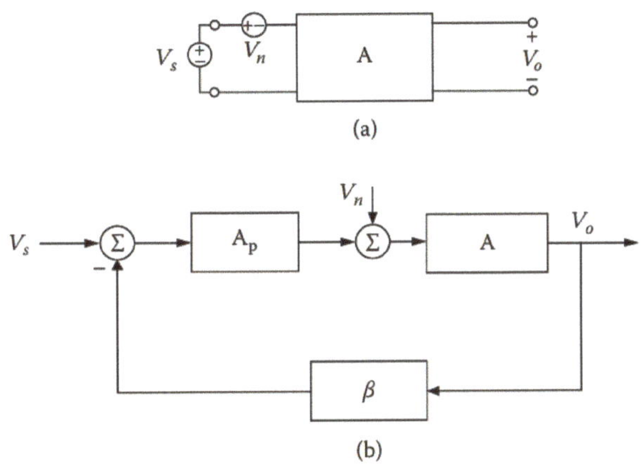

Figure 7.3. *Concept of increasing the SNR of a circuit: (a) basic amplifier indicating the case of input referred noise; (b) with an additional noise-free amplifier at the front end for SNR improvement.*

Source:http://library.navoiy-uni.uz/files/nihal%20kularatna%20-%20electronic%20circuit%20design-%20from%20concept%20to%20implementation%20(2008)(504s).pdf.

Thus, the modified SNR becomes:

$$SNR_{new} = \frac{V_S}{V_n} A_p. \quad (14)$$

This is an improvement by a factor of A_p.

- **Reduction in Nonlinear Distortion, Bandwidth Extension, and Input/Output Impedance Modification by Feedback:** The linearization of an amplifier's transfer characteristics is another practical use for feedback. Likewise, in the case of a low-pass filter, feedback can be used to boost the circuit's bandwidth at the price of a decreased mid-band gain. Consider a single-pole amplifier with a 3-dB upper cutoff frequency of ω_H and a mid-band gain of A_M. The transfer function of this amplifier is given by. This serves as an example of the situation.

$$A(s) = \frac{A_M}{1+s/\omega_H}. \quad (15)$$

With a feedback circuit with a feedback factor of β added to the circuit, commencing from the relationships in Equation (11) and Equation (15), we can show that the new transfer function is given by

$$A_f(s) = \frac{A_M / (1 + A_M\beta)}{1 + s / \omega_H (1 + A_M\beta)} . \tag{16}$$

This merely demonstrates the rationale behind altering the bandwidth at the price of changing the mid-band gain. Other kinds of amplifiers could be the subject of similar discussions. It also serves as a reminder that a single-pole low-pass circuit's gain bandwidth product is constant. Amplifier stages' input and output impedances can be changed via feedback. It might be appropriate to talk about the four common types of amplifiers at this point: transconductance, transresistance, voltage, and current. To facilitate the design process, we can include a feedback stage in each case along with appropriate practical simplifications like (1) an open-loop amplifier, which functions as a unilateral circuit (no reverse feedback passes through the amplifier), and (2) a β circuit, which prevents any forward signal transfer from the amp's input side to its output side. The four topologies with the feedback networks are displayed in Figure 7.4. Depending on how we combine and sample the signals at the input and output sides, respectively, we assign distinct names in each instance. For instance, when using a voltage amplifier, we sample the output voltage to combine a voltage sample with the input (resulting in a series configuration at the input) (hence a shunt sampling case). This situation is known as "series-shunt" feedback. It might be important to note that the cases stem from the fundamental need to combine signals, voltages, or currents of the same kind. (While two ideal current sources can only be mixed in parallel, two ideal voltage sources can only be mixed in series.).

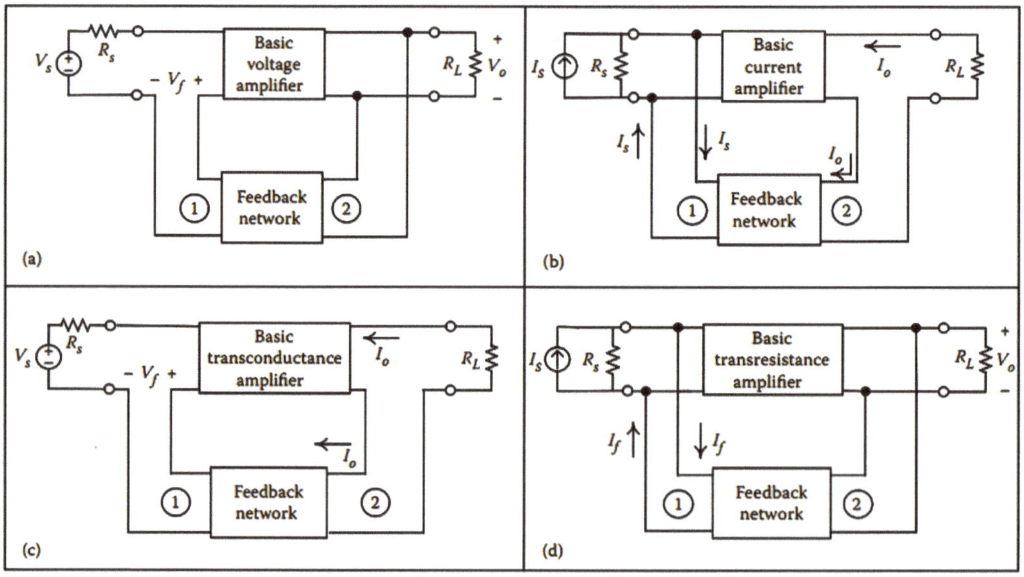

Figure 7.4. *The four basic feedback topologies: (a) voltage-sampling series mixing (series-shunt) topology; (b) current-sampling shunt mixing (shunt-series) topology; (c) current-sampling series-mixing (series-series) topology; (d) voltage-sampling shunt-mixing (shunt-shunt) topology.*

Source: http://library.navoiy-uni.uz/files/nihal%20kularatna%20-%20electronic%20circuit%20design-%20from%20concept%20to%20implementation%20(2008)(504s).pdf.

The utilization of a voltage amplifier in conjunction with a two-port h parameter feedback circuit is demonstrated in Figure 7.5, along with the derivation of the A circuit—a modified version of the basic amplifier—and the β circuit for the series-shunt feedback. A unilateral β circuit, as depicted in Figure 7.5b, can only be achieved in a practical design by simplifying the case and ignoring the parameter h_{21}. Similar to this, Figure 7.6 for the transconductance amplifier shows how the A and β circuits, which employ series-series feedback, were derived.

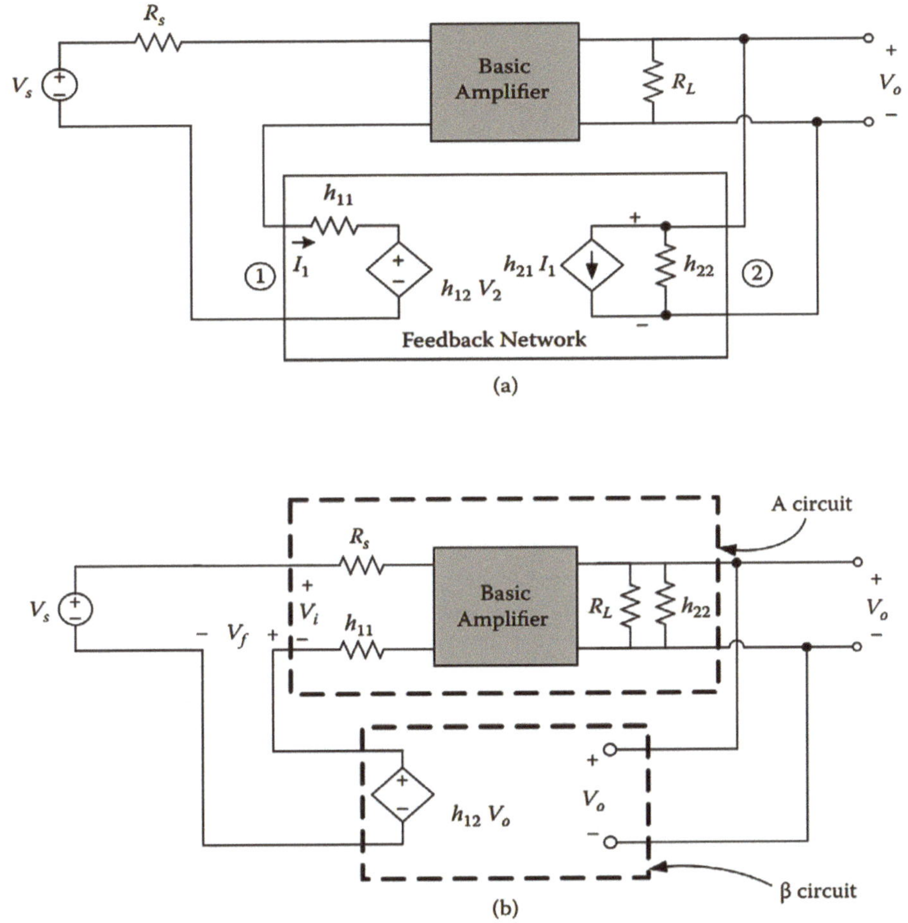

Figure 7.5. *Derivation of the A circuit and β circuit for the series-shunt feedback amplifier: (a) the circuit in Figure 7.4a with the feedback represented by h parameters; (b) the circuit in Figure 7.5a, neglecting h_{21} and modifying the basic amplifier to form the A circuit.*

Source: http://library.navoiy-uni.uz/files/nihal%20kularatna%20-%20electronic%20circuit%20design-%20from%20concept%20to%20implementation%20(2008)(504s).pdf.

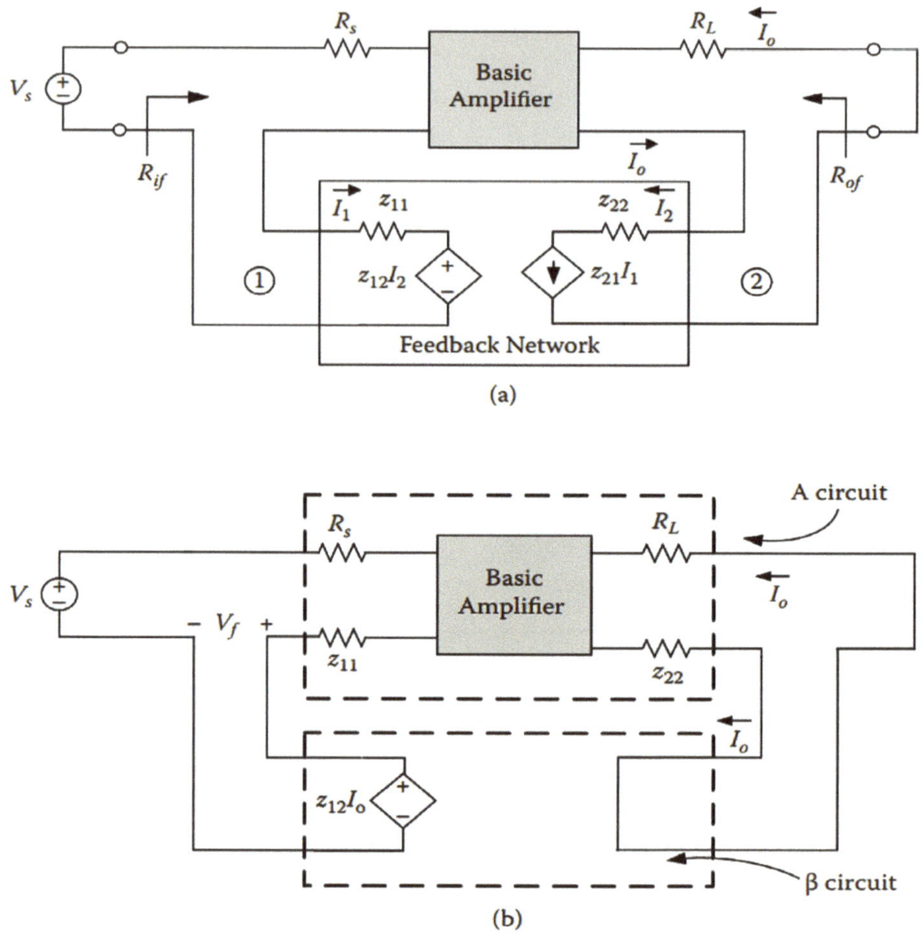

Figure 7.6. *Derivation of the A circuit and the β circuit for a series-series feedback amplifier: (a) the circuit in Figure 7.4c with the feedback represented by z parameters; (b) the same circuit neglecting z_{21} and modifying the basic amplifier to form the A circuit.*

Source: http://library.navoiy-uni.uz/files/nihal%20kularatna%20-%20electronic%20circuit%20design-%20from%20concept%20to%20implementation%20(2008)(504s).pdf

Figure 7.7 depicts the ideal configuration of a series-shunt feedback amplifier, which consists of an ideal voltage sampling series mixing feedback network (the β circuit) and a unilateral open-loop amplifier (the A circuit). The voltage gain A, the output resistance R_o, and the input resistance R_i make up the A circuit. Presumably, the source and load resistances are part of the A circuit. Moreover, take note of the fact that the β circuit does not load the A circuit, meaning that connecting it will not alter A's value, which is determined by V_o/V_i. Similarly, Figure 7.8 depicts the ideal structure and the equivalent circuit for the series-series feedback case.

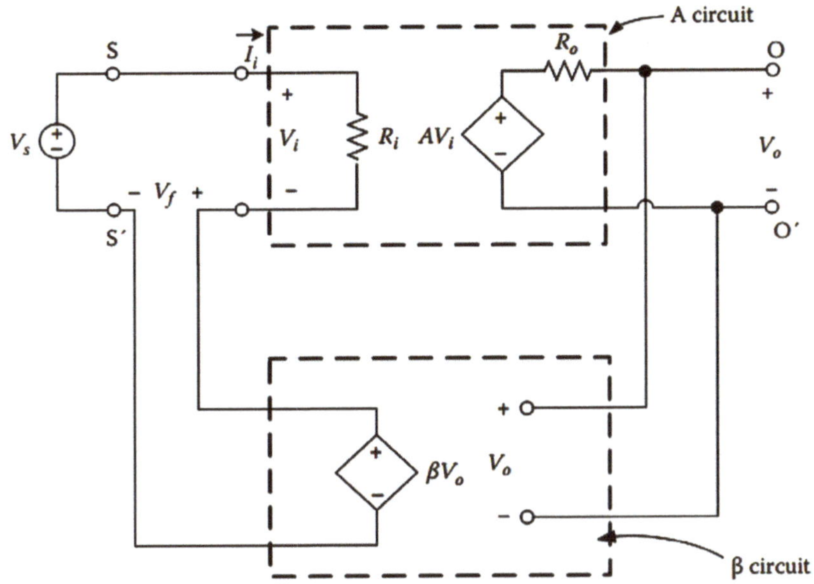

Figure 7.7. *The series-shunt feedback amplifier.*

Source: http://library.navoiy-uni.uz/files/nihal%20kularatna%20-%20electronic%20circuit%20design-%20from%20concept%20to%20implementation%20(2008)(504s).pdf.

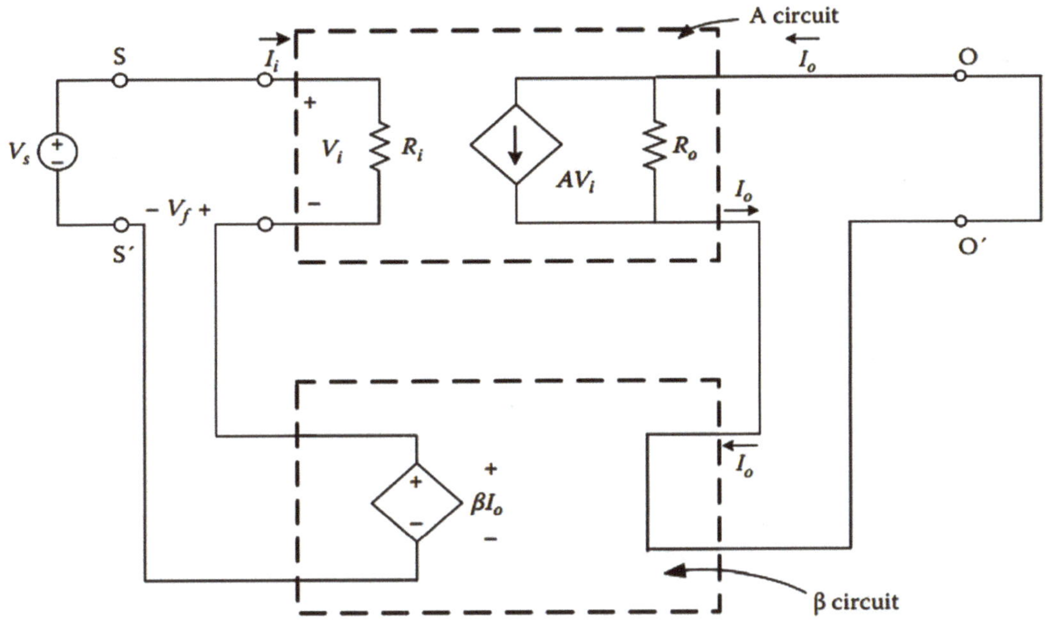

(a)

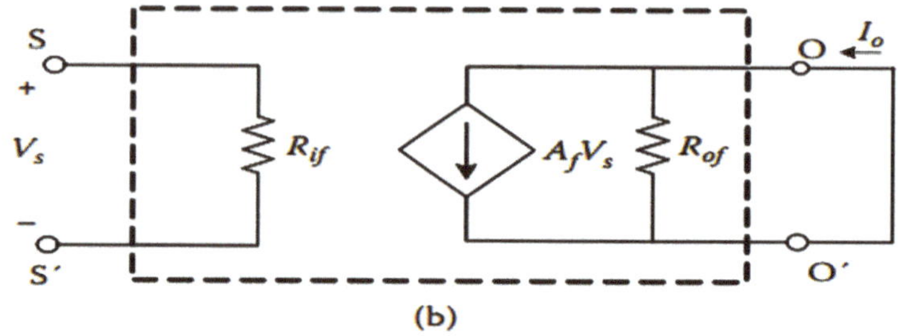

Figure 7.8. *The series-series feedback amplifier: (a) ideal structure; (b) equivalent circuit with modified input and output impedances.*

Source:http://library.navoiy-uni.uz/files/nihal%20kularatna%20-%20electronic%20circuit%20design-%20from%20concept%20to%20implementation%20(2008)(504s).pdf.

7.2.5. Loop Gain and the Stability Problem

The open-loop gain is a function of frequency, and the feedback factor β is dependent on frequency in a general case. This situation can be indicated by the generalized case for the feedback with a closed-loop transfer function A_f (s) as in Equation (17):

$$A_f(s) = \frac{A(s)}{1+A(s)\beta(s)}. \quad (17)$$

If the amplifier is assumed to be a direct-coupled case with a constant DC gain of A_0 with the poles and zeros occurring in the high-frequency band, the loop gain A(s) β(s) becomes a constant ($A_0\beta_0$) at low frequencies when the feedback factor β(s) at low frequencies reduces to a constant value (β_0). This is a common situation we assume in design work for simplicity and convenience and to achieve a negative feedback condition. However, at higher frequencies this need not be the case, and by substituting s = j_ω,

$$A_f(j\omega) = \frac{A(j\omega)}{1+A(j\omega)\beta(j\omega)}. \quad (18)$$

This clearly indicates that $A(j_\omega)\beta(j_\omega)$ is a complex number $L(j_\omega)$ that can be represented by

$$L(j\omega) \equiv |A(j\omega)\beta(j\omega)| e^{j\phi(\omega)} \quad (19)$$

This indicates that at higher frequencies, loop gain $L(j_\omega)$ can be positive or negative, depending on the phase angle ϕ. To appreciate this fact, consider the frequency at which ϕ(ω) becomes 180°. At this frequency, ω_{180}, $A(j_\omega)\beta(j_\omega)$, or the loop gain will be a negative real number. As per Equation (18), at this frequency the feedback will be positive. If at

$\omega = \omega_{180}$ the loop gain is less than unity, the closed-loop gain will be greater than the open-loop gain $A(j_{\omega})$. Nevertheless, the feedback amplifier will be stable.

In the special case where the loop gain becomes −1 at $\omega = \omega_{180}$, the amplifier closed-loop gain will be infinite, with the practical situation of output rising to a very high value for no appreciable input signal, which represents the case of an oscillator with a frequency of oscillation equal to ω_{180}.

When the magnitude of the loop gain is greater than unity at ω180 it is not obvious from Equation (18), but the circuit will oscillate with the amplitude growing gradually until the circuit nonlinearities limit the amplitude and the loop gain becomes unity.

- *The Nyquist Plot:* The Nyquist plot, which is essentially a polar plot of loop gain with frequency as a parameter, is a formalized method for testing the stability based on the discussion that came before it. The Nyquist plot displays the phase angle ⊠ and the radial distance as |Aβ|. A mirror image of the plot for the positive and negative ω is formed by the solid line representing positive frequencies and the dotted line representing negative frequencies. The Nyquist plot intersects the negative real axis at frequency ω_{180}. This indicates that the amplifier becomes unstable if the intersection occurs to the left of the point (−1,0), as loop gain |Aβ| > 1. The amplifier will be stable, though, if the plot intersects the negative real axis to the right of (−1,0). In short, the amplifier will be unstable if the Nyquist plot, which is a condensed form of the Nyquist criterion, encircles the point (−1,0).
- *Poles and Zeros, S-Domain, and Bode Plots:* An analysis of an amplifier's frequency response can be done by plotting the gain as a function of the complex frequency s. In the s domain analysis, an equivalent impedance of 1/sC represents a capacitor of value C, and L_s represents an inductance of value L. T(s) ≡ Vo(s)/Vi (s) is the voltage transfer function that can be obtained using standard circuit analysis methods. We can assess this function's frequency behavior after it has been derived by substituting jω for s. Generally speaking, the transfer function T(s) in its unaltered form (without changing s = jω) can provide a wealth of insightful information regarding the circuit's stability. There are numerous ways to express function T(s), including.

$$T(s)=a_m\frac{(s-z_1)(s-z_2)(s-z_3)\ldots(s-z_m)}{(s-p_1)(s-p_2)(s-p_3)\ldots(s-p_n)} \quad (20)$$

and,

$$T(s)=\frac{a_m s^m+a_{m-1}s^{m-1}+a_{m-2}s^{m-2}+\ldots+a_0}{s^n+b_{n-1}s^{n-1}+b_{n-2}s^{n-2}\ldots+b_0} \quad (21)$$

Equation (20) refers to $p_1, p_2, \ldots, p_n$ as transfer function poles or the network's natural modes, and $z_1, z_2, \ldots, z_m$ as transmission zeros or transfer function zeros. The network order is denoted by the n in the transfer function. The condition that the poles or zeros must occur as conjugate pairs is given by the equivalent form of the transfer function in Equation (21), where the coefficients a and b are real numbers. This means that if a pole or zero occurs at 3 + 2j, for example, there should be another pole or zero at 3 – 2j. At $\omega = \omega_z$, the transfer function is exactly zero due to a pure imaginary zero ($\pm j\omega_z$). A system's poles must be located in the left half of the s plane in order for it to be stable. On the jω axis, two complex-conjugate poles produce long-lasting oscillations. In a practical system, nonlinearities will limit the growing oscillations caused by poles on the right-hand side of the s plane. The three options are shown in Figure 7.9. Equation (17)'s closed-loop transfer function indicates that the equation's poles can be found by solving it for the feedback amplifier:

$$1 + A(s)\beta(s) = 0, \tag{22}$$

which is called the characteristic equation. This indicates that feedback to a system changes its poles. In an amplifier with an open-loop transfer function characterized by a single pole, the open-loop transfer function is:

$$A(s) = \frac{A_o}{1 + \dfrac{s}{\omega_p}}. \tag{23}$$

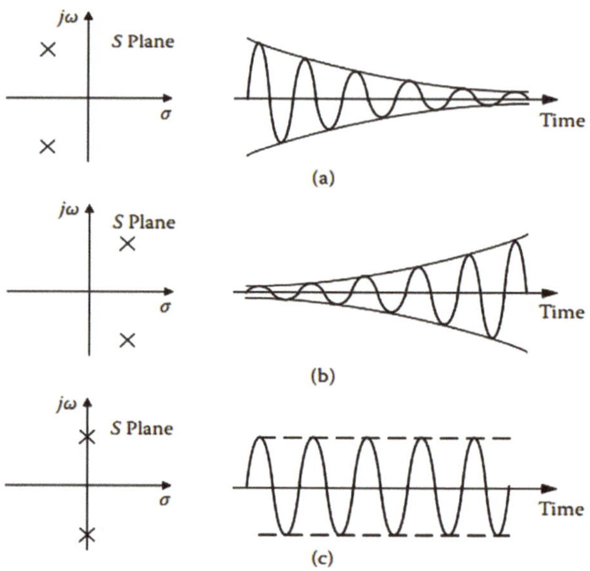

Figure 7.9. *Relationship between pole location and transient response: (a) poles on the left-hand side of the imaginary axis; (b) poles on the right-hand side of the imaginary axis; (c) poles on the imaginary axis.*

Source: http://library.navoiy-uni.uz/files/nihal%20kularatna%20-%20electronic%20circuit%20design-%20from%20concept%20to%20implementation%20(2008)(504s).pdf.

The closed-loop transfer function is given by

$$A_f(s) = \frac{A_o / (1 + A_o\beta)}{1 + s / \omega_p(1 + A_o\beta)}. \quad (24)$$

Feedback moves the pole along the negative real axis to a frequency ω_{Pf}

$$\omega_{Pf} = \omega_P(1 + A_o\beta) \quad (25)$$

With feedback, the new pole shifts to the left-hand side of the original pole. Its effect on the Bode plot, where at low frequencies the gain drops by an amount equal to $20\log(1 + A_o\beta)$ and the two curves coincide at high frequencies. Applying feedback extends the bandwidth of the amplifier at the expense of loop gain. Because the pole of the closed-loop amplifier never enters the right-hand side of the s plane, the single-pole amplifier is unconditionally stable.

7.3. NOISE IN ELECTRONIC CIRCUITS

Any undesired disturbance that masks or obstructs a desired signal can be broadly characterized as noise. The primary cause of noise is that electrical charge is carried in discrete amounts equivalent to the electron charge rather than being continuous. The frequency components that make up noise are completely random in terms of both amplitude and phase. The precise amplitude at any given time cannot be predicted, even though the long-term root mean square (RMS) value of noise can be measured. Noise is significant in real-world design settings because it can reduce a sensor's resolution or a system's dynamic range. The instantaneous amplitude with time distribution of most noise is either normal or Gaussian. Common electrical noise is, to a good engineering approximation, located within three times the noise waveform's root mean square (RMS) value. Device noise emitted or radiated noise, and conducted noise are the three main types of noise in circuits. Three types of noise can be found in devices: shot noise, low-frequency (1/f) noise, and thermal noise.

7.3.1. Noise in Passive Components

The random vibration of thermally excited charge carriers in a conductor is the source of thermal noise, also known as Johnson noise. The electrons in any conductor vibrate randomly, and their motion is temperature-dependent, above absolute zero. The absolute temperature and the measuring system's bandwidth are proportional to the available noise power (P_n). P_n is supplied by:

$$P_n = kT\Delta f, \tag{26}$$

where k is the Boltzmann constant (1.38×10^{-23}), T is the absolute temperature, and Δf is the noise bandwidth of the measuring system. There is equal noise power in each hertz of the bandwidth, regardless of the center frequency. For example, power in the bandwidth from 1 to 2 Hz is the same as from 1000 to 1001 Hz. This property characterizes thermal noise as "white noise." From the relationship for a resistance of R, the equivalent RMS noise voltage, E_n, can be written as

$$E_n = \sqrt{4kTR\Delta f} \tag{27}$$

The above relationship defines a useful parameter, the noise-voltage spectral density, which is the equivalent noise voltage per 1-Hz bandwidth. Some useful facts in designing circuits are that the noise voltage spectral density of a 50-Ω resistor is approximately 0.9 nV/√Hz and for a 1-kΩ resistor is approximately 4 nV/√Hz. Figures 7.10(a) and 7.10(b) represent alternatives for circuit equivalents for thermal noise.

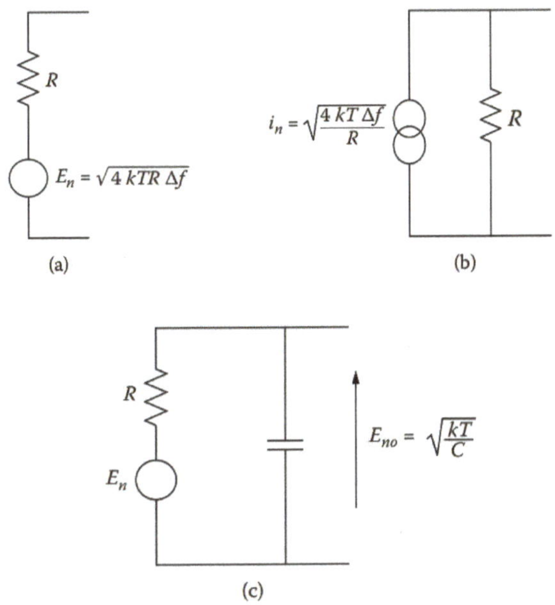

Figure 7.10. *Thermal noise in a resistor: (a) noise as a voltage source; (b) noise as a current source; (c) noise in a practical resistor shunted by its parasitic capacitance.*

Source:http://library.navoiy-uni.uz/files/nihal%20kularatna%20-%20electronic%20circuit%20 design-%20from%20concept%20to%20implementation%20(2008)(504s).pdf.

A resistor generates excess noise in addition to thermal noise (or Johnson noise) when a DC passes through it. Current noise is another name for this. This happens when current passes through a discontinuous medium, like a resistor made of carbon composition, which combines carbon granules with a binding medium. Because of the differences in conductivity, this manufacturing process may result in microarcs. Thermal noise predominates at higher frequencies, while excess noise typically has a 1/f power spectrum and is more prominent at lower frequencies. The excess noise voltage (E_{ex}) for a resistor can be found experimentally as follows:

$$E_{ex}^2 = \frac{K_{ex} I_{DC}^2 R^2}{f}, \tag{28}$$

where K_{ex} is a constant dependent on the manufacturing process and IDC is the DC through the resistor. Excess noise in a resistor becomes significant at low frequencies only when there is significant voltage drop across the resistor. Excess noise in a resistor can be measured in terms of a noise index, expressed in decibels. The noise index is the RMS noise in microvolts in the resistor per volt of DC drop across the resistor in every 10 units of frequency.

7.3.2. Effect of Circuit Capacitance

Inductors and capacitors typically do not produce any device noise. According to Eq. (27), an infinite resistance or an open circuit will produce an infinite noise voltage. Since there is always some shunt capacitance limiting the voltage, this is not observed in practice. The case of a resistor shifted by its capacitance is depicted in Figure 7.10. It is evident from this example that the total RMS noise at the output is provided by:

$$E_{no} = \sqrt{\int_0^\infty [E_{no}^2(f)]df} = \sqrt{\frac{kT}{C}} \tag{29}$$

This relationship indicates that in practical circumstances, total noise energy of a resistor is limited by temperature and capacitance.

7.3.3. Noise in Semiconductors and Amplifiers

Semiconductor devices contain various noise sources. Shot noise, flicker noise, burst (or popcorn) noise, and avalanche noise are notable types. Shot noise can be found in diodes, bipolar transistors, and MOS transistors and is always connected to a DC flow. It can be demonstrated that the resulting noise current, i_{ns}, has a mean square value of because the DC in a semiconductor is always based on electrons and holes.

$$\overline{i_{ns}^2} = 2qI_D\Delta f \text{ ,} \tag{30}$$

where I_D is the DC. Equation (30) is valid until the frequency becomes comparable to $1/\tau$, where τ is the transit time through the depletion region. For most devices, τ is extremely small, and Eq. (30) is accurate well into the gigahertz region. Note that shot noise current is independent of temperature. The junction diode small-signal equivalent circuit with noise is shown in Figure 7.11. The equivalent resistance, r_d, is considered a noiseless element, as it represents the dynamic resistance of the diode.

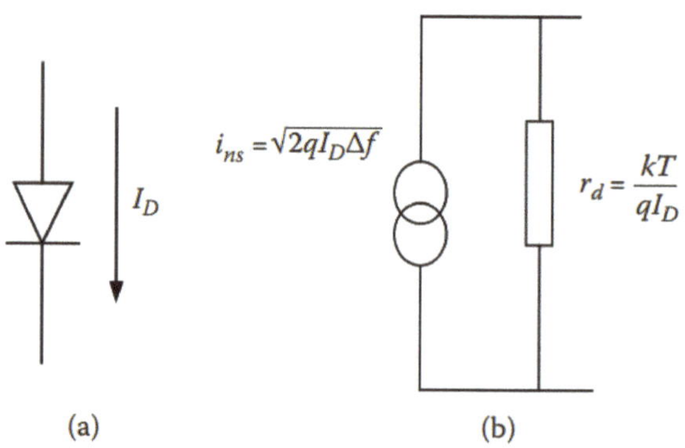

Figure 7.11. *Small-signal equivalent of junction diode with noise.*

Source:http://library.navoiy-uni.uz/files/nihal%20kularatna%20-%20electronic%20circuit%20design-%20from%20concept%20to%20implementation%20(2008)(504s).pdf.

Another type of noise in semiconductors is flicker noise, which is comparable to the situation of excessive noise in resistors. Flicker noise comes from a variety of sources, but its primary causes are contamination-related traps and crystal defects. Randomly capturing and releasing carriers is how these traps work, and the time constants involved in the process create a noise signal with low frequency energy concentration. This is always connected to a DC flow and shows the form's spectral density.

$$\overline{i_{nf}^2} = K_f \frac{I^a}{f^b} \Delta f, \tag{31}$$

In this case, I represent the DC, K_f denotes a constant specific to a given device, a lies in the interval 0–2, and b is a constant that approximates unity. Since b is nearly equal to unity, the frequency dependence of the noise spectral density is roughly 1/f. Hence, flicker noise is also referred to as 1/f noise occasionally. Fringer noise has an unknown constant K_f, in contrast to shot noise and thermal noise, which have well-defined mean square values that can be expressed in terms of current flow, resistance, and well-known physical constants. The variation of this constant can be significant for distinct transistors or integrated circuits from the same wafer, in addition to variations ranging from orders of magnitude across different device types. For ICs from a given process, this means that the average or typical flicker noise performance is predicted using statistically derived process values. Burst noise, sometimes known as popcorn noise, is another kind of low-frequency noise found in integrated circuits (ICs). This type of noise is superimposed on multiple discrete levels, usually two or more, and falls within the audio range. This contains the spectral density and wave shape depicted in Figure 7.12 and may make a popping sound when played through a loudspeaker. By, we can find the burst noise's spectral density.

$$i_{nb} = K_b \frac{I^c}{1+\left(\frac{f}{f_{cb}}\right)^2} \Delta f, \tag{32}$$

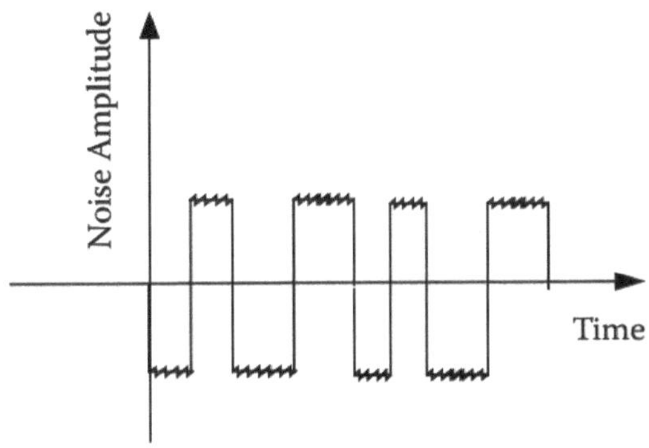

(a)

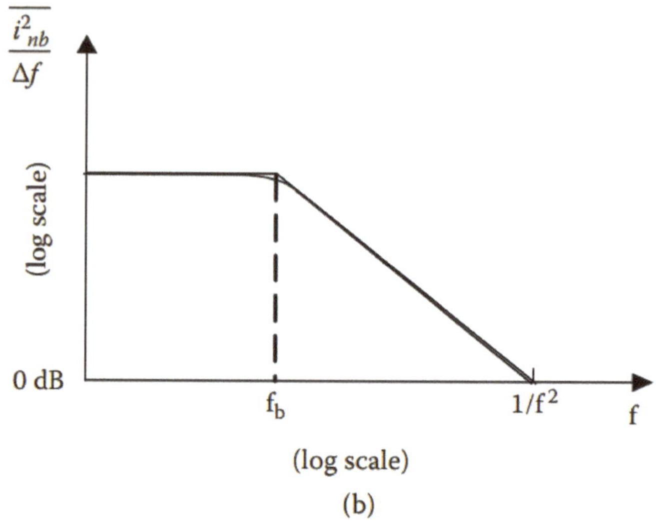

Figure 7.12. *Burst noise: (a) typical waveform; (b) spectral density.*

Source: http://library.navoiy-uni.uz/files/nihal%20kularatna%20-%20electronic%20circuit%20design-%20from%20concept%20to%20implementation%20(2008)(504s).pdf.

where K_b is the constant for a particular device carrying current of I, c is a constant in the range of 0.5 to 2, and f_{cb} is a particular frequency for a given noise process. Burst noise can occur with different time constants and can produce multiple humps in the spectrum. Figure 7.13 shows a combination of burst noise and flicker noise.

Avalanche noise is another form of a noise produced in zener or avalanche diodes. This is a cumulative effect of a large series of noise spikes and is associated with the DC through the device. The magnitude is difficult to predict, and it depends on the device structure and the uniformity of the silicon crystal. A typical value for $\overline{v_n^2}/\Delta f$ could be 10–14 V2 /Hz, and this could be equivalent to the thermal noise voltage of a 600-kΩ resistor and hence significant in value.

One can create small-signal equivalent circuits for diodes, BJTs, and MOS transistors using basic semiconductor components, as discussed in the above discussion of noise sources. Since r_c is a relatively tiny series resistor in the BJT noise model, only r_b's significant contribution has been included. r_b and r_c are physical resistors. Because r_π and r_o are imaginary resistors, they don't contribute any noise to the model. Since avalanche noise in BJTs is insignificant if VCE is maintained at least 5V below the breakdown voltage, the model does not include it. Figure 7.14 displays the base current noise spectrum (absent the burst noise component). Depending on the processing, the intersection of the shot noise and flicker noise asymptotes can occur at a corner frequency f_a, which ranges from 100 Hz to 10 MHz.

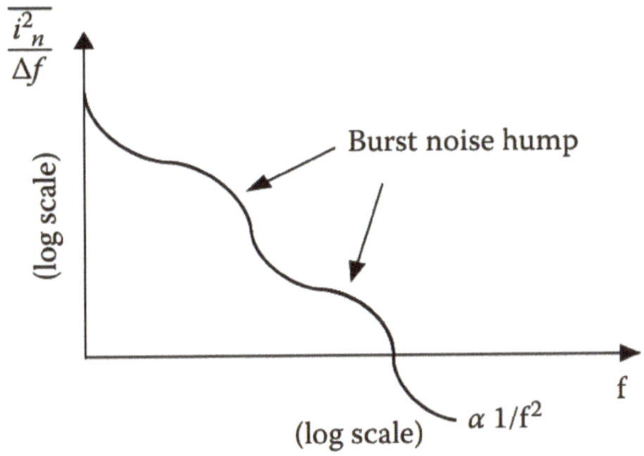

Figure 7.13. *Combined spectral density due to burst and flicker noise.*

Source:http://library.navoiy-uni.uz/files/nihal%20kularatna%20-%20electronic%20circuit%20 design-%20from%20concept%20to%20implementation%20(2008)(504s).pdf.

Due to the resistive nature of the channel material, flicker noise and thermal noise are important factors in MOS devices. For short-channel devices (less than 1-µm channel length), thermal noise is two to five times larger, and the relationship is typically valid only for long-channel devices. Another correlated gate noise component is added to high-frequency long-channel MOS devices, with a magnitude of correlation of approximately 0.39. This component may be greater for short-channel devices. For operational amplifiers' and transistor combinations' noise performance.

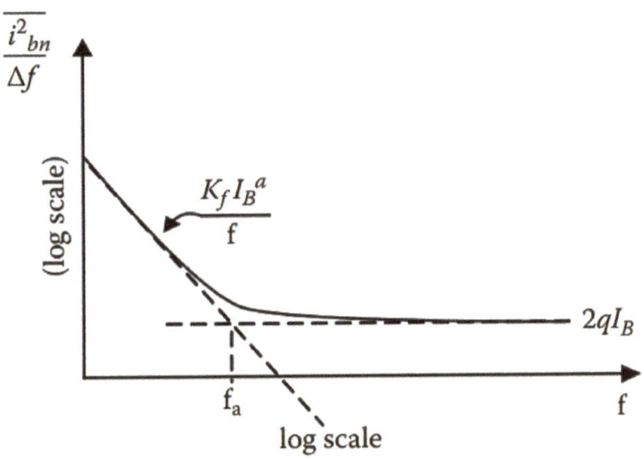

Figure 7.14. *Base current noise density of a BJT.*

Source: http://library.navoiy-uni.uz/files/nihal%20kularatna%20-%20electronic%20circuit%20 design-%20from%20concept%20to%20implementation%20(2008)(504s).pdf.

7.3.4. Circuit Noise Calculations and Noise Bandwidth

Analyzing a circuit's different noise sources to ascertain how much each one contributes to the overall noise performance is known as circuit noise calculations. This usually includes, depending on the components and operating conditions, shot noise, flicker noise, thermal noise, and other sources. Quantifying the noise at the circuit output and confirming that it satisfies the intended performance requirements are the objectives. The noise bandwidth, or the frequency range over which the noise power is integrated, is an important parameter in noise analysis. For a circuit's noise performance to be accurately characterized, it is imperative to consider the noise bandwidth, which considers the fact that noise power is distributed across a range of frequencies. In the event that there is a mean square value noise current source, it should be included in the analysis:

$$\overline{v_n^2} = S(f)\Delta f\ , \tag{33}$$

this gives the equivalent RMS amplitude of:

$$v_n = \sqrt{S(f)\Delta f}\ . \tag{34}$$

We can handle noise sources mathematically by treating them as well-known sinusoidal waveforms, using values comparable to the scenario in Equation (34). In the context of a single-pole amplifier, where frequency rolls off with a 3-dB amplitude at the corner frequency, noise bandwidth is defined as a range of frequencies where the noise amplitude is flat for a given noise source.

$$f_N = \frac{1}{A_{v0}^2} \int_0^{\infty} \left|A_v(jf)\right|^2 df\ , \tag{35}$$

where $A_v(jf)$ is the circuit's voltage gain magnitude at a given frequency f and A_{v0} is the circuit transfer function's low-frequency gain. The output noise voltage spectral density and the circuit transfer function, in that order. The notion of noise bandwidth, characterized by a level noise voltage spectral density throughout the f_N bandwidth. The value of f_N for a low-pass filter with a 3-dB corner frequency of f_1 is $1.57f_1$, according to the relationship in Equation (35).

7.3.5. Noise Figure and Noise Temperature

The noise figure is a commonly used method of specifying the noise performance of a circuit or device. The definition of the noise figure (F) of a circuit is:

$$F = [\mathrm{SNR}]_{\mathrm{input}}/[\mathrm{SNR}]_{\mathrm{output}} \tag{36}$$

F is usually expressed in decibels. This parameter approximates the level of degradation of the SNR caused by the circuit. For an ideal noiseless amplifier, this figure is unity or 0 dB. A useful alternative representation of Equation (32) is:

F = total output noise/output noise due to source resistance. (37)

The temperature at which the source resistance R_s must be maintained in a circuit in order for the noise output resulting from R_s to equal the noise output resulting from the circuit itself is known as the noise temperature (T_n) of that circuit. Determining the cumulative impact of all noise sources is necessary in real-world circuits with various noise sources. The noise voltage that results from connecting two noise sources in series, v_{n1} and v_{n2}, with equivalent noise voltages is as follows.

$$V_{total} = \sqrt{v_{n1}^2 + v_{n2}^2 + 2C_n v_{n1} v_{n2}}, \tag{38}$$

where C_n is a correlation coefficient between −1 and 1, including 0. When the noise sources are completely independent (such as in the case of two series resistances), the value of this coefficient is zero. This relationship indicates that for totally independent noise sources connected in series, an RMS value can be estimated.

CASE STUDY

BUILDING AUTOMATION SYSTEM CONTROL CIRCUIT

A commercial building management company is tasked with upgrading the control system for a large office complex. The goal is to create an efficient and integrated Building Automation System (BAS) that monitors and controls various building systems such as lighting, HVAC, security, and energy management. The engineering team is responsible for designing the control circuits to ensure seamless operation and energy efficiency.

- **Challenges:**

1. **Multi-System Integration:** The control circuit must integrate diverse systems like lighting, heating, cooling, ventilation, access control, and fire detection.
2. **Energy Efficiency:** The BAS aims to optimize energy usage by controlling systems based on occupancy, time schedules, and ambient conditions.
3. **Interactivity:** The control circuit should allow manual user control while also responding to automated inputs for optimal system performance.
4. **Emergency Scenarios:** The circuit must include provisions for emergency shutdown and response during fire alarms or power outages.
5. **User Interface:** Designing an intuitive user interface for building occupants to control and monitor various systems is crucial.

- **Solution:** The engineering team employs control circuit design to create an intelligent Building Automation System that enhances comfort, safety, and energy efficiency.

1. **System Mapping:** The team identifies all building systems that need to be integrated into the BAS, including sensors, actuators, switches, and control panels.
2. **Control Logic:** Control logic is designed to manage system operations based on occupancy, time schedules, temperature thresholds, and user preferences.

3. **User Interface:** An intuitive Human-Machine Interface (HMI) is developed, enabling building occupants to adjust settings, monitor conditions, and receive alerts.
4. **Sensor Integration:** Sensors like motion detectors, temperature sensors, and light sensors are integrated into the control circuit to provide real-time data for decision-making.
5. **Actuator Control:** The control circuit activates actuators to control lighting levels, adjust temperature setpoints, and manage ventilation based on occupancy and environmental conditions.
6. **Emergency Scenarios:** Emergency shutdown procedures are integrated into the control circuit, allowing for quick response to fire alarms or power disruptions.
7. **Networking:** The control circuit uses networking protocols to facilitate communication between different systems, allowing centralized monitoring and control.
8. **Energy Management:** The control circuit optimizes energy usage by dimming lights, adjusting HVAC settings, and turning off non-essential systems during unoccupied hours.
9. **Testing and Validation:** The control logic is thoroughly tested in simulation before deployment to identify and address any issues or conflicts.

- **Results:** The effective use of control circuit design in the Building Automation System leads to several benefits:

1. **Energy Efficiency:** Smart control of lighting, HVAC, and other systems results in reduced energy consumption and lower utility costs.
2. **Comfort and Productivity:** Occupants experience improved comfort due to optimized temperature, lighting, and ventilation conditions.
3. **Safety:** Integrated fire detection and emergency shutdown capabilities enhance occupant safety in emergency situations.
4. **Remote Monitoring:** Centralized monitoring through the control circuit enables real-time tracking of system performance and quick response to anomalies.
5. **Sustainability:** The efficient use of resources contributes to the building's environmental sustainability and reduces its carbon footprint.
6. **Cost Savings:** The automation of system controls leads to operational cost savings and a more efficient use of maintenance resources.

In this case study, control circuit design plays a vital role in creating an intelligent Building Automation System that improves occupant comfort, safety, and energy efficiency while streamlining building management operations.

CLASS ACTIVITY

DIY PRINTED CIRCUIT BOARD (PCB) FABRICATION WORKSHOP

- Objective: To provide students with hands-on experience in designing, fabricating, and assembling their own PCBs, enhancing their understanding of electronic circuit fabrication techniques.
- Materials Needed: Copper-clad boards, etching chemicals, photocopier, iron, drill press, soldering equipment, multimeters, etc.
- Assessment: Evaluate students based on the accuracy and functionality of their fabricated PCBs, adherence to safety protocols during fabrication, and ability to troubleshoot and resolve assembly issues. Assess students' understanding of PCB fabrication techniques through written reflections or presentations.

SUMMARY

- An electronic circuit is a path or network of interconnected electronic components that allows the flow of electric current to perform a specific function. These circuits are fundamental to modern technology and are found in virtually every electronic device, from simple gadgets like flashlights and calculators to complex systems like computers and communication networks.
- Electronic circuits can be broadly classified into two types: analog and digital. Analog circuits process continuous signals, making them essential for tasks involving sound, light, and radio waves.
- Integrated circuits (ICs) are a significant advancement in electronic circuitry, miniaturizing and combining multiple components into a single chip. This innovation has revolutionized the electronics industry, enabling the creation of compact, efficient, and powerful devices.
- Electronic circuit design consists of analysis and synthesis of electronic circuits. While designing an analog circuit or digital circuit the designer should be able to predict the voltage and current at every node in the circuit.
- Analogue electronic circuit designs are those in which current or voltage varies with time to correspond to the information being represented. Diodes, capacitors, resistors, transistors and wires are the major components of an analog circuit.
- Digital electronic circuit design takes the electrical signals in the form of discrete values. The data are represented in the form of zeros and ones. Digital circuits extensively use transistors, interconnected to give create logic gates that provide the function of Boolean logic.
- A schematic circuit diagram is the representation of components and interconnections in a circuit using standardized symbols without using the actual image of the component. Circuit diagrams are used for design, construction and maintenance of the electrical and electronic equipment.
- Circuits need to be compact and have reliable interconnections among their components. As devices get smaller, overlapping wires cause interferences, which can cause them to malfunction.
- Passive components are essential to electronic circuits and devices. They can reduce the amount of electricity flowing through a circuit, store energy, and produce inductance. Passice components can also help increase voltage and current.
- All electrical signals can be described as a function of either time or frequency. When we observe signals as a function of time, they are called time domain measurements. Sometimes we observe the frequencies present in signals, in which case they are called frequency domain measurements.

REVIEW QUESTIONS

1. What are the electronic circuit development strategies?
2. List the factors to consider when creating a circuitry design.
3. Discuss about feedback and frequency response.
4. What is the effect of circuit capacitance?
5. Describe the circuit noise calculations and noise bandwidth?

REFERENCES

1. Bhattacharya, S. (2005). Template-driven parasitic-aware optimization of analog/RF IC layouts. Thesis, University of Washington.
2. Bor, J.-C., & Wu, C.-Y. (1996). Analog electronic cochlea design using multiplexing switched-capacitor circuits. *IEEE Transactions on Neural Networks, 7*(1), 155–166.
3. Castejón, F., & Carmona, E. J. (2018). Automatic design of analog electronic circuits using grammatical evolution. *Applied Soft Computing, 62*, 1003–1018.
4. Dieste-Velasco, M. I. (2021). Application of a fuzzy inference system for optimization of an amplifier design. *Mathematics, 9*(17), 2168.
5. Felgueiras, M. C., Macedo, J., Fidalgo, A., Petry, C., & Alves, G. (2016). How to use remote labs for enhancing e-learning on PSoCs. *International Journal of Online Engineering (iJOE), 12*(4), 61.
6. Flynn, M. P., Park, S., & Lee, C. C. (2005). Achieving analog accuracy in nanometer CMOS. *International Journal of High-Speed Electronics and Systems, 15*(2), 255–275.
7. Jiao, S. M., Wang, C. H., & Wang, X. M. (2012). Large-scale analog circuit evolutionary design using a real-coded scheme. *Applied Mechanics and Materials, 220–223*, 2036–2039.
8. Lima, E. C. de O., Cavalcante, A. B., & Da Fonseca Neto, J. V. (2020). Optimization of amplifier circuits by using gradient boosted trees and probability annealing policy. *Journal of Integrated Circuits and Systems, 15*(3), 1–5.
9. Lui, S. (2007). Analog circuit design by nonconvex polynomial optimization: Two design examples. Retrieved from: HKUTO.
10. Mhiri, N., Alahdal, A., Ghulman, H., & Ammous, A. (2017). A novel analog circuit design for maximum power point tracking of photovoltaic panels. *Advances in Power Electronics, 2017*, 1–9.
11. Odame, K. (2008). Exploiting device nonlinearity in analog circuit design. Dissertation, Georgia Institute of Technology.
12. Shana'a, O. K. (1996). Circuit implementation of a high-speed continuous-time current-mode field programmable analog array (FPAA). Retrieved from: PDXScholar.
13. Wang, W., & Jin, D. (2005). CMOS design of analog fuzzy system. *Journal of Circuits, Systems and Computers, 14*(6), 1101–1112.

CHAPTER

8

Operational Amplifiers

LEARNING OBJECTIVES

After studying this chapter, you will be able to:

- Give an overview of operation;
- Discuss the internal circuitry of 741-type op-amp;
- Focus in operational amplifier applications.

KEY TERMS FROM THIS CHAPTER

Amplifiers
Digital circuits
Oscillators
Voltage
Circuit
Magnitude
Transfer function

INTRODUCTION

Op-amps, or operational amplifiers, are essential parts of analog electronic circuits. These are multipurpose, high-gain voltage amplifiers that usually have a single-ended output and differential inputs. An op-amp's main job is to increase the voltage differential applied to its non-inverting (+) and inverting (-) terminals. Op-amps can handle a wide range of signal processing tasks, including filtering, oscillation, and voltage regulation, thanks to their high gain, which frequently exceeds 100,000. Op-amps are distinguished by a number of important characteristics, such as bandwidth, slew rate, input and output impedances, and gain. While the low output impedance enables the op-amp to drive large loads, the high input impedance guarantees that there is little current drawn from the source. An op-amp's bandwidth is the range of frequencies over which it can efficiently amplify signals, and its slew rate is the fastest rate at which the output voltage can fluctuate. Op-amps, short for operational amplifiers, are essential components in electronic circuits due to their versatility and wide array of uses. These devices are commonly found in integrated circuits and are capable of performing various mathematical functions like addition, subtraction, integration, and differentiation. Op-amps are defined by their high input impedance, low output impedance, and high open-loop gain, making them well-suited for tasks involving amplification and signal conditioning. An op-amp is made up of a single-ended output stage and a differential input stage that amplifies the difference between two input voltages. The output is a function of the voltage difference between the two inputs, multiplied by the amplifier's gain. The inputs are labeled as non-inverting (+) and inverting (-). Op-amps can be employed in a variety of linear applications, including differential amplifiers, summing amplifiers, and voltage followers, thanks to this configuration. Furthermore, their stability and linearity can be improved by precisely controlling their gain through the incorporation of feedback networks. Analog signal processing is one of the main applications for op-amps. Op-amps, for instance, amplify audio signals in audio equipment with little distortion, guaranteeing high-fidelity sound reproduction. They amplify low-level signals from sensors in instrumentation and measurement systems, enabling the detection and measurement of minute physical changes. Op-amps are also essential for analog computing, as they enable real-time mathematical operations that facilitate the creation of waveform generators, oscillators, and analog filters, among other applications.

Op-amps are used in nonlinear circuits in addition to linear ones. In digital circuits, they perform the function of comparators, comparing an input voltage to a reference voltage and producing a digital signal that indicates the outcome. They create intricate waveforms for signal processing and communication systems through waveform generation. Furthermore, op-amps are essential components of power supply circuits because they regulate voltage and convert DC to DC in order to maintain stable output voltages under a range of load scenarios. Contemporary op-amps come in low-power, high-speed, and precision varieties, all of which are tailored to specific application needs. Advances in semiconductor technology have resulted in the creation of op-amps that exhibit improved performance attributes, such as reduced noise, increased bandwidth, and enhanced dependability. These developments guarantee that op-amps will always be essential parts of analog and digital electronic systems, supporting a wide range of uses from industrial automation to consumer electronics and more.

8.1. AN OVERVIEW OF OPERATION

An op-amp, short for operational amplifier, is an electronic device that amplifies the difference between two input voltages. It is a versatile amplifier that is commonly used in various electronic circuits, such as filters, oscillators, and voltage regulators.

Because of their low output impedance, high gain, and high input impedance, op-amps are effective at amplifying signals with extremely small amplitudes. They can amplify minuscule variations in input voltage because they are usually built with an extremely high open-loop gain. A differential amplifier stage is typically used in the construction of op-amps, followed by one or more amplifier stages. The signal is further amplified and given the required output impedance by the subsequent amplifier stages, after the differential amplifier stage has amplified the difference between the two input voltages. Op-amps can be used as integrators, differential amplifiers, inverting amplifiers, and non-inverting amplifiers, among other configurations. They can also be employed as switches, oscillators, and comparators. Op-amps' affordability, high gain, and adaptability make them a popular choice for electronic circuits. They can be found widely as integrated circuits (ICs) and are employed in a variety of settings, such as industrial control systems and audio amplifiers.

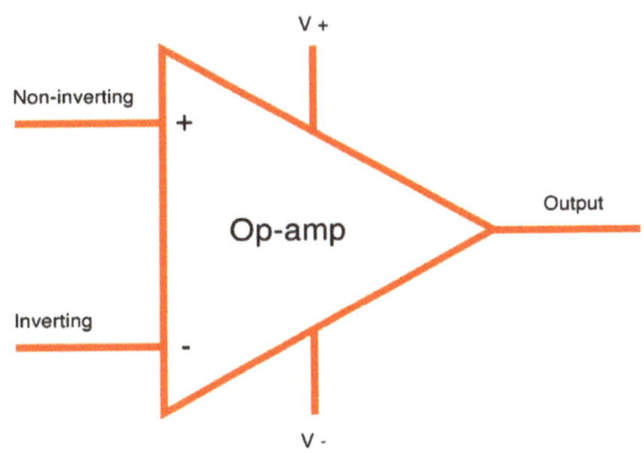

Figure 8.1. *The op amp is one type of differential amplifier.*

Source: https://www.allaboutcircuits.com/uploads/articles/op-amp-in-schematics.jpg.

Op-amps are commonly employed in a wide range of electronic devices nowadays, spanning from consumer gadgets to industrial and scientific instruments. While many typical integrated circuit op-amps are budget-friendly, costing only a few cents each, there are specialized models that can be priced at over US$100 in small quantities. These op-amps can be found as standalone components or as part of more intricate integrated circuit designs.

One kind of differential amplifier is the op-amp. Differential amplifiers can also be built as fully differential amplifiers, instrumentation amplifiers, isolation amplifiers, and negative-feedback amplifiers. The former is constructed from one or more op amps and a resistive feedback network, while the latter is typically built from three op amps and has a differential rather than a single-ended output.

Operational amplifiers, or op-amps, are versatile linear devices perfect for DC amplification and commonly utilized in signal conditioning, filtering, or calculations. They are considered one of the most valuable components in analog electronic circuits, as they can perform a multitude of signal processing functions with just a few external parts. Additionally, op-amps are cost-effective, with many general-purpose models available for less than a dollar each. Modern op-amps are designed for durability, with some models able to withstand direct short-circuits on their outputs without being damaged.

The engineering principle of feedback, especially negative feedback, which forms the basis of nearly all automatic control processes, holds a key to the utility of these small circuits. A non-inverting input (+) with voltage V_+ and an inverting input (–) with voltage V_- make up the differential inputs of the amplifier. Ideally, the op amp amplifies only the voltage difference between the two, which is known as the differential input voltage. The equation provides the output voltage of the op amp V_{out}.

$$V_{out} = A_{OL}(V_+ - V_-),$$

where A_{OL} is the open-loop gain of the amplifier (the term "open-loop" refers to the absence of an external feedback loop from the output to the input).

8.1.1. Open-Loop Amplifier

AOL's magnitude is usually very high (typically over 100,000 for integrated circuit op-amps) leading to even a slight variance between V+ and V- causing the amplifier to clip or saturate. As manufacturing does not tightly regulate AOL's magnitude, using an open-loop amplifier as a standalone differential amplifier is impractical.

If an op-amp does not receive negative feedback, and potentially positive feedback for regeneration, it will function as a comparator. When the inverting input is grounded and the non-inverting input receives a positive input voltage (V_{in}), the output will be maximum positive. Conversely, when V_{in} is negative, the output will be maximum negative. Since there is no feedback from the output to either input, this open-loop circuit operates solely as a comparator.

8.1.2. Closed-Loop Amplifier

Negative feedback is applied by applying a portion of the output voltage to the inverting input if consistent operation is required. The circuit's gain is significantly decreased by the closed-loop feedback. When negative feedback is employed, the feedback network—rather than the op-amp characteristics—determines the circuit's overall gain and response. The value of the op amp's open-loop response (A_{OL}) has little effect on the circuit's performance if the feedback network is composed of components whose values are small in relation to the op-amp's input impedance. High input impedance at the input terminals and low output impedance at the output terminal or terminals are particularly helpful characteristics of an op amp in this situation.

Electrical engineering is the field that designs an op-amp circuit to have a desired transfer function. A transfer function is a mathematical representation of the response of the op-amp circuit, which consists of input, output, and feedback circuits, to an input. Most op-amp applications, including analog computers, depend on the transfer functions.

In the non-inverting amplifier on the right, the presence of negative feedback via the voltage divider R_f, R_g determines the closed-loop gain $A_{CL} = V_{out}/V_{in}$. Equilibrium will be established when V_{out} is just sufficient to pull the inverting input to the same voltage as V_{in}. The voltage gain of the entire circuit is thus $1 + R_f/R_g$. As a simple example, if $V_{in} = 1$ V and $R_f = R_g$, V_{out} will be 2 V, exactly the amount required to keep V_- at 1 V. Because of the feedback provided by the R_f, R_g network, this is a closed-loop circuit.

Another way to analyze this circuit proceeds by making the following (usually valid) assumptions:

- When an op amp operates in linear (i.e., not saturated) mode, the difference in voltage between the non-inverting (+) and inverting (–) pins is negligibly small.
- The input impedance of the (+) and (–) pins is much larger than other resistances in the circuit.

The input signal V_{in} appears at both (+) and (–) pins per assumption 1, resulting in a current i through: R_g equal to V_{in}/R_g:

$$i = \frac{V_{in}}{R_g}$$

Since Kirchhoff's current law states that the same current must leave a node as enter it, and since the impedance into the (–) pin is near infinity per assumption 2, we can assume practically all of the same current i flows through R_f, creating an output voltage

$$V_{out} = V_{in} + iR_f = V_{in} + \left(\frac{V_{in}}{R_g}R_f\right) = V_{in} + \frac{V_{in}R_f}{R_g} = V_{in}\left(1 + \frac{R_f}{R_g}\right)$$

By combining terms, we determine the closed-loop gain A_{CL}:

$$A_{CL} = \frac{V_{out}}{V_{in}} = 1 + \frac{R_f}{R_g}$$

8.1.3. Op-Amp Characteristics

There are many different important characteristics and parameters related to op-amps.

- **Ideal op-Amps**: An ideal op-amp is usually considered to have the following characteristics:
  - o Infinite open-loop gain $G = v_{out}/v_{in}$
  - o Infinite input impedance R_{in}, and so zero input current
  - o Zero input offset voltage
  - o Infinite output voltage range
  - o Infinite bandwidth with zero phase shift and infinite slew rate
  - o Zero output impedance Rout, and so infinite output current range
  - o Zero noise
  - o Infinite common-mode rejection ratio (CMRR)
  - o Infinite power supply rejection ratio

These ideals can be summarized by the two golden rules:

- In a closed loop the output does whatever is necessary to make the voltage difference between the inputs zero
- The inputs draw zero current

The first rule typically applies when the operational amplifier (op-amp) is utilized in a closed-loop configuration, specifically in negative feedback circuits where there is a feedback loop from the output to the inverting input. These rules are commonly used as a preliminary approximation for analyzing and designing op-amp circuits. However, it is important to note that achieving these ideals is not always possible in reality. A real op-amp can be represented using equivalent resistors and capacitors in the model to account for non-infinite or non-zero parameters. By incorporating these effects into the circuit design, the designer can better understand the overall performance. Some parameters may have minimal impact on the final design, while others may pose actual limitations that need to be addressed during evaluation.

- **Real op Amps:** Real op-amps differ from the ideal model in various aspects.

1. *Finite Gain:* In real operational amplifiers, open-loop gain is finite. Common devices have open-loop DC gain greater than 100,000. Provided that the loop gain (i.e., the sum of the open-loop and feedback gains) is very high, and the quantity of negative feedback will be the only factor influencing the closed-loop gain (i.e., it won't depend on the open-loop gain. Applications where the closed-loop gain needs to be extremely high (nearing the open-loop gain) will have very low feedback gains, and the circuit will behave less than optimally in these situations due to the lower loop gain.
2. *Non-zero Output Impedance:* Low-impedance loads benefit from low output impedance because the voltage drop across the output impedance effectively lowers the open-loop gain for these loads. Op-amp circuits typically have a very low output impedance in linear applications because configurations with voltage-sensing negative feedback effectively lower the amplifier's output impedance. Usually, low-impedance outputs call for high quiescent (i.e., Low-power designs may deliberately forgo low output impedance to reduce power consumption due to idle (i.e., current) in the output stage.
3. *Finite Input Impedances:* The operational amplifier's differential input impedance is the impedance between its two inputs, while the common-mode input impedance is the impedance from each input to ground. MOSFET-input operational amplifiers may have protection circuits that short circuit input differences above a certain threshold, making the input impedance seem low in certain tests. However, in high-gain negative feedback applications, these protection circuits are usually inactive. Instead, the input bias and leakage currents play a more crucial role in typical operational amplifier designs.
4. *Input Capacitance:* Additional input impedance due to parasitic capacitance can be a critical issue for high-frequency operation where it reduces input impedance and may cause phase shifts.

5. *Input Current:* A small amount of current enters the inputs as a result of leakage or biasing requirements. These tiny currents can cause voltage drops in circuits that use high resistances or sources with high output impedances. The voltages generated at each input will be equal if the input currents and the impedance coming out of both inputs match. These matched voltages have no effect because the operational amplifier operates on the difference between its inputs. A slight mismatch in the input currents is more typical. Even with matched resistances, a tiny offset voltage may be generated. The difference is known as the input offset current. The operational amplifier may experience drifting or offsets as a result of this offset voltage.
6. *Input Offset Voltage:* The voltage needed across the op amp's input terminals to drive the output voltage to zero is known as input offset voltage. In a perfect amplifier, there would be no input offset voltage, but in reality, it exists due to imperfections in the differential amplifier input stage of op amps. Input offset voltage causes two issues: when operating the amplifier without negative feedback, the high voltage gain can lead to saturation of the output, even with the input terminals connected. Additionally, in a closed loop configuration with negative feedback, the input offset voltage is amplified along with the signal, which can be problematic for high precision DC amplification or when dealing with very small input signals.
7. *Common-Mode Gain:* An ideal operational amplifier rejects all voltages that are common to both of its inputs and only amplifies the voltage difference between them. These common voltages are amplified to some extent because operational amplifiers' differential input stage is never flawless. The common-mode rejection ratio (CMRR) is the commonly used metric to quantify this defect. When operating at high gain, non-inverting amplifiers must minimize common-mode gain.
8. *Power-Supply Rejection:* An ideal operational amplifier's output will be unaffected by changes in the voltage of the power supply. The power supply rejection ratio (PSRR) of each real operational amplifier is finite and indicates how well the op amp can prevent noise from entering its power supply and reaching the output.
9. *Temperature Effects:* Performance and properties of the amplifier typically changes, to some extent, with changes in temperature. Temperature drift of the input offset voltage is especially important.
10. *Drift:* Real op-amp parameters are subject to slow change over time and with changes in temperature, input conditions, etc.
11. *Finite Bandwidth:* Amplifications all have a limited bandwidth. The operational amplifier (op-amp) approximates the frequency response of an integrator with gain. In other words, an ordinary operational amplifier's gain is directly proportional to frequency and can be identified by its gain–bandwidth product (GBWP). An

op- amp with a GBWP of 1 MHz, for instance, would have a gain of 1 at 1 MHz and 5 at 200 kHz. The op- amp has the characteristics of a first-order low-pass filter with a very high DC gain and low cutoff frequency determined by the GBWP divided by the DC gain because of its dynamic response and extremely high DC gain.

The finite bandwidth of an op-amp can be the source of several problems, including:

a. *Stability:* A phase difference between the input signal and the amplifier output that is linked to the bandwidth limitation may cause oscillation in some feedback circuits. For instance, if a sinusoidal output signal is delayed by 180 degrees, it can interfere constructively with an input signal of the same frequency, creating positive feedback. In these situations, frequency compensation—which raises the open-loop circuit's gain or phase margin—can stabilize the feedback circuit. This compensation can be externally implemented by the circuit designer using a different circuit component. As an alternative, a dominant pole that adequately attenuates the operational amplifier's high-frequency gain can be added to the operational amplifier to implement the compensation. The manufacturer may set this pole's location internally, or the circuit designer may configure it using op amp-specific techniques. Generally speaking, dominant-pole frequency compensation narrows the operational amplifier's bandwidth even more. Op- amps with higher bandwidths can be used in applications with high closed-loop gain because op-amp frequency compensation is often unnecessary when the desired closed-loop gain is high because the required open-loop gain is low enough.

b. *Frequency Response and Bandwidth (BW):* No matter the signal frequency, an ideal op-amp would be able to maintain a high gain and have an infinite bandwidth (BW). All op-amps, however, have a limited bandwidth that is commonly referred to as the "-3dB point," at which point the gain starts to roll off as frequency rises. Then, as the frequency rises, the amplifier's gain drops at a rate of -20dB/decade. Because they maintain higher gains at higher frequencies, op-amps with a higher BW perform better; however, this higher gain comes at the expense of greater power consumption or cost.

c. *Distortion and Other Effects:* Reduced bandwidth also leads to decreased feedback at higher frequencies, which raises output impedance and distortion as the frequency rises. The GBWP of typical low-cost general-purpose op-amps is a few megahertz. High-speed and specialty op-amps are available with GBWPs reaching hundreds of megahertz. In circuits with extremely high frequencies, current-feedback operational amplifiers are frequently employed.

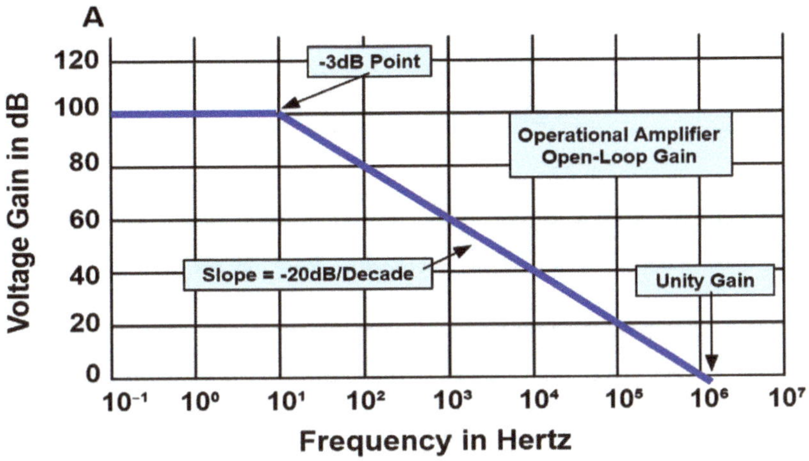

Figure 8.2. *Operational amplifier open-loop frequency response curve.*

Source: https://media.monolithicpower.com/wysiwyg/2_23.png.

d. *Noise:* Noise is an inherent output of amplifiers, even in the absence of a signal. The device's flicker noise and internal thermal noise may be to blame for this. Noise becomes a significant factor in applications with high gain or high bandwidth, and in order to meet performance requirements, a low-noise amplifier—which is specifically made for minimum intrinsic noise—might be needed.

e. *Power-Supply Rejection:* With increasing frequency, the power-supply rejection usually gets worse. So, it can be important to keep the supply clean of higher frequency ripples and signals, e.g., by the use of bypass capacitors.

8.2. INTERNAL CIRCUITRY OF 741-TYPE OP AMP

The 741 integrated circuit, created in 1968 by David Fullagar at Fairchild Semiconductor following Bob Widlar's LM301 integrated circuit design, is an example of a bipolar transistor operational amplifier. It is sourced by numerous manufacturers and found in numerous similar products. We describe the small-signal, grounded emitter properties of a transistor using the parameters of the hybrid-pi model. In this model, the current gain of a transistor is denoted h_{fe}, more commonly called the β.

8.2.1. Architecture

A small-scale integrated circuit, the 741 op-amp shares with most op-amps an internal structure consisting of three gain stages:

- **Differential amplifier (outlined dark blue):** provides high differential amplification (gain), with rejection of common-mode signal, low noise, high input impedance, and drives a
- **Voltage amplifier (outlined magenta):** provides high voltage gain, a single-pole frequency roll-off, and in turn drives the
- **Output amplifier (outlined cyan and green):** provides high current gain (low output impedance), along with output current limiting, and output short-circuit protection.

Additionally, it contains current mirror (outlined red) bias circuitry and compensation capacitor (30 pF).

1. **Differential amplifier:** A current-mirror active load is placed after a cascaded differential amplifier (highlighted in dark blue) in the input stage in Figure 8.4. By converting a differential voltage signal at the bases of Q1, Q2, and Q15 into a current signal at the base of Q15, this functions as a transconductance amplifier. Two cascaded transistor pairs are needed to satisfy competing requirements. The matched NPN emitter follower pair Q1, Q2, which offer high input impedance, make up the first stage. The second drives an active load Q7 plus matched pair Q5, Q6, and eliminates the unwanted Miller effect. It is the matched PNP common-base pair Q3, Q4. That active load is implemented as a modified Wilson current mirror, and its purpose is to increase the open-loop gain of the op amp

by 3 dB by converting the (differential) input current signal to a single-ended signal without the associated 50 percent losses. As a result, at the base of Q15, the voltage gain stage's input, a small-signal differential current in Q3 versus Q4 appears summed (doubled).

2. **Voltage amplifier:** The output side of the current mirror created by Q12 and Q13 serves as the collector (dynamic) load for the (class-A) voltage gain stage (shown in magenta), which is made up of the two NPN transistors Q15 and Q19 connected in a Darlington configuration. This allows the stage to achieve a high voltage gain.

The common collectors of Q15 and Q19 supply base drive to the output sink transistor Q20, while the output source transistor Q14 receives base drive from the level-shifter Q16. The output sink current is limited because transistor Q22 stops this stage from giving Q20 an excessive amount of current.

8.2.2. Varactor Bridge Op Amps

George Philbrick promoted the varactor bridge amplifier, a cutting-edge operational amplifier type that turned a profit for GAP/R. In this circuit, the op amp error voltage is processed as a phase-sensitive AC carrier in an input stage using voltage variable capacitors, also known as varactors. The op amp input terminals are made to see only very small DC leakage currents by carefully placing the bridge components, i.e., as little as 1 pA (or much less in some cases). Consequently, of all the op amps available during the solid-state era, a varactor bridge op amp achieved the lowest input current. Less than common tubes, in fact! Moreover, the permissible input CM voltage of a varactor bridge op amp could reach extremely high levels—up to ±200V—because there was no input DC path to common. A varactor bridge op amp is shown in block diagram form in the figure below. There are four main parts: an AC amplifier to boost the bridge output error voltage, a synchronous phase detector to convert the amplified AC error to a corresponding DC error, and an output amplifier to provide additional DC gain and load drive comprise the front end, which consists of the bridge circuit and a high frequency oscillator.

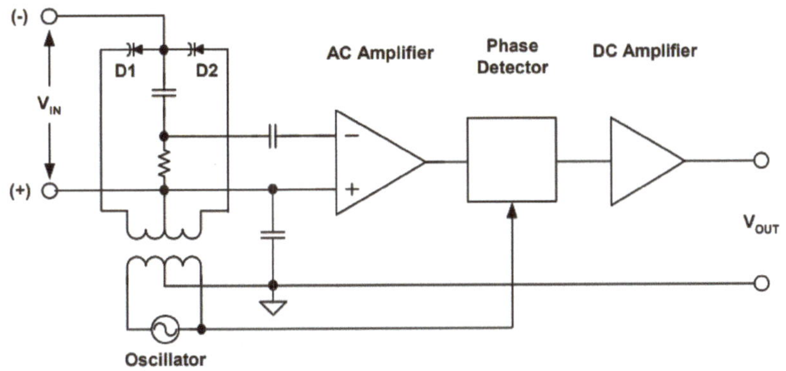

Figure 8.3. *Generalized block diagram for a varactor bridge solid-state op amp.*

Source: https://ars.els-cdn.com/content/image/3-s2.0-B9780750678445501594-f08-10-9780750678445.gif.

The circuit operates as follows: an AC bridge imbalance is produced by applying a small DC error voltage V_{IN} to the matched varactor diodes D1 and D2, which is then fed into the AC amplifier. Phase sensitivity will be present in this AC voltage, contingent on the DC error. The loop's remaining components detect and amplify the DC error. As with conventional op amps, an external feedback loop is closed from V_{OUT} back to the inverting input terminal in order to apply the amplifier. Due to the existence of two unique degrees of freedom for both bias current and common-mode (CM) voltage, the varactor bridge op- amp case differs from the others.

8.2.2.1. Output Amplifier

Q14, Q20, the output stage, is a Class AB complementary-symmetry amplifier; it is indicated in cyan. It offers a current gain output drive with an impedance of about 50 Ω. The quiescent current for the output transistors is provided by transistor Q16 (highlighted in green), and the output source current is limited by transistor Q17.

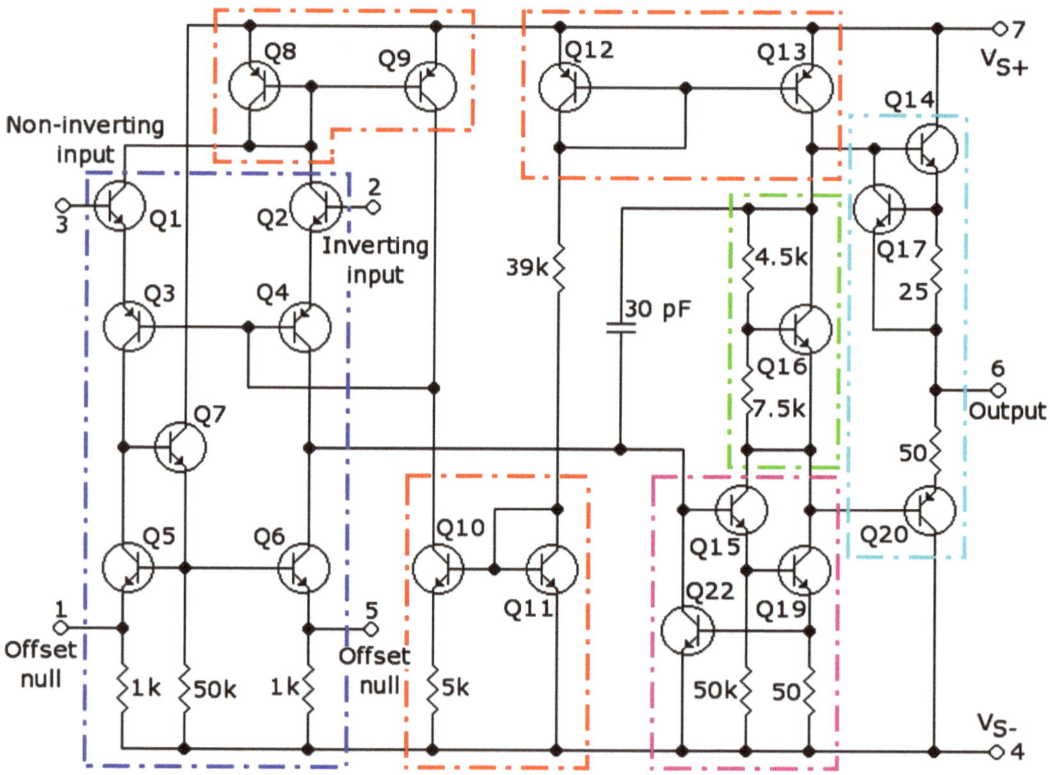

Figure 8.4. *A component-level diagram of the common 741 op amp.*

Source: https://upload.wikimedia.org/wikipedia/commons/2/2d/Opamptransistorlevelcoloredlabeled.png?20050831013336.

8.2.3. Biasing Circuits

Provide appropriate quiescent current for each stage of the op amp. The resistor (39 kΩ) connecting the (diode-connected) Q11 and Q12, and the given supply voltage ($V_{S+} - V_{S-}$), determine the current in the current mirrors, (matched pairs) Q10/Q11 and Q12/Q13. The collector current of Q11, $i_{11} \times 39$ kΩ $= V_{S+} - V_{S-} - 2\ V_{BE}$. For the typical $V_S = \pm 20$ V, the standing current in Q11/Q12 (as well as in Q13) would be ~1 mA. A supply current for a typical 741 of about 2 mA agrees with the notion that these two bias currents dominate the quiescent supply current.

Transistors Q11 and Q10 form a Widlar current mirror, with quiescent current in Q10 i_{10} such that $\ln(i_{11}/i_{10}) = i_{10} \times 5$ kΩ/28 mV, where 5 kΩ represents the emitter resistor of Q10, and 28 mV is V_T, the thermal voltage at room temperature. In this case $i_{10} \approx 20$ µA.

- *Differential Amplifier:* A feedback loop that forces the collector currents of Q10 and Q9 to (nearly) match sets the biasing circuit of this stage. The common base of Q3/Q4 is driven by the slight difference in these currents (keep in mind that the base drive of input transistors Q1/Q2 is the input bias current and needs to be sourced externally). Mirrored from Q8 into Q9, the summed quiescent currents of Q1/Q3 plus Q2/Q4 are added to the collector current in Q10, and the resulting value is applied to the bases of Q3/Q4.

The quiescent currents of Q1/Q3 (resp., Q2/Q4) i_1 will thus be half of i_{10}, of order ~10 µA. Input bias current for the base of Q1 (resp. Q2) will amount to i_1/β; typically, ~50 nA, implying a current gain $h_{fe} \approx 200$ for Q1(Q2).

This feedback circuit tends to draw the common base node of Q3/Q4 to a voltage $V_{com} - 2\ V_{BE}$, where V_{com} is the input common-mode voltage. At the same time, the magnitude of the quiescent current is relatively insensitive to the characteristics of the components Q1–Q4, such as h_{fe}, that would otherwise cause temperature dependence or part-to-part variations.

Transistor Q7 drives Q5 and Q6 into conduction until their (equal) collector currents match that of Q1/Q3 and Q2/Q4. The quiescent current in Q7 is V_{BE}/50 kΩ, about 35 µA, as is the quiescent current in Q15, with its matching operating point. Thus, the quiescent currents are pairwise matched in Q1/Q2, Q3/Q4, Q5/Q6, and Q7/Q15.

- *Voltage Amplifier*: Quiescent currents in Q16 and Q19 are set by the current mirror Q12/Q13, which is running at ~1 mA. Through some mechanism, the collector current in Q19 tracks that standing current.
- *Output Amplifier*: The Q16 (also known as the rubber diode or V_{BE} multiplier) circuit requires the 4.5 kΩ resistor to conduct at approximately 100 µA, and the Q16 VBE to be at approximately 700 mV. Then, the V_{CE} and V_{CB} must be at approximately 1.0 and 0.4 V, respectively. Regardless of the common-mode voltage of Q14/Q20 base, the Q16 transistor establishes a voltage difference of approximately 1 V between Q14 base and Q20 base because the Q16 emitter drives into the Q19 collector current sink and the Q16 collector is driven by a current source. In Q14/Q20, the standing current will be a factor exp (100 mV/VT) ≈ 36 less than the class

A quiescent current of 1 mA in the op amp. The output stage in class AB operation is established and its crossover distortion is decreased by this (small) standing current in the output transistors.

8.2.4. Small-Signal Differential Mode

A small differential input voltage signal gives rise, through multiple stages of current amplification, to a much larger voltage signal on output.

- *Input Impedance:* When Q2 and Q4 are added, the input stage with Q1 and Q3 is similar to an emitter-coupled pair (long-tailed pair). Due to the low current passing through Q1–Q4, the input impedance is comparatively high. The differential input impedance of a typical 741 op amp is approximately 2 MΩ. Because the input stage operates at nearly constant current, the common mode input impedance is even higher.
- *Differential Amplifier:* A differential voltage V_{in} at the op-amp inputs (pins 3 and 2, respectively) gives rise to a small differential current in the bases of Q1 and Q2 $i_{in} \approx V_{in}/(2h_{ie}h_{fe})$. This differential base current causes a change in the differential collector current in each leg by $i_{in}h_{fe}$. Introducing the transconductance of Q1, $g_m = h_{fe}/h_{ie}$, the (small-signal) current at the base of Q15 (the input of the voltage gain stage) is $V_{in}g_m/2$.

In order to avoid needlessly discarding the signal in either leg, this part of the op-amp cleverly converts a differential signal at the op amp inputs to a single-ended signal at the base of Q15. To understand how, observe that, a tiny negative voltage change at the inverting input (Q2 base) causes it to stop conducting. This incremental current drop then flows directly from Q4's collector to its emitter, causing Q15's base drive to drop. However, this transistor is driven into conduction by a slight positive voltage change at the non-inverting input (Q1 base), which is reflected in an increase in current at Q3's collector. By forcing Q7 into further conduction, this current activates current mirror Q5/Q6. As a result, an increase in Q6 collector current mirrors the increase in Q3 emitter current. Consequently, the base drive current for Q15 decreases as the increased collector current shunts farther from the collector node. This method reduces feed-through of power supply noise and common-mode gain in addition to preventing the loss of 3 dB of gain.

- *Voltage Amplifier*: A current signal i at Q15's base gives rise to a current in Q19 of order $i\beta^2$ (the product of the h_{fe} of each of Q15 and Q19, which are connected in a Darlington pair). This current signal develops a voltage at the bases of output transistors Q14/Q20 proportional to the h_{ie} of the respective transistor.
- *Output Amplifier:* Since output transistors Q14 and Q20 are set up as emitter followers, there is no voltage gain at that stage; instead, current gain equal to the h_{fe} of Q14 is provided. Instead of being zero, as it would be in a perfect op amp, the output impedance is almost zero at low frequencies thanks to negative feedback.
- *Overall Open-Loop Voltage Gain*: The product of the current gain $_{hfe}$ of approximately 4 transistors determines the net open-loop small-signal voltage gain of the

operational amplifier. A typical 741-style op-amp has a voltage gain of approximately 200,000 in practice. Additionally, the current gain is further increased (power) by the ratio of input impedance (~2–6 MΩ) to output impedance (~50 Ω).

8.2.5. Other Linear Characteristics

- **Small-Signal Common Mode Gain:** Zero common-mode gain, or infinite common-mode rejection ratio, characterizes the perfect op-amp. The current circuit's negative feedback causes the base voltage of Q3/Q4 to follow input voltage variations (with two V_{BE} drops below) if the input voltages change in the same direction. Currently, despite fluctuating voltage, the output portion (Q10) of the Q10-Q11 current mirror maintains a constant common current through Q9/Q8. The output current at Q15's base is unaffected by the Q3/Q4 collector currents. The common-mode rejection ratio of 90 dB in a typical 741 op amp indicates an open-loop common-mode voltage gain of roughly 6.
- **Frequency Compensation:** The Fairchild μA741 introduced a novel feature: an on-chip (monolithic) capacitor for frequency compensation. This eliminated the need for external components, simplifying the op-amp's application. The 30 pF capacitor works similarly to an op-amp integrator circuit and uses Miller compensation to stabilize the amplifier. Also referred to as "dominant pole compensation," this technique adds a pole to the open loop frequency response that masks (dominates) the effects of other poles. In a 741 op amp, this pole can be as low as 10 Hz, which results in a –3 dB loss of open- loop voltage gain.

When the feedback network is non-reactive and the closed-loop gain is unity or greater in negative feedback configurations, this internal compensation is meant to provide unconditional stability for the amplifier. On the other hand, amplifiers like the μA748 that need external compensation might need closed-loop gains that are much higher than unity or external compensation.

- **Input offset voltage:** To modify the balance of the Q5/Q6 current mirror, external resistors (usually in the form of a potentiometer with the slider connected to VS–) can be placed in parallel with the emitter resistors of Q5 and Q6 using the offset null pins. When the inputs are shorted together, the potentiometer is set so that the output is null (midrange).

8.2.6. Non-Linear Characteristics

- **Input Breakdown Voltage:** The transistors Q3, Q4 help to increase the reverse V_{BE} rating: the base-emitter junctions of the NPN transistors Q1 and Q2 break down at around 7 V, but the PNP transistors Q3 and Q4 have V_{BE} breakdown voltages around 50 V.
- **Output-Stage Voltage Swing and Current Limiting:** Crossover distortion and quiescent current may vary significantly because temperature and part-to-part variations in quiescent current are

frequent occurrences. The V_{BE} of the output transistors Q14 and Q20 is partially to blame for the amplifier's output range being roughly one volt lower than the supply voltage.

Q17 and the 25 Ω resistor at the Q14 emitter work together to restrict Q14 current to around 25 mA; in the absence of any current, Q17 conducts nothing. The voltage gain stage is where Q22 senses the voltage across Q19's emitter resistor (50 Ω) and limits Q20's current by reducing the drive current to Q15 base as it turns on. Subsequent iterations of this amplifier schematic might display an output current limitation technique that is slightly different.

8.2.7. Applicability Considerations

Although the 741 was formerly employed in sensitive equipment such as audio systems, its use in these applications is now uncommon due to the enhanced noise performance of more recent op amps. In addition to producing audible hiss, 741s and other older operational amplifiers might have low common-mode rejection ratios, which means they frequently cause cable-borne mains hum and other common-mode interference, like switch 'clicks,' into equipment that is sensitive. The "741" has come to often mean a generic op-amp IC (such as µA741, LM301, 558, LM324, TBA221—or a more modern replacement such as the TL071). The description of the 741 output stage is qualitatively similar for many other designs (that may have quite different input stages), except:

- Some devices (µA748, LM301, LM308) are not internally compensated (require an external capacitor from output to some point within the operational amplifier, if used in low closed-loop gain applications).
- Some modern devices have rail-to-rail output capability, meaning that the output can range from within a few millivolts of the positive supply voltage to within a few millivolts of the negative supply voltage.

8.2.8. Classification

Op-amps may be classified by their construction:

- Discrete, built from individual transistors or tubes/valves.
- Hybrid, consisting of discrete and integrated components.
- Fully integrated circuits — most common, having displaced the former two due to low cost.

IC op-amps may be classified in many ways, including:

- Device grade, including acceptable operating temperature ranges and other environmental or quality factors. For example: LM101, LM201, and LM301 refer to the military, industrial, and commercial versions of the same component. Military and industrial grade components offer better performance in harsh conditions than their commercial counterparts but are sold at higher prices.
- Classification by package type may also affect environmental hardiness, as well as manufacturing options; DIP, and other through-hole packages are tending to be replaced by surface-mount devices.
- Classification by internal compensation: op-amps may suffer from high frequency instability in some negative feedback circuits unless a small compensation

capacitor modifies the phase and frequency responses. Op-amps with a built-in capacitor are termed compensated, and allow circuits above some specified closed-loop gain to operate stably with no external capacitor. In particular, op-amps that are stable even with a closed loop gain of 1 are called unity gain compensated.

- Single, dual and quad versions of many commercial op-amp IC are available, meaning 1, 2 or 4 operational amplifiers are included in the same package.
- Rail-to-rail input (and/or output) op-amps can work with input (and/or output) signals very close to the power supply rails.
- CMOS op-amps (such as the CA3140E) provide extremely high input resistances, higher than JFET-input op amps, which are normally higher than bipolar-input op amps.
- Other varieties of op amp include programmable op amps (simply meaning the quiescent current, bandwidth and so on can be adjusted by an external resistor).
- Manufacturers often tabulate their op-amps according to purpose, such as low-noise pre-amplifiers, wide bandwidth amplifiers, and so on.

8.3. OPERATIONAL AMPLIFIER APPLICATIONS

The equivalent circuit of a non-ideal operational amplifier has a finite gain, a non-zero output impedance, and a finite input impedance. The diagram illustrates many of the less-than-ideal characteristics of a real op-amp; however, since it uses simplified schematic notation, many details, including power supply connections and device selection, are not displayed.

8.3.1. Practical Considerations

1. **Operational Amplifiers Parameter Requirements**

For a particular device to be used in an application, it must satisfy certain requirements. The operational amplifier must

- Have large open-loop signal gain (voltage gain of 200,000 is obtained in early integrated circuit exemplars), and
- Have input impedance large with respect to values present in the feedback network.
- With these requirements satisfied, the op-amp is considered ideal, and one can use the method of virtual ground to quickly and intuitively grasp the 'behavior' of any of the op-amp circuits below.

2. *Component Specification:* Practical solid-state op-amp circuits usually use resistors in the kΩ range. Much larger resistors than 1 MΩ result in excessive thermal noise and increase the likelihood of bias or leakage current errors affecting the circuit's operation.
3. *Input Bias Currents and Input Offset:* Because of leakage (in the case of MOSFET-based inputs) or bias requirements (in the case of bipolar junction transistor-based inputs), practical operational amplifiers require a small amount of current from each of their inputs. Small voltage drops are created across the resistances as a result of these currents passing through them and connecting to the inputs. A well-designed feedback network can mitigate issues related to common-mode gain and input bias currents. Ensuring that each input terminal's impedance is the same is the heuristic rule.

If the input bias currents do not match, an input offset voltage may arise, causing potential issues with circuit performance. Many commercial op-amp options offer a way to adjust the operational amplifier to equalize the inputs, such as using offset null or balance pins that can interact with an external voltage source connected to a potentiometer. Alternatively, a tunable external voltage can be applied to one of the inputs to counteract the offset effect. In situations where a design requires one input to be grounded, a variable resistance can replace the short circuit to help alleviate the offset issue. Operational amplifiers with MOSFET-based input stages typically have input leakage currents that are negligible in many designs.

4. *Power supply effects:* Although power supplies are not indicated in the (simplified) operational amplifier designs below, they are nonetheless present and can be critical in operational amplifier circuit design.

a. *Supply noise*: Imperfections in the power supply (e.g., power signal ripple, and non-zero source impedance) can cause discernible departures from the optimal behavior of an operational amplifier. An operational amplifier's power supply rejection ratio, for instance, measures how well the output can reject signals that show up on the power supply inputs. Due to inductance effects that prevent current from being instantly delivered to every component at once and the fact that almost every component in the design uses the power supply, power supply inputs are frequently noisy in large designs. Therefore, when a component needs significant current injections (e.g., nearby components may sag at their connection to the power supply, such as a digital component that alternates between states frequently. bypass capacitors that are appropriately placed across each pin of the power supply and connected to ground can help to mitigate this issue. When a component needs sudden bursts of current, it can get that current straight from the adjacent capacitor (which the power supply then gradually recharges). This allows the component to avoid using the power supply.

b. *Using power supply currents in the signal path:* Furthermore, the operational amplifier's power supply current can be utilized as an input for external circuitry to increase the operational amplifier's capabilities. For instance, if an operational amplifier's output is needed to produce signals outside of the amplifier's safe range, it might not be suitable for a certain high-gain application. In this instance, the current into and out of the operational amplifier can be used to control an external push-pull amplifier. As a result, the operational amplifier can function within the parameters set by the manufacturer while still permitting a sizable output signal to be included in the negative feedback path that is well outside of those parameters.

8.3.2. Amplifiers

The differential amplifier is the first example, from which many other applications, such as the voltage follower, integrator, differentiator, gyrator, and inverting, non-inverting, and summing amplifier, can be derived.

8.3.2.1. Differential Amplifier (Difference Amplifier)

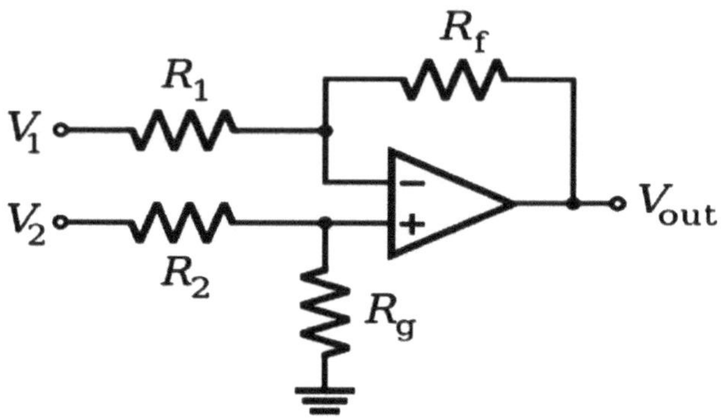

Amplifies the difference in voltage between its inputs.

- The name "differential amplifier" must not be confused with the "differentiator".
- The "instrumentation amplifier," which is also shown on this page, is a modification of the differential amplifier that also provides high input impedance.
- The circuit shown computes the difference of two voltages, multiplied by some gain factor. The output voltage.

$$V_{out} = \frac{(R_f + R_1)R_g}{(R_g + R_2)R_1}V_2 - \frac{R_f}{R_1}V_1 = \left(\frac{R_1 + R_f}{R_1}\right)\cdot\left(\frac{R_g}{R_g + R_2}\right)V_2 - \frac{R_f}{R_1}V_1.$$

Or, expressed as a function of the common-mode input V_{com} and difference input V_{dif}:

$$V_{com} = (V_1 + V_2)/2; V_{dif} = V_2 - V_1,$$

the output voltage is,

$$V_{out}\frac{R_1}{R_f} = V_{com}\frac{R_1/R_f - R_2/R_g}{1 + R_2/R_g} + V_{dif}\frac{1 + (R_2/R_g + R_1/R_f)/2}{1 + R_2/R_g}.$$

In order for this circuit to produce a signal proportional to the voltage difference of the input terminals, the coefficient of the V_{com} term (the common-mode gain) must be zero, or

$$R_1/R_f = R_2/R_g.$$

With this constraint in place, the common-mode rejection ratio of this circuit is infinitely large, and the output

$$V_{out} = \frac{R_f}{R_1}V_{dif} = \frac{R_f}{R_1}(V_2 - V_1),$$

where the simple expression $R_{f/}R_1$ represents the closed-loop gain of the differential amplifier.

The special case when the closed-loop gain is unity is a differential follower, with

$V_{out} = V_2 - V_1$.

8.3.2.2. Inverting Amplifier

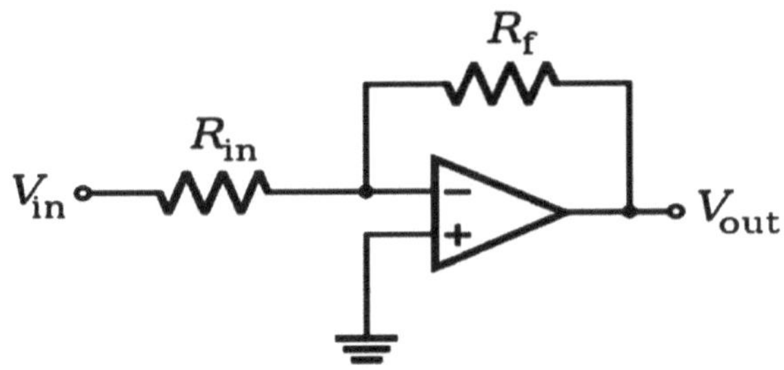

An inverting amplifier is a special case of the differential amplifier in which that circuit's non-inverting input V_2 is grounded, and inverting input V_1 is identified with V_{in} above. The closed-loop gain is R_f/R_{in}, hence

$$V_{out} = -\frac{R_f}{R_{in}} V_{in}$$

The simplified circuit above is like the differential amplifier in the limit of R_2 and R_g very small. In this case, though, the circuit will be susceptible to input bias current drift because of the mismatch between R_f and R_{in}.

To intuitively see the gain equation above, calculate the current in R_{in}:

$$i_{in} = \frac{V_{in}}{R_{in}}$$

then recall that this same current must be passing through R_f, therefore (because $V_- = V_+ = 0$):

$$V_{out} = -i_{in} R_f = -V_{in} \frac{R_f}{R_{in}}$$

A mechanical analogy is a seesaw, with the V_- node (between R_{in} and R_f) as the fulcrum, at ground potential. V_{in} is at a length R_{in} from the fulcrum; V_{out} is at a length R_f. When V_{in} descends "below ground," the output V_{out} rises proportionately to balance the seesaw, and vice versa.

As the negative input of the op-amp acts as a virtual ground, the input impedance of this circuit is equal to R_{in}.

8.3.2.3. Non-Inverting Amplifier

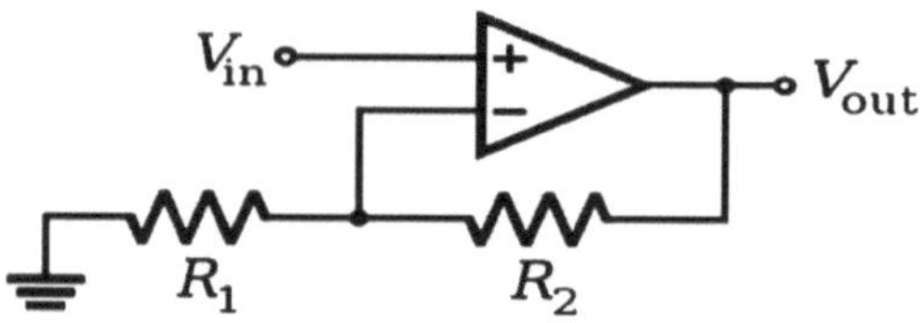

A non-inverting amplifier is a special case of the differential amplifier in which that circuit's inverting input V_1 is grounded, and non-inverting input V_2 is identified with V_{in} above, with $R_1 \gg R_2$. Referring to the circuit immediately above,

$$V_{out} = \left(1 + \frac{R_2}{R_1}\right) V_{in}$$

To intuitively see this gain equation, use the virtual ground technique to calculate the current in resistor R_1:

$$i_1 = \frac{V_{in}}{R_1},$$

then recall that this same current must be passing through R_2, therefore:

$$V_{out} = V_{in} + i_1 R_2 = V_{in}\left(1 + \frac{R_2}{R_1}\right)$$

Unlike the inverting amplifier, a non-inverting amplifier cannot have a gain of less than 1.

A mechanical analogy is a class-2 lever, with one terminal of R_1 as the fulcrum, at ground potential. V_{in} is at a length R_1 from the fulcrum; V_{out} is at a length R_2 further along. When V_{in} ascends "above ground," the output V_{out} rises proportionately with the lever.

The input impedance of the simplified non-inverting amplifier is high:

$$Z_{in} = (1 + A_{OL} B) Z_{dif}$$

where Z_{dif} is the op-amp's input impedance to differential signals, and A_{OL} is the open-loop voltage gain of the op-amp (which varies with frequency), and B is the feedback factor (the fraction of the output signal that returns to the input). In the case of the ideal op-amp, with A_{OL} infinite and Z_{dif} infinite, the input impedance is also infinite. In this case, though, the circuit will be susceptible to input bias current drift because of the mismatch between the impedances driving the V_+ and V_- op-amp inputs.

The feedback loop similarly decreases the output impedance:

$$Z_{out} = \frac{Z_{OL}}{1 + A_{OL}B}$$

where Z_{out} is the output impedance with feedback, and Z_{OL} is the open-loop output impedance.

8.3.2.4. Voltage Follower (Unity Buffer Amplifier)

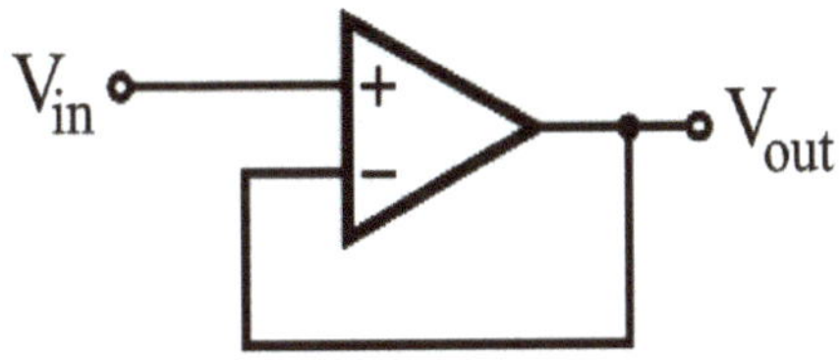

Used as a buffer amplifier to eliminate loading effects (e.g., connecting a device with a high source impedance to a device with a low input impedance).

$$V_{out} = V_{in}$$

$Z_{in} = \infty$(realistically, the differential input impedance of the op-amp itself (1 MΩ to 1 TΩ), multiplied by the open-loop gain of the op-amp)

Due to feedback (unity gain) and some imperfect features of actual operational amplifiers, this feedback system is prone to having narrow stability margins. Consequently, when connected to sufficiently capacitive loads, the system might become unstable. Under certain circumstances, a lag compensation network (e.g., stability can be restored by using a resistor to connect the load to the voltage follower. The operational amplifier's manufacturer data sheet might offer suggestions for selecting the parts of external compensation networks. As an alternative, an operational amplifier with a better internal compensation scheme can be selected.

8.3.3. Filters

Active filters that provide band-pass, band-reject, delay, high-pass, and low-pass functions can be built using operational amplifiers. An op-amp's high input impedance and gain make it simple to calculate element values, which enabling precise implementation of any desired filter topology with little regard for the loading effects of filter stages or of stages that come after. Nevertheless, there is a limit to the frequencies at which active filters can be used; filter performance deteriorates when amplifier behavior deviates considerably from the ideal behavior assumed in the basic filter design.

8.3.4. Comparator

It is possible to force an operational amplifier to function as a comparator if needed. The output will swing to almost the supply voltage because of the slightest difference between the input voltages being greatly amplified. But since a dedicated comparator's output can reach either power supply rail and has a higher slew rate, it is usually preferable to use one for this purpose. Clamping diodes on the input of some op-amps prohibit them from being used as comparators.

8.3.5. Integration and Differentiation

1. **Inverting Integrator**: The integrator is mostly used in analog computers, analog-to-digital converters and wave-shaping circuits.

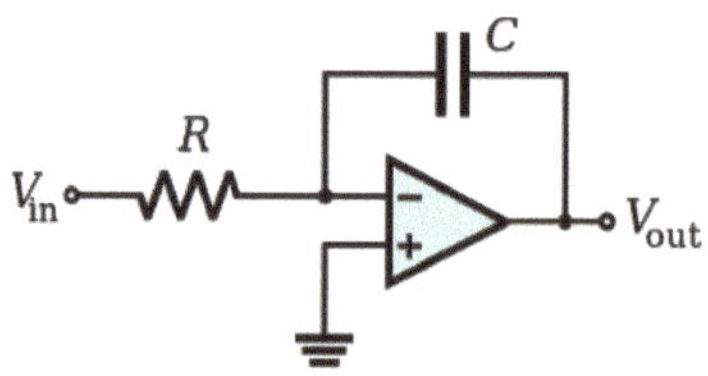

Integrates (and inverts) the input signal $V_{in}(t)$ over a time interval t, $t_0 < t < t_1$, yielding an output voltage at time $t = t_1$ of:

$$V_{out}(t_1) = V_{out}(t_0) - \frac{1}{RC}\int_{t_0}^{t_1} V_{in}(t)dt,$$

where $V_{out}(t_0)$ represents the output voltage of the circuit at time $t = t_0$. This is the same as saying that the output voltage changes over time $t_0 < t < t_1$ by an amount proportional to the time integral of the input voltage:

$$-\frac{1}{RC}\int_{t_0}^{t_1} V_{in}(t)dt.$$

This circuit can be viewed as a low-pass electronic filter, one with a single pole at DC (i.e., where $\omega = 0$) and with gain. In a practical application one encounters a significant difficulty: unless the capacitor C is periodically discharged, the output voltage will eventually drift outside of the operational amplifier's operating range. This can be due to any combination of:

- The input V_{in} has a non-zero DC component,
- Input bias current is non-zero,

- Input offset voltage is non-zero.

A slightly more complex circuit can ameliorate the second two problems, and in some cases, the first as well.

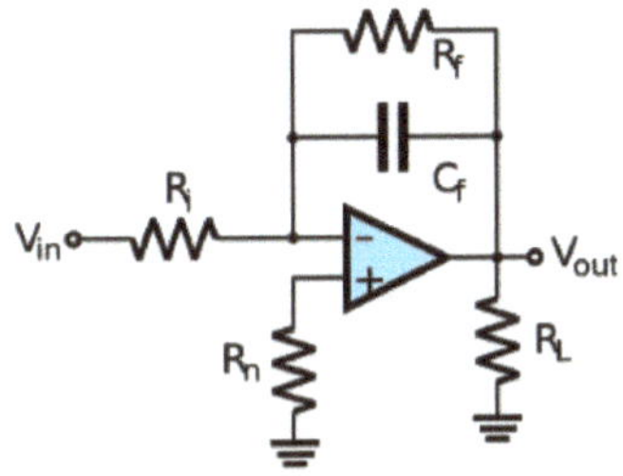

Here, the feedback resistor R_f provides a discharge path for capacitor C_f, while the series resistor at the non-inverting input R_n, when of the correct value, alleviates input bias current and common-mode problems. That value is the parallel resistance of R_i and R_f, or using the shorthand notation ||:

$$R_n = \frac{1}{\frac{1}{R_i} + \frac{1}{R_f}} = R_i || R_f.$$

The relationship between input signal and output signal is now

$$V_{out}(t_1) = V_{out}(t_0) - \frac{1}{R_i C_f} \int_{t_0}^{t_1} V_{in}(t)dt.$$

2. **Inverting Differentiator:** Differentiates the (inverted) signal over time:

$$V_{out} = -RC\frac{dV_{in}}{dt},$$

where V_{in} and V_{out} are functions of time.

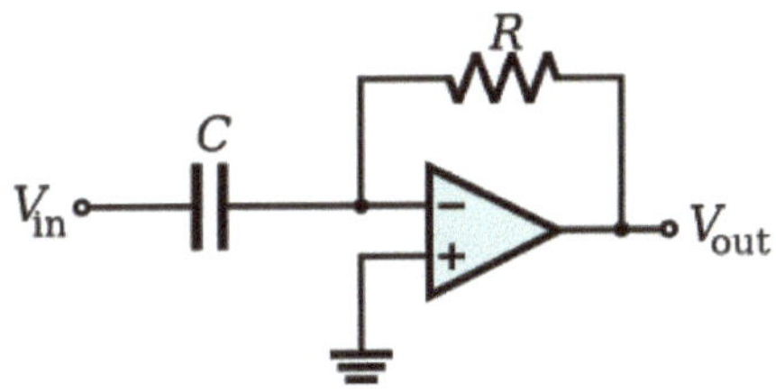

The transfer function of the inverting differentiator has a single zero in the origin (i.e., where angular frequency $\omega = 0$). The high-pass characteristics of a differentiating amplifier can lead to stability challenges when the circuit is used in an analog servo loop (e.g., in a PID controller with a significant derivative gain). As a root locus analysis would show, increasing feedback gain will drive a closed-loop pole toward marginal stability at the DC zero introduced by the differentiator.

ROLE MODEL

HAROLD STEPHEN BLACK: DISCOVERED THE NEGATIVE FEEDBACK AMPLIFIER IN 1927

Harold Stephen Black was an American electrical engineer, who revolutionized the field of applied electronics by discovering the negative feedback amplifier in 1927. To some, his discovery is considered the most important breakthrough of the twentieth century in the field of electronics, since it has a wide area of application. This is because all electronic devices (vacuum tubes, bipolar transistors and MOS transistors) are inherently nonlinear, but they can be made substantially linear with the application of negative feedback. Negative feedback works by sacrificing gain for higher linearity (or in other words, smaller distortion/intermodulation). By sacrificing gain, it also has an additional effect of increasing the bandwidth of the amplifier. However, a negative feedback amplifier can be unstable such that it may oscillate. Once the stability problem is solved, the negative feedback amplifier is extremely useful in the field of electronics. Black published a famous paper, Stabilized feedback amplifiers, in 1934.

After years of work Black invented the negative feedback amplifier which uses negative feedback to reduce the gain of a high-gain, non-linear amplifier and makes it act as a low-gain, linear amplifier with much lower noise and distortion. The Negative feedback amplifier allowed Bell system to reduce overcrowding of lines and extend its long-distance network by means of carrier telephony. It enabled the design of accurate fire-control systems in World War II, and it formed the basis of early operational amplifiers, as well as precise, variable-frequency audio oscillators.

CLASS ACTIVITY

An amplifier is an electronic device that increases the power of a signal, making it stronger without altering its original content. It is commonly used in audio equipment to boost sound signals, allowing for clearer and louder audio output. Amplifiers function by taking a weak input signal and using power from a power supply to enhance it, delivering a more powerful output signal. They are essential in various applications, including musical instruments, home entertainment systems, and broadcasting. The effectiveness of an amplifier is measured by its gain, which is the ratio of output signal strength to input signal strength.

SUMMARY

- An operational amplifier (often op-amp or opamp) is a DC-coupled high-gain electronic voltage amplifier with a differential input and, usually, a single-ended output.
- Op-amps are used widely in electronic devices today, including a vast array of consumer, industrial, and scientific devices.
- The amplifier's differential inputs consist of a non-inverting input (+) with voltage V_+ and an inverting input (–) with voltage V_-; ideally the op-amp amplifies only the difference in voltage between the two, which is called the differential input voltage.
- The magnitude of AOL is typically very large (100,000 or more for integrated circuit op-amps), and therefore even a quite small difference between V+ and V– drives the amplifier into clipping or saturation.
- Open-loop gain is finite in real operational amplifiers. Typical devices exhibit open-loop DC gain exceeding 100,000.
- A perfect operational amplifier amplifies only the voltage difference between its two inputs, completely rejecting all voltages that are common to both.
- The input stage consists of a cascaded differential amplifier (outlined in dark blue) followed by a current-mirror active load. This constitutes a transconductance amplifier, turning a differential voltage signal at the bases of Q1, Q2 into a current signal into the base of Q15.
- A differential voltage V_{in} at the op-amp inputs (pins 3 and 2, respectively) gives rise to a small differential current in the bases of Q1 and Q2 $i_{in} \approx V_{in}/(2h_{ie}h_{fe})$.
- While the 741 was historically used in audio and other sensitive equipment, such use is now rare because of the improved noise performance of more modern op-amps.
- A non-ideal operational amplifier's equivalent circuit has a finite input impedance, a non-zero output impedance, and a finite gain.
- Practical operational amplifiers draw a small current from each of their inputs due to bias requirements (in the case of bipolar junction transistor-based inputs) or leakage (in the case of MOSFET-based inputs).

REVIEW QUESTIONS

1. What are the key characteristics of an ideal operational amplifier? Discuss how these ideal characteristics influence op-amp circuit design.
2. Explain the concept of negative feedback in op-amp circuits. How does negative feedback contribute to the performance and stability of op-amp circuits? Provide an example.
3. Describe the different configurations of op-amp circuits (inverting amplifier, non-inverting amplifier, summing amplifier, difference amplifier, integrator, differentiator). Provide schematic diagrams and explain their operational principles.

4. Discuss the significance of input impedance and output impedance in op-amp circuits. How are these parameters defined, and how do they affect circuit performance?
5. Explain the concept of slew rate in op-amps. What factors affect the slew rate, and how does it impact the performance of the op-amp in practical applications?

REFERENCES

1. Allen, P. E., & Holberg, D. R. (2002). *CMOS Analog Circuit Design* (2nd ed.). Oxford University Press.
2. Alzaher, H., & Ismail, M. (2001). A CMOS fully balanced differential difference amplifier and its applications. *IEEE Transactions on Circuits and Systems II: Analog and Digital Signal Processing, 48*(6), 614–620. https://doi.org/10.1109/82.962408.
3. Amaudov, D., & Dudin, E. O. (2020). Distance learning and remote laboratory exercises on power supply systems. In *2020 XI National Conference with International Participation (ELECTRONICA)* (pp. 1–4). Sofia. https://doi.org/10.1109/ELECTRONICA50363.2020.9251161.
4. Chan, P. K., Ng, K. A., & Zhang, X. L. (2004). A CMOS chopper-stabilized differential difference amplifier for biomedical integrated circuits. In *Midwest Symposium on Circuits and Systems* (Vol. 3, pp. III33–III36). https://doi.org/10.1109/MWSCAS.2004.1356684.
5. Comer, J., & Comer, D. T. (2004). Using the weak inversion region to optimize input stage design of CMOS op amps. *Circuits and Systems II: Express Briefs, 51*(1), 8–14. https://doi.org/10.1109/TCSII.2003.822328.
6. Degen, T., & Jäckel, H. (2006). A pseudodifferential amplifier for bioelectric events with DC-offset compensation using two-wired amplifying electrodes. *IEEE Transactions on Biomedical Engineering, 53*(2), 300–310. https://doi.org/10.1109/TBME.2005.862577.
7. Duque-Carrillo, J. F., Ausín, J. L., Torelli, G., Valverde, J. M., & Domínguez, M. A. (2000). 1-V rail-to-rail operational amplifiers in standard CMOS technology. *IEEE Journal of Solid-State Circuits, 35*(1), 33–44. https://doi.org/10.1109/4.818918.
8. Fu, Q., Xiao, Y., Tan, K., Liu, X., & Shan, Q. (2010). Analysis and design of instrumentation amplifier based on chopper technology. In *2010 Academic Symposium on Optoelectronics and Microelectronics Technology and 10th Chinese-Russian Symposium on Laser Physics and Laser Technology, RCSLPLT/ASOT 2010* (pp. 318–321). https://doi.org/10.1109/RCSLPLT.2010.5740573.
9. Lindberg, E. (2004). Is the quadrature oscillator a multivibrator? *IEEE Circuits & Devices Magazine, 20*(6), 23–28. https://doi.org/10.1109/MCD.2004.1374222.
10. Raydovska, V. (2020). Building a laboratory for simulation and experimental research in electrical engineering. In *ICERI2020 Conference* (pp. 2687–2689). ISBN: 978-84-09-24232-0.
11. Taur, Y. (2002). CMOS design near the limit of scaling. *IBM Journal of Research and Development, 46*(2/3), 213–222. https://doi.org/10.1147/rd.462.0213.

12. Tokmakov, S., & Asenov, S. (2020). Operational amplifier open loop gain simulation in electronics engineering education. In *2020 XI National Conference with International Participation (ELECTRONICA)*. Sofia, Bulgaria: Electronic.

13. Zhang, L., Zhaohui, L., & He, L. (2007). System design of a low noise, low offset instrumentation amplifier with chopper stabilization. In *ASIC, 2007. ASICON '07. 7th International Conference* (pp. 627–630). https://doi.org/10.1109/ASICON.2007.4420717.

CHAPTER

9 Oscillators

LEARNING OBJECTIVES

After studying this chapter, you will be able to:

- Understand the oscillator
- Define Armstrong oscillator and tuned collector oscillator
- Explain the Hartley oscillator
- Discuss crystal oscillator

KEY TERMS FROM THIS CHAPTER

Carrier wave
Frequency stability
Mechanical resonance
Phase-locked loop (PLL)
Positive feedback
Signal generation
Square wave
Crystal oscillator
Harmonic oscillator
Modulation
Piezoelectric material
Relaxation oscillator
Sine wave
Timekeeping

INTRODUCTION

Oscillators are fundamental components in both analog and digital electronics, playing a critical role in a variety of applications ranging from timekeeping to signal generation. Essentially, an oscillator is a circuit that generates a continuous, oscillating electrical signal, typically in the form of a sine wave, square wave, or other periodic waveform. This ability to produce a stable and repetitive signal makes oscillators indispensable in numerous electronic devices, including clocks, radios, computers, and communication systems.

At the heart of an oscillator is the principle of positive feedback. This involves feeding a portion of the output signal back to the input in such a way that it reinforces the oscillations. The circuit is designed such that it can sustain oscillations once they are initiated, without the need for any external input signal. Common types of oscillators include the harmonic oscillator, which produces sinusoidal waves, and the relaxation oscillator, which generates non-sinusoidal waves like square or triangular waves. Each type has its specific applications and characteristics, making them suitable for different uses.

One of the most widespread applications of oscillators is in clock generation for digital systems. In microprocessors and other digital circuits, oscillators provide clock signals that synchronize operations and ensure that data is processed at the correct times. Crystal oscillators, which use the mechanical resonance of a vibrating crystal of piezoelectric material to create an electrical signal with a very precise frequency, are particularly valued for their high stability and accuracy. They are found in almost every digital device that requires precise timing, from wristwatches to global positioning systems (GPS).

In communication systems, oscillators are equally crucial. They generate carrier waves that are modulated with information signals for transmission. The stability and purity of these oscillators directly affect the quality of the transmitted signal and the efficiency of the communication system. Advances in oscillator design, including phase-locked loops (PLLs) and voltage-controlled oscillators (VCOs), have significantly enhanced the performance and reliability of modern communication equipment.

Overall, oscillators are key building blocks in modern electronics, providing the necessary signals for timing, control, and communication. Their ability to produce stable and reliable waveforms underpins the operation of countless devices and systems, highlighting their importance in both everyday technology and advanced scientific applications.

9.1. WHAT IS OSCILLATOR?

A circuit that generates alternating current (AC) at a desired frequency is known as an oscillator. Fundamentally, an oscillator can create a recurring AC signal by converting direct current (DC) from a power source. By supplying positive feedback to offset circuit losses, it maintains oscillations. The initial oscillations caused by noise or signals are amplified and continuously fed back when power is applied, producing output waves that are sustained (Snelgrove, 2011).

Oscillations can occur at very low audio frequencies or very high radio waves, depending on the components used. Oscillators are widely used in all electronic devices that need signals at frequencies or clock pulses.

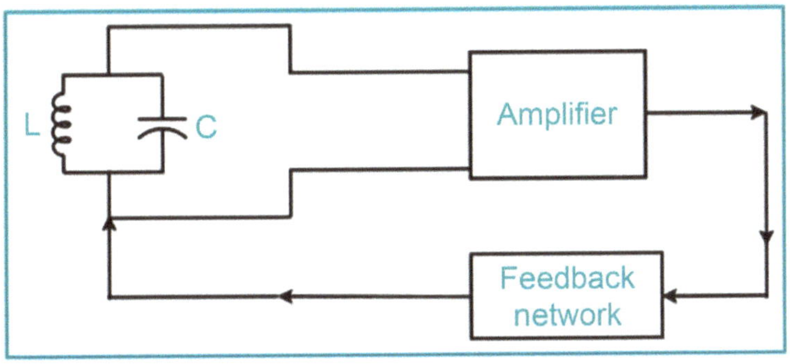

Figure 9.1. Practical oscillator circuit.

Source: https://testbook.com/electrical-engineering/oscillators.

9.1.1. Positive Feedback Oscillators

The oscillator consists of an amplifier and a Frequency Selective Network. The amplifier output is applied to the frequency selective network, the output of which is applied to the amplifier input. For the analysis a switch can be imagined opening the closed loop signal path. If V_2 is exactly the same as V_1, both in amplitude and phase then steady state oscillations result. If the magnitude of $V_2 > V_1$ and the phase is the same, then the oscillations will grow. If $V_2 < V_1$, then the oscillations will decay (Figure 9.1).

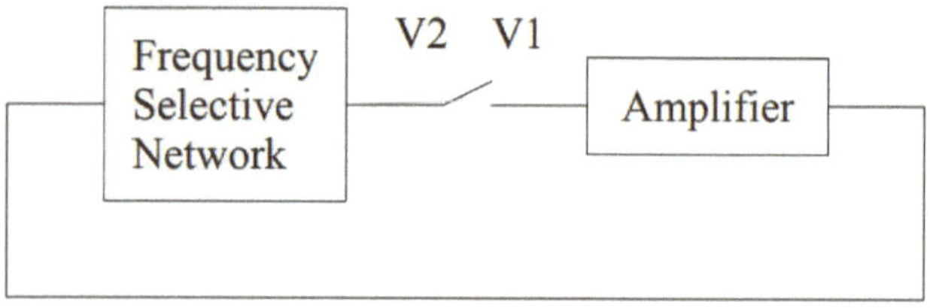

Figure 9.2. Feedback oscillator block diagram.

Source: http://mwl.diet.uniroma1.it/people/pisa/SISTEMI_RF/MATERIALE%20INTEGRATIVO/Kikkert_RF_Electronics_Course/08-RF_Electronics_Kikkert_Ch6_Oscillators.pdf

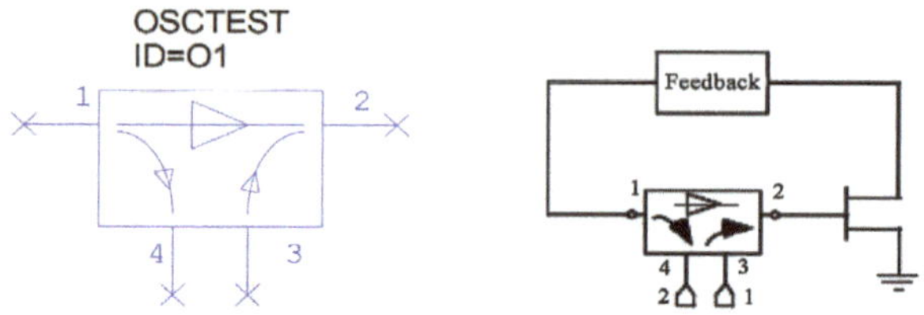

Figure 9.3. The use of OSCTEST for linear analysis of oscillators.

Source: http://mwl.diet.uniroma1.it/people/pisa/SISTEMI_RF/MATERIALE%20INTEGRATIVO/Kikkert_RF_Electronics_Course/08-RF_Electronics_Kikkert_Ch6_Oscillators.pdf

9.1.2. *Oscillator Design Process*

1. At the necessary operating frequency, design the feedback network to exhibit the proper frequency-selective behavior. An oscillator with a high Q network will have less phase noise because of its quick phase shift with frequency.
2. Select or design the amplifier to have the required gain at the operating frequency. The input and output impedances should also match the expected loads to be seen by the frequency selective network.
3. To provide the necessary gain and phase shift at the operating frequency, simulate the amplifier and make sure the biasing is correct (Chattopadhyay, 2006).
4. Connect the frequency selective network and amplifier and do a linear analysis to ensure that the required gain (>1) and phase (V2 in-phase with V1) conditions are satisfied. In MWO the OSCTEST element is specially designed to perform this function and is used in place of the switch shown in Figure 9.3.
5. Replace the OSCTEST probe with the Oscaprobe to alter the linear circuit after the circuit has been adjusted to have the proper gain and phase for oscillation. As indicated in Figure 9.3, the Oscaprobe is connected to ground at the location of the switch.

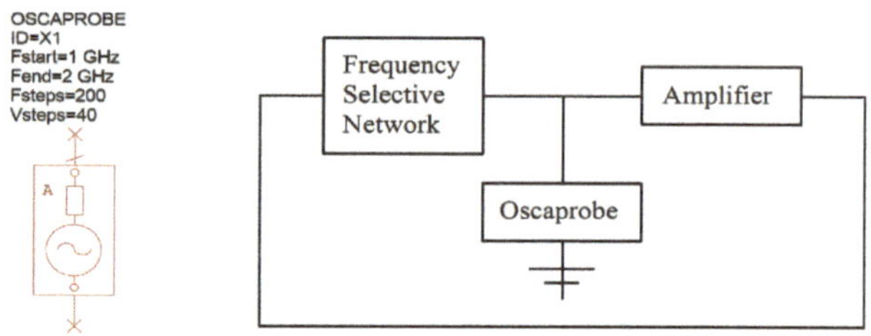

Figure 9.4. The use of Oscaprobe for nonlinear analysis of oscillators.

Source: http://mwl.diet.uniroma1.it/people/pisa/SISTEMI_RF/MATERIALE%20INTEGRATIVO/Kikkert_RF_Electronics_Course/08-RF_Electronics_Kikkert_Ch6_Oscillators.pdf

The Oscaprobe is a network in which the voltage amplitude and phase are changed until no current passes through the probe, at which point the proper oscillating amplitude is found (Figure 9.4). The network is a short circuit at the designated oscillation frequency and an open circuit at all other frequencies. Next, a non-linear (spice) simulation of the oscillator circuit is performed using this amplitude. This then makes it possible to determine the oscillator's phase noise, harmonics, and output waveforms. The design examples that adhere to this procedure are the Colpitts oscillator design example found in these lecture notes and the MWO oscillator example found in the MWO directory: Examples/Circuit Design Types/Oscillators/BJT Oscillator.

9.1.3. Oscillator Design Procedure

Step 1: Select a Transistor, FET or Amplifier

The oscillator's active device is crucial, with key factors to consider including output power, oscillating frequency, supply voltage, and cost. The phase noise at low frequencies is influenced by the flicker noise of the amplifier, making the noise figure of the amplifier important for low-noise oscillators.

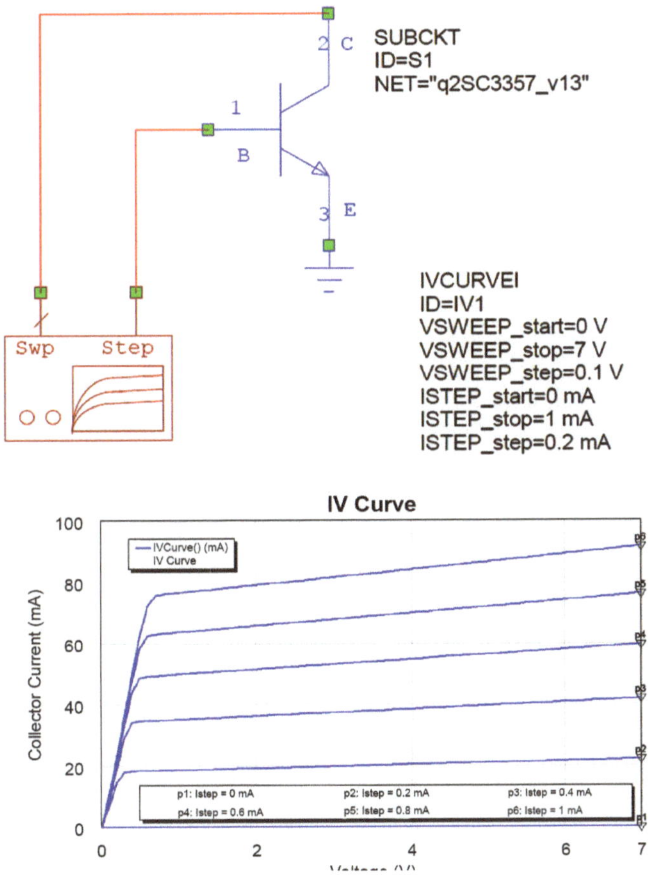

Figure 9.5. Circuit for determining the I-V characteristic of a transistor.

Source: http://mwl.diet.uniroma1.it/people/pisa/SISTEMI_RF/MATERIALE%20INTEGRATIVO/Kikkert_RF_Electronics_Course/08-RF_Electronics_Kikkert_Ch6_Oscillators.pdf

The transistor's biasing is the most crucial design parameter. It was demonstrated in EE2300 and EE3300 that the IV curve is highly helpful in figuring out the appropriate biasing conditions. It should be understood that there can be a 5:1 ratio variation in some of the critical transistor parameters, such as ⊠. Typical values are among the transistor parameters used in the computer simulation software's models. The oscillator must function under all permissible transistor parameter variations, according to the design. It is simple to obtain the I-V curve in MWO, as Figure 9.5 illustrates.

For linear biasing, a 7V supply and quiescent collector current of between 40 and 5 mA is required. 15 mA is a good compromise to minimize the power dissipation in the transistor and still have sufficient output power.

Step 2: Design an Amplifier

A typical amplifier configuration and one that is used for this design is shown in Figure 9.3. The first step is to select the bias voltages. For a 7 Volt supply, and having an emitter follower configuration for the amplifier, the emitter voltage should be close to half of the supply voltage, minus one volt.

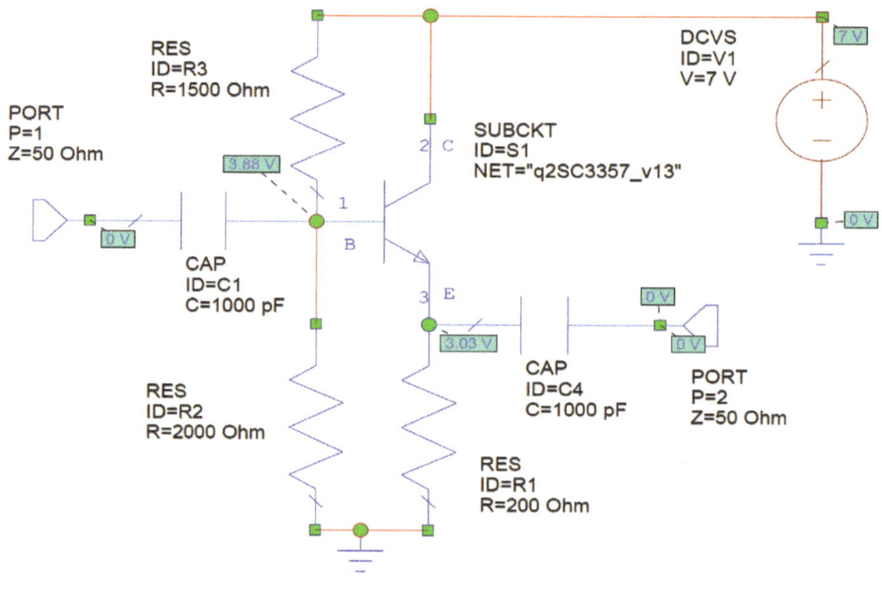

Figure 9.6. Amplifier for the Colpitts oscillator.

Source: http://mwl.diet.uniroma1.it/people/pisa/SISTEMI_RF/MATERIALE%20INTEGRATIVO/Kikkert_RF_Electronics_Course/08-RF_Electronics_Kikkert_Ch6_Oscillators.pdf

This results in a large swing without the transistor saturating. For a quiescent current of 15 mA, the current through the biasing chain should be of the order of 1.5 mA to 3 mA, for good quiescent current stability with temperature. For an emitter voltage of 3 Volt, a base voltage of 3.5. Volt is desired, so that a biasing resistor, between 2.4 kΩ and 1.2 kΩ is required for R3, when the values are rounded to the nearest available resistor values. Since a relatively high input impedance is required, a 2 kΩ resistor is used. A 1.5 kΩ resistor is required to then obtain the 3.5. V at the base of the transistor using a 7 V supply. The annotation setting allows the circuit voltages to be indicated on the circuit diagram in Microwave Office and allows the biasing and voltages to be checked.

Keep in mind that the power gain of 4dB accounts for the various input and output impedances. The Collpits oscillator's resonance will be affected if the large reactive component is not considered during the design phase (Figure 9.6). The input impedance is 638 (and the output impedance is 17 Ω).

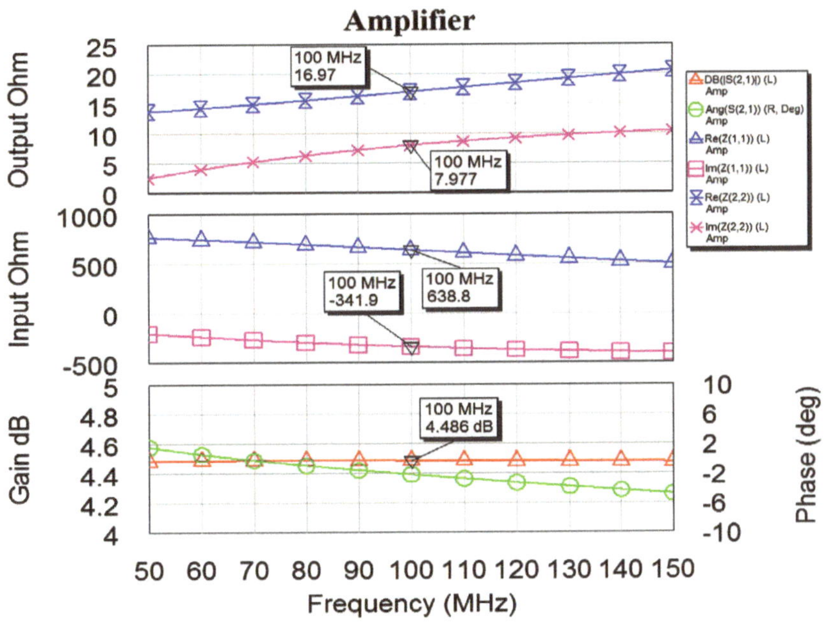

Figure 9.7. Amplifier performance.

Source: http://mwl.diet.uniroma1.it/people/pisa/SISTEMI_RF/MATERIALE%20INTEGRATIVO/Kikkert_RF_Electronics_Course/08-RF_Electronics_Kikkert_Ch6_Oscillators.pdf

Step 3: Design a Resonator

To guarantee that the conditions for oscillation only hold true at a single frequency, the resonator needs to be a frequency-selective network. The network must also have a high Q and be matched to the amplifier's input and output impedances (Figure 9.7). The components can be adjusted to guarantee a zero phase shift at the needed operating frequency, or a 180 degree phase shift if necessary. Since the network in Figure 9.8 matches the one actually used in the oscillator and permits phase changes brought about by the amplifier, the imaginary part of the transfer function is not precisely zero.

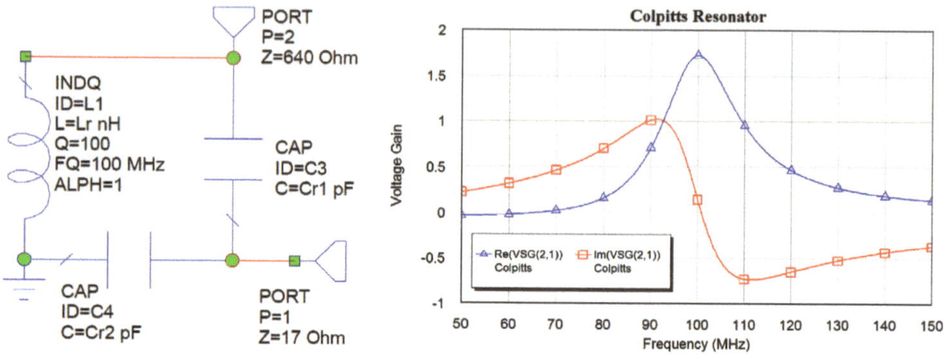

Figure 9.8. Colpitts resonator network.

Source: http://mwl.diet.uniroma1.it/people/pisa/SISTEMI_RF/MATERIALE%20INTEGRATIVO/Kikkert_RF_Electronics_Course/08-RF_Electronics_Kikkert_Ch6_Oscillators.pdf

Step 4: Linear Oscillator Analysis

Now, the amplifier and resonator are integrated into a single circuit. The phase shift and loop gain are found using the OSCTEST element. At the necessary operating frequency, the resonator network is now adjusted to provide a phase shift of zero degrees between ports 1 and 2. To make sure that it is only marginally above one at the operating frequency, the loop gain is adjusted as necessary. Thus, that matches the ideal circumstances for oscillations.

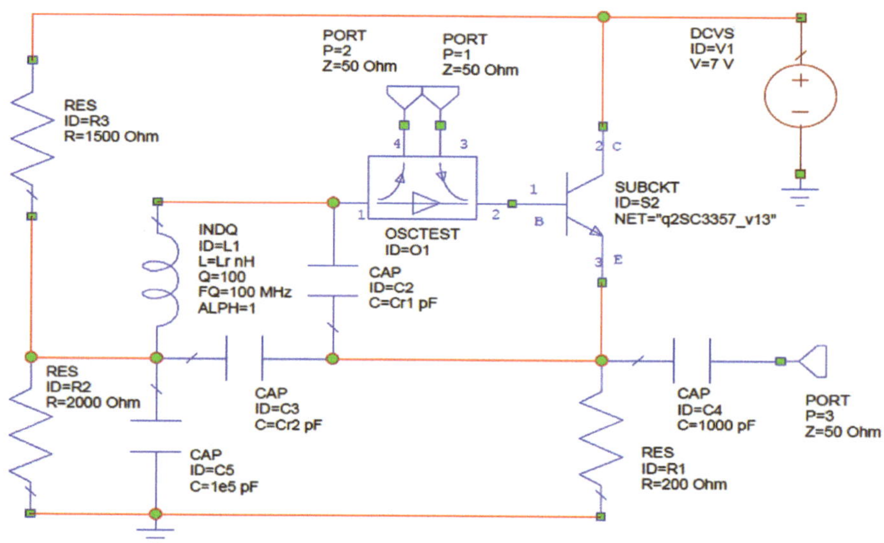

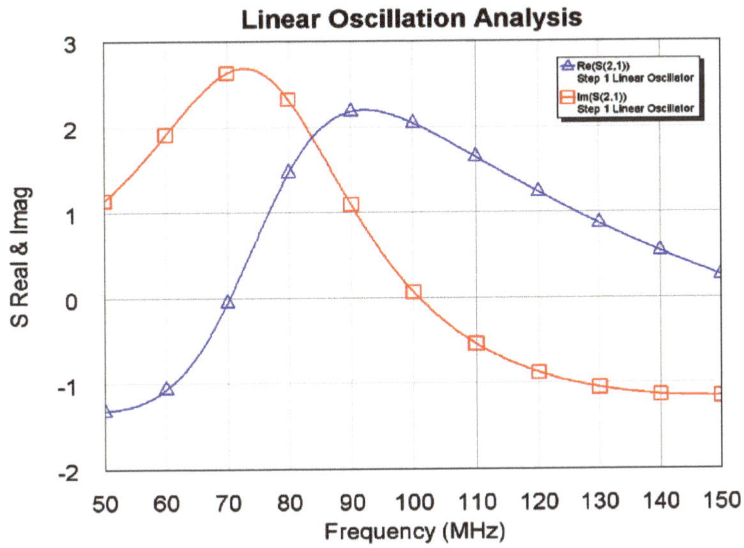

Figure 9.9. Linear analysis using OCSTEST.

Source: http://mwl.diet.uniroma1.it/people/pisa/SISTEMI_RF/MATERIALE%20INTEGRATIVO/Kikkert_RF_Electronics_Course/08-RF_Electronics_Kikkert_Ch6_Oscillators.pdf

The gain in Figure 9.9 is too large and the 17 Ω output impedance of the amplifier, causes a severe reduction of the Q of the resonator network, resulting in a high phase noise. The gain can be reduced by inserting a 47 Ω resistor in series with the output of the amplifier. That also increases the Q if the resonator and reduces the phase noise of the oscillator, since the output impedance dominates the load to the network.

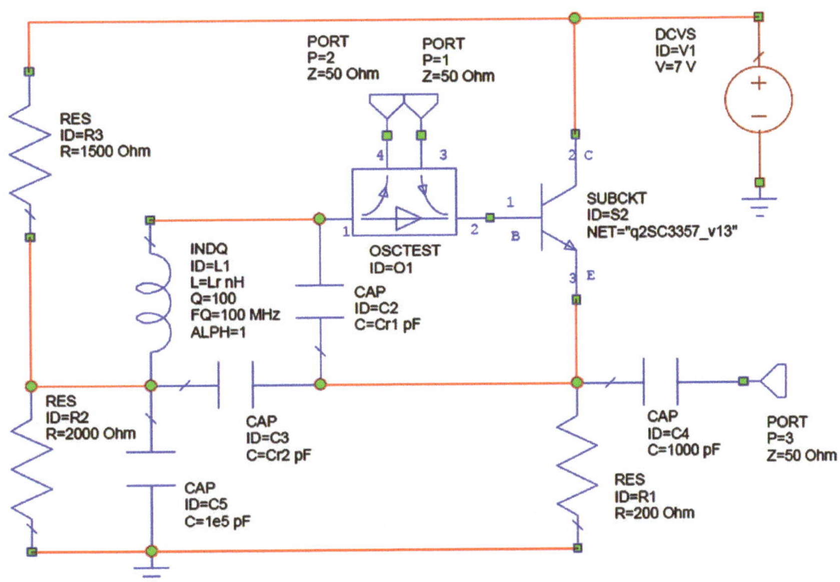

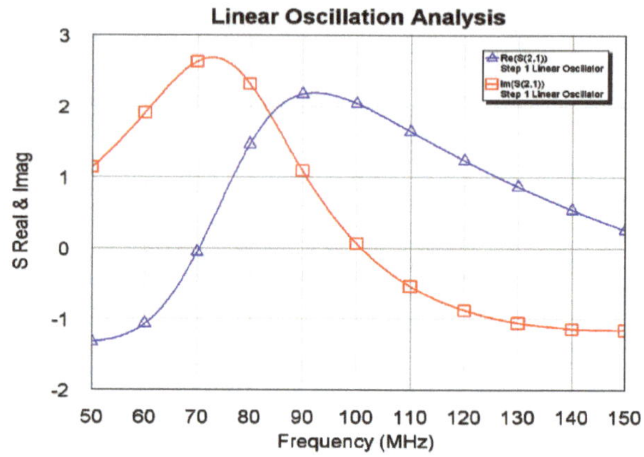

Figure 9.10. Linear analysis using OCSTEST Resistor added to increase Q and reduce gain.

Source: http://mwl.diet.uniroma1.it/people/pisa/SISTEMI_RF/MATERIALE%20INTEGRATIVO/Kikkert_RF_Electronics_Course/08-RF_Electronics_Kikkert_Ch6_Oscillators.pdf

Step 5: Nonlinear Analysis

As illustrated in Figure 9.11, replace OSCTEST for OSCPROBE to perform a harmonic balance analysis. This will make it possible to determine the oscillation frequency, output power, and output waveforms, which will include the amplitude of each harmonic. Plotting the required inductance against oscillating frequency allows one to investigate the tuning frequency range, and Figure 9.12 displays the output waveform. The output spectrum obtained with the harmonic balance.

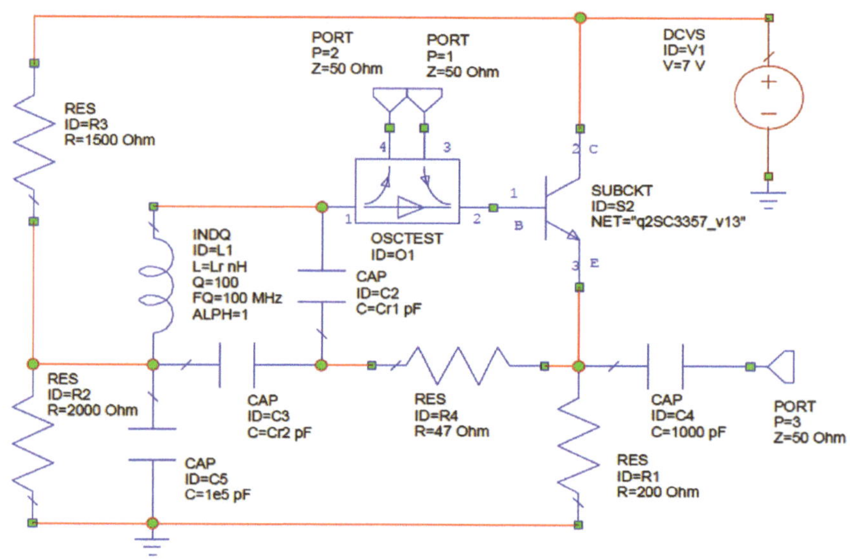

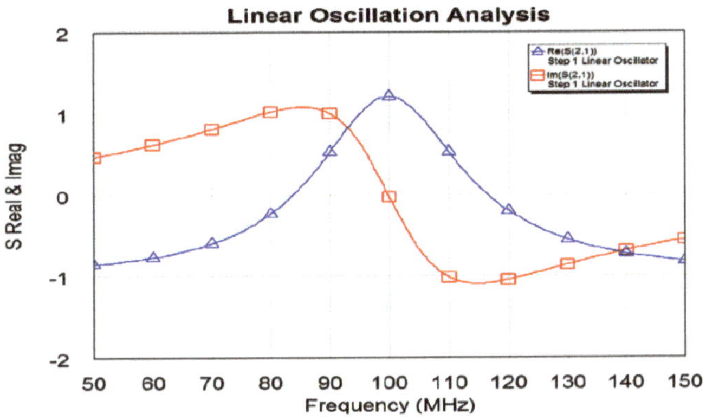

Figure 9.11. Non-linear analysis using OSCAPROBE, circuit and output spectrum.

Source: http://mwl.diet.uniroma1.it/people/pisa/SISTEMI_RF/MATERIALE%20INTEGRATIVO/Kikkert_RF_Electronics_Course/08-RF_Electronics_Kikkert_Ch6_Oscillators.pdf

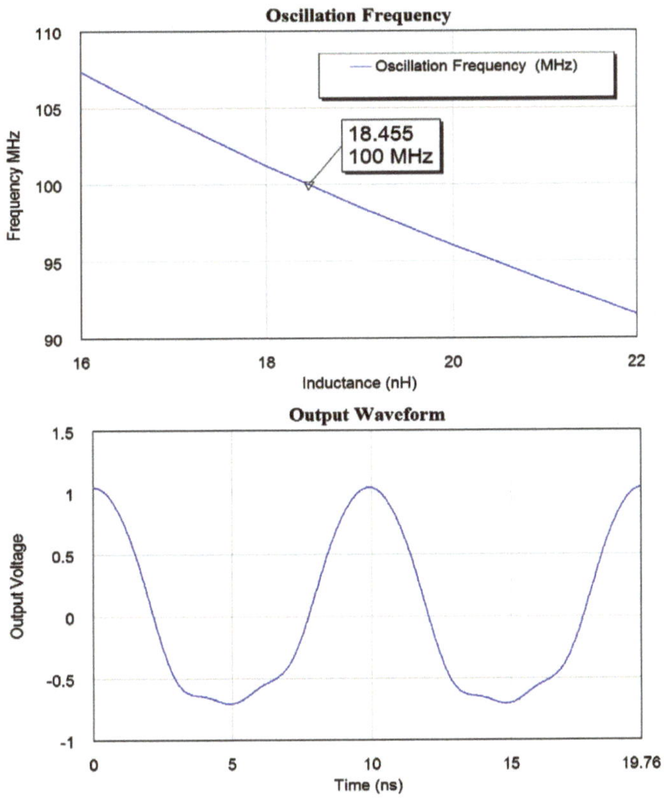

Figure 9.12. Non-linear analysis using OSCAPROBE, Tuning range and output waveform.

Source: http://mwl.diet.uniroma1.it/people/pisa/SISTEMI_RF/MATERIALE%20INTEGRATIVO/Kikkert_RF_Electronics_Course/08-RF_Electronics_Kikkert_Ch6_Oscillators.pdf

The SWPVAR simulation control (Element ⇒ Simulation Control ⇒ SWPVAR) shown in the schematic of Figure 9.11 allows the plotting of the sweep of inductor values and their corresponding frequency. Note that the output waveform is smaller than expected, since the amplifier should have a linear output swing of at least ±2 Volt. The waveform can then be used to investigate the suitability of the biasing (Carr, 2002).

MicroWave Office makes it possible to assess an oscillator's performance before any hardware is manufactured. This significantly lowers the oscillator design cost. The production of the non-linear analysis is contingent upon the availability of a complete SPICE model for the active device. The necessary models for their transistors are offered by numerous transistor manufacturers. Unfortunately, not all MMICs can use the Spice models.

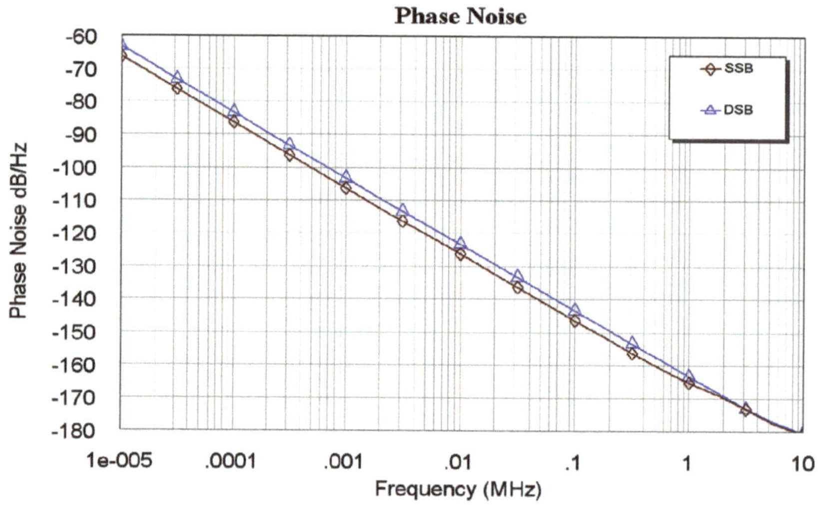

Figure 9.13. Phase noise of the oscillator using non-linear (harmonic balance) analysis.

Source: http://mwl.diet.uniroma1.it/people/pisa/SISTEMI_RF/MATERIALE%20INTEGRATIVO/Kikkert_RF_Electronics_Course/08-RF_Electronics_Kikkert_Ch6_Oscillators.pdf

9.2. ARMSTRONG OSCILLATOR

There are various varieties of resonant LC feedback oscillators (LC electronic oscillators), including an Armstrong oscillator, a Colpitts, Clapp, Hartley, and Crystal-controlled oscillators. An LC feedback oscillator with an inductor and capacitor in its feedback network commonly referred to as an Armstrong oscillator, or Meissner oscillator. One can construct the Armstrong oscillator circuit using a transistor, operational amplifier, tube, or other active (amplifying) components. Generally, the oscillators consist of three basic parts:

- An amplifier: This will typically be a class A, B, or C biased voltage amplifier.
- A wave shaping network: Passive components like filter circuits are used to shape the waveform and control the frequency of the output signal.
- A POSITIVE feedback path: In order to regenerate and re-amplify the feedback signal, a portion of the output signal is fed back into the amplifier input. Without requiring an external input signal, this signal was fed back once more to maintain a steady output signal.

The below given two conditions for the oscillation. Every oscillator must satisfy these conditions in order to make proper oscillations.

- The oscillations should be taken place at one particular frequency. The oscillation frequency F is determined by the tank circuit (L and C) and is approximately given by

$$F = \frac{1}{2\pi\sqrt{LC}}$$

- The amplitude of the oscillations should be constant.

9.2.1. Armstrong Oscillator Circuit and Its Working

The output of the Armstrong oscillator is a sinusoidal wave with a consistent frequency and amplitude within the specified radio frequency range. It is typically employed in receivers as a local oscillator, but it can also be used as a medium- and high-frequency radio frequency oscillator, as well as a source in signal generators.

The identifying characteristics of the Armstrong oscillator

- It uses an LC tuned circuit to establish the frequency of oscillation.

- Feedback is accomplished by mutual inductive coupling between the tickler coil and the LC tuned circuit.
- Its frequency is stable, and the output amplitude is relatively constant.

The diagram displayed above illustrates a common Armstrong circuit utilizing an NPN BJT transistor. The inductor L2, also known as the Trickler Coil, provides feedback (regeneration) to the BJT input by coupling with L1 separately. Certain signals within the output circuit are inductively linked to the input circuit through L2. The transistor's base circuit features a parallel tuned tank circuit composed of L1 and C1. This tank circuit is responsible for determining the oscillator circuit's oscillation frequency.

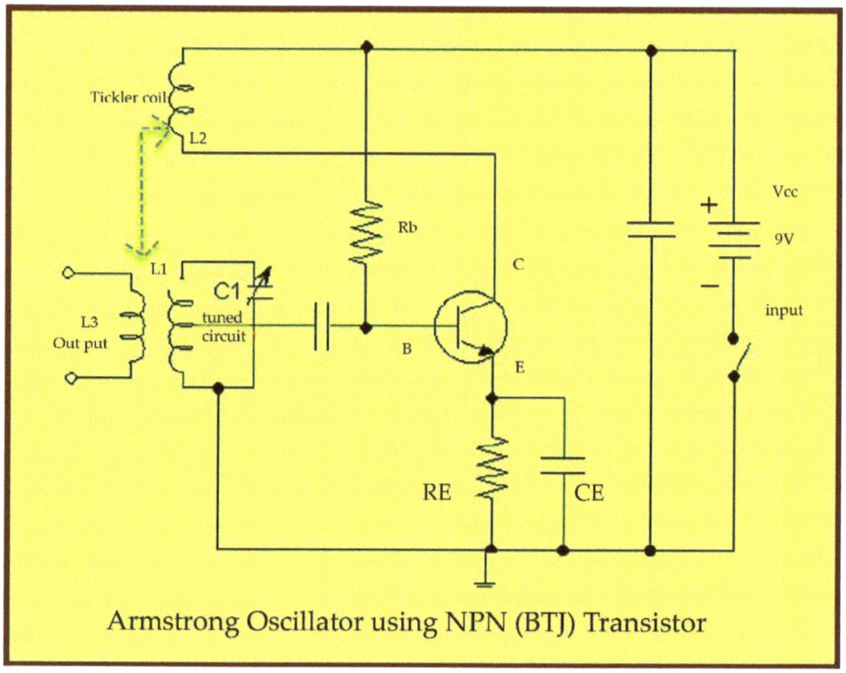

Figure 9.14. Armstrong oscillator circuit and its working.

Source: https://www.elprocus.com/armstrong-oscillator-circuit-working-application/

Here C1 is a variable capacitor to change the frequency of oscillation. The resistor Rb provides foe=r the correct amount of bias current. DC bias current flows from ground to emitter via Re, out of the base, via Rb and then back to the positive. The value of Rb and Re determines the amount of bias current (generally Rb> Re). The resistor Re provides emitter stabilization to prevent thermal runaway and capacitor CE is the emitter bypass capacitor.

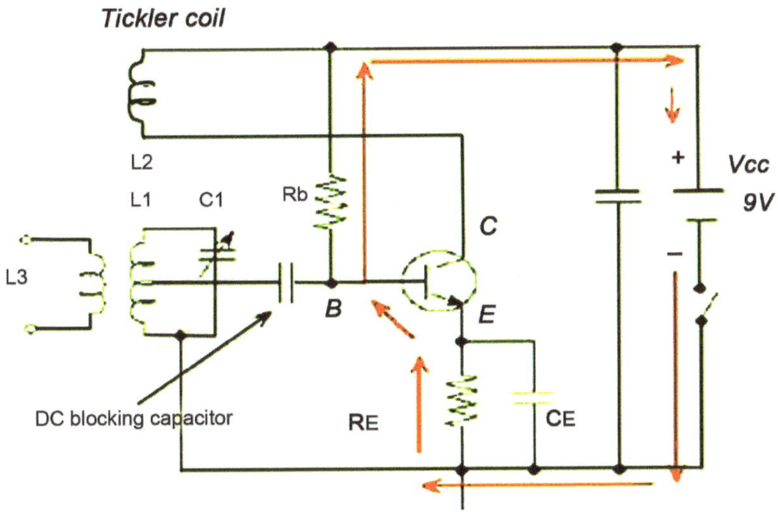

(A)

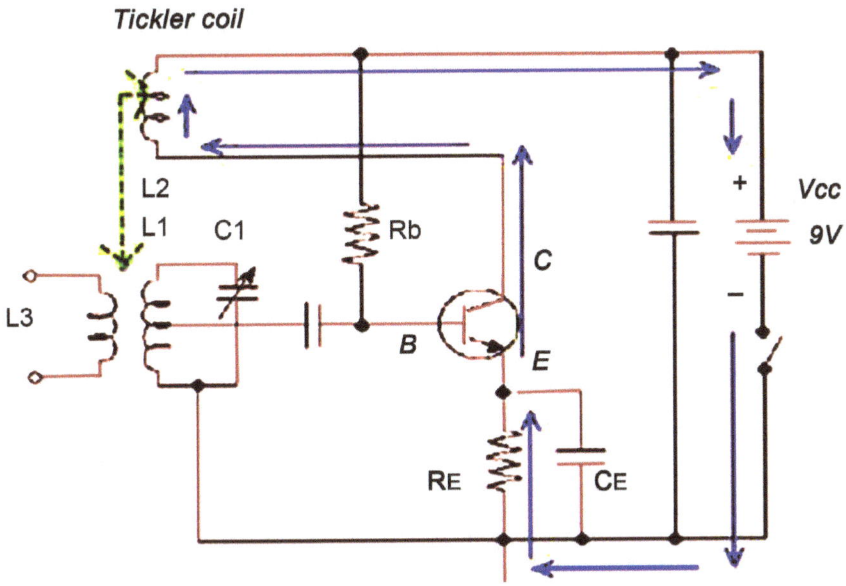

(B)

Figure 9.15. A-DC bias is determined primarily by Rb, B-DC collector emitter circuit current.

Source: https://www.elprocus.com/armstrong-oscillator-circuit-working-application/

The resistor Rb's value, as shown in circuit Figure 9.15(a), determines the amount of DC biased current. A DC blocking capacitor is the capacitor, C, connected in series with the base (B). This will stop the DC bias current from entering L1, but it will still permit the signal from L1–C1 to reach the Base. Figure 9.15(b) displays the emitter-collector current in DC output.

In this case, the transistor's emitter-base circuit is forward biased. It will then be subjected to the emitter-collector current. Thus, the signal current occurs when the circuit is oscillating, according to the circuits in Figures 9.15 (a and b). Thus, we would only have the previously mentioned DC currents if the oscillations were stopped, which would entail opening the tickler coil.

Figure 9.15(b) above illustrates the DC output emitter-collector current. In this instance, the transistor's emitter-base circuit is forward biased. It will then be subjected to the emitter-collector current. Thus, the signal current happens when the circuit is oscillating, according to the circuits in Figures 9.15(a and b) above. Thus, we would only have the previously mentioned DC currents if the oscillations were stopped, which would entail opening the tickler coil.

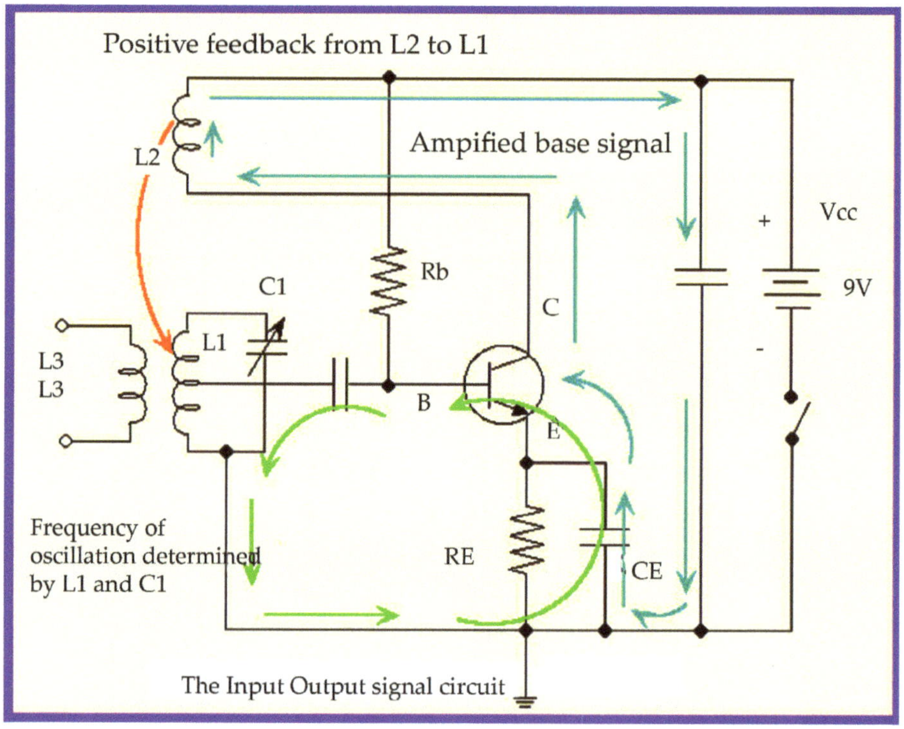

Figure 9.16. The input and output signal circuit.

Source: https://www.elprocus.com/armstrong-oscillator-circuit-working-application/

The signals would go in this oscillator is depicted in the schematic above. Assume that a sine wave at 1MHz is what the oscillator is intended to produce. This will vary the DC, not the AC, using a sine wave. Since many current devices are incompatible with AC power. L1 and C1 begin oscillating at 1MHz when the Armstrong oscillator is turned on. Normally, losses in the tank circuit would cause this oscillation to decrease (L1 andC1). Overlying the DC bias current in the base circuit is the oscillating voltage across L1 and C1. Thus, as can be shown in the green line above, a 1MHz signal current flow in the base circuit.

Here, the capacitive resistance of CE at 1MHz would be one-tenth the value of RE, indicating that the current flowing through the resistor Re is negligible. As a result, the base circuit's 1MHz signal now generates an aqua blue 1MHz signal in the collector circuit. The signal surrounding the supply is avoided by the capacitor across the battery. The tickler coil is where the amplified signal travels. L1 and L3 are both inductively coupled to the tickler coil (L2). Thus, we can use L3's amplified output signal.

9.2.1.1. Advantages and Disadvantages

- The main advantage is that, the construction of Armstrong-type tube oscillators using a tuning capacitor where one side is earthed. It produces a stable frequency and stably amplified output waveform.
- The main disadvantage of this circuit is that the resulting electromagnetic vibrations may contain interfering harmonics very light, which are undesirable in most cases.

9.2.1.2. Applications of Armstrong Oscillator

- It is used to generate the sinusoidal output signals with a very high frequency.
- It is generally used as a local oscillator in receivers.
- It is used in the radio and mobile communications.
- Used as a source in signal generators and as a radio-frequency oscillator in the medium- and high-frequency range.

9.3. TUNED COLLECTOR OSCILLATOR

One type of transistor LC oscillator is the tuned collector oscillator, in which the transistor's collector terminal is connected to a transformer and capacitor in the tank circuit. The most basic and straightforward type of LC oscillator is the tuned collector oscillator circuit. The oscillator frequency is determined by the tank circuit, which is connected to the collector circuit and operates at resonance much like a basic resistive load. This circuit can be used in frequency demodulators, mixers, signal generators, RF oscillator circuits, and other general applications. The following section describes the circuit diagram and operation of a tuned collector oscillator.

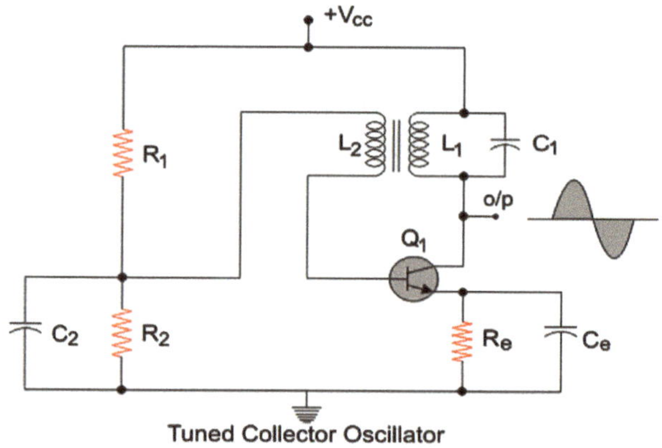

Figure 9.17. Tuned collector oscillator circuit.

Source: https://www.tutorialspoint.com/sinusoidal_oscillators/images/base_oscillator.jpg

9.3.1. Tuned Collector Oscillator Circuit Working

Let's acknowledge the fact that a transistor amplifies an input voltage and induces a 180-degree phase shift before delving into the oscillator's operation. The oscillations originate from the two elements that make up the tank circuit, L_1 and C_1. Positive feedback is provided by the transformer, and the output is amplified by the transistor. Now that this is established, let's move on to comprehend how the circuit functions.

Upon turning on the power supply, the capacitor C_1 begins to charge. It begins to discharge through the inductor L_1 once it is fully charged. Electrostatic energy stored in the capacitor is transformed into electromagnetic energy and stored in inductor L_1. The inductor begins recharging the capacitor after it has fully discharged. This is because inductors do not experience rapid changes in current; as a result, they will alter the polarity across themselves to maintain the current's direction of flow. The cycle keeps going in this way as the capacitor begins to charge once more. An oscillating signal is produced as a result of periodic changes in the polarity across the inductor and capacitor.

Through electromagnetic induction, the Coil L_2 receives a charge, which it then transfers to the transistor. The signal is amplified by the transistors and used as the output. Positive feedback occurs when a portion of the output is returned to the system.

Positive feedback is when the input and the feedback are in sync. Two phase shifts of 180 degrees are introduced: one by the transistor and the other by the transformer. Thus, a 360-degree phase shift is obtained, which is then fed back into the tank circuit. Sustained oscillations require positive feedback. The oscillation frequency is determined by calculating the values of the inductor and capacitor in the tank circuit.

$$F = \frac{1}{[2\pi\sqrt{(L_1C_1)}]}$$

where,

F = Frequency of the oscillation.

L_1 = value of the inductance of primary of the transformer L_1.

C_1 = value of capacitance of capacitor C_1.

9.4. THE HARTLEY OSCILLATOR

In the RF range (30 kHz to 30 MHz), the Hartley oscillator is a very helpful circuit for generating high-quality sine wave signals; however, at higher frequencies in this range and beyond, the Colpitts oscillator is typically recommended. Although the frequency of both oscillators is controlled by an LC tuned (tank) circuit, the Hartley design is distinguished by the use of a tapped inductor (L1 and L2 in Figure 9.18).

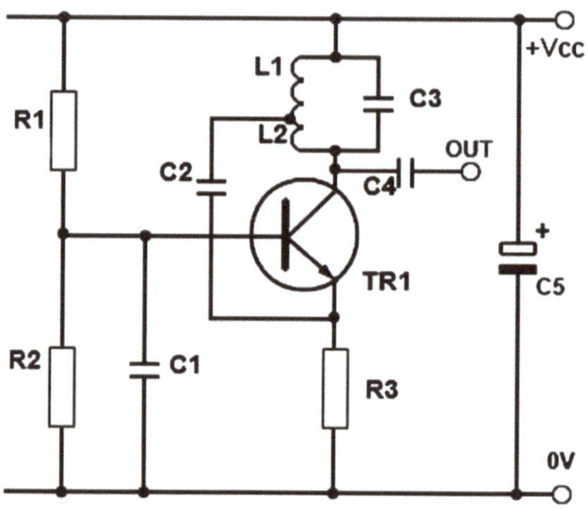

Figure 9.18. The Hartley Oscillator.

Source: https://learnabout-electronics.org/Oscillators/osc21.php

The frequency of oscillation can be calculated in the same way as any parallel resonant circuit, using:

$$F_r = \frac{1}{2\pi\sqrt{LC}}$$

where, $L = L_1 + L_2$.

This basic formula is adequate where the mutual inductance between L_1 and L_2 is negligible but needs to be modified when the mutual inductance between L_1 and L_2 is considerable.

9.4.1. Mutual Inductance in Hartley Oscillators

When a current is induced into one inductor (or one portion of a tapped inductor) by the magnetic field surrounding it, the other inductor experiences mutual inductance, which is an additional effective amount of inductance. As illustrated in Figure 9.19, when both inductors are wound around a single core, the total inductance is determined by the following formula, and the effect of mutual inductance (M) can be significant.

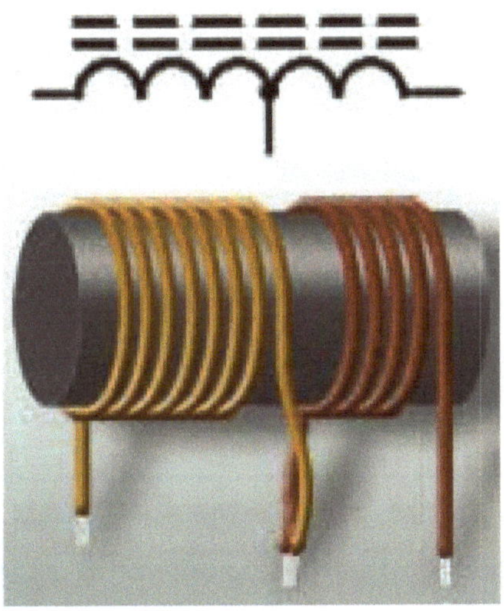

Figure 9.19. Centre tapped inductors on a common core.

Source: https://learnabout-electronics.org/Oscillators/osc21.php

$$L_{TOT} = L_1 + L_2 \pm 2M$$

The effective magnetic coupling between the two inductors determines the actual value of M. This coupling is dependent on several factors, including the distance between the inductors, the number of turns on each inductor, the dimensions of each coil, and the composition of the common core.

Using independent fixed inductors, as illustrated in Figure 9.20a, the considerations of mutual inductance are reduced to the physical distance between the inductors and the direction of their magnetic fields. The mutual inductance effect is only apparent when the inductors are positioned within a few millimeters of one another, as illustrated in Figure 9.20, because in practice the tiny values of inductance of the inductors required at RF create very little magnetic field outside the component (Razavi, Behzad, 2001).

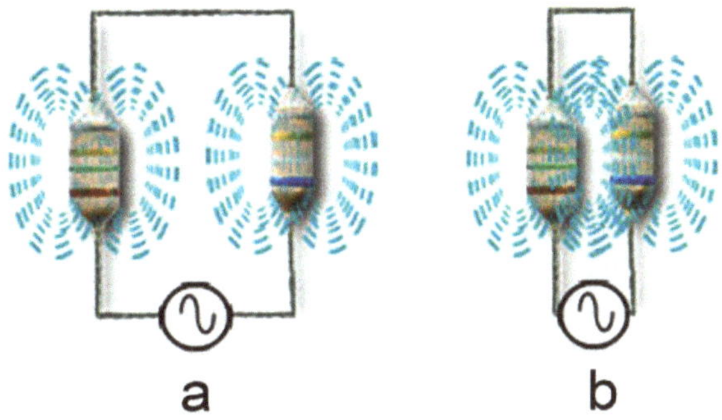

Figure 9.20. Inductors with adding fields.

Source: https://learnabout-electronics.org/Oscillators/osc21.php

It is dependent on the North-South polarity of the fields surrounding the individual coils L1 and L2 as to whether M, measured in Henrys or, more likely, microhenrys (μH), adds to or subtracts from the total inductance of very closely mounted tapped inductors. If the magnetic fields are oriented in the same direction, the mutual inductance will add to the total inductance; however, if the magnetic fields are arranged to oppose each other, as in Figure 9.21 Mutual inductance has the effect of lowering total inductance, which raises the oscillator's actual working frequency.

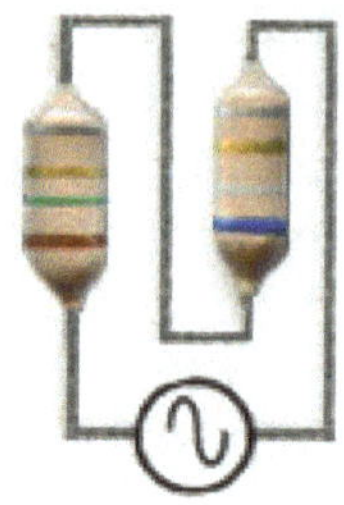

Figure 9.21. Inductors with opposing fields.

Source: https://learnabout-electronics.org/Oscillators/osc21.php

Whether the mutual inductance between tiny inductors of a few microhenrys increases or decreases the computed oscillator frequency of an RF Hartley oscillator, this will typically only alter the operating frequency by a factor comparable to that which could be brought about by the oscillator component tolerances' normal component value variation. The mutual induction effect, however, can be much larger in real-world Hartley oscillators that use inductors with a shared core. This is dependent on the coefficient of coupling (k), which ranges from 0 when there is no coupling between the two inductors to 1 when the mutual inductance is roughly equal to 100% magnetic coupling.

Because there are so many variables influencing the mutual coupling, calculating a theoretical value for k requires some fairly complex math. Ultimately, the calculation comes down to determining whether there is little mutual coupling—that is, whether less than half of the magnetic flux generated by one coil affects the other coil. If this is the case, the inductors are referred to as "loosely coupled," and it is assumed that k has a value less than 0.5. As an alternative, if there is no spacing between the inductors that share a common core, they are referred to as "tightly coupled," and k is taken to be between 0.5 and 1. In actual use, it is frequently discovered that the shared core of a tapped inductor operating at radio frequency is of a variable type, as illustrated in Figure 9.22 so that the oscillator frequency can be corrected by adjusting the core's position, thereby compensating for any frequency shift caused by mutual inductance.

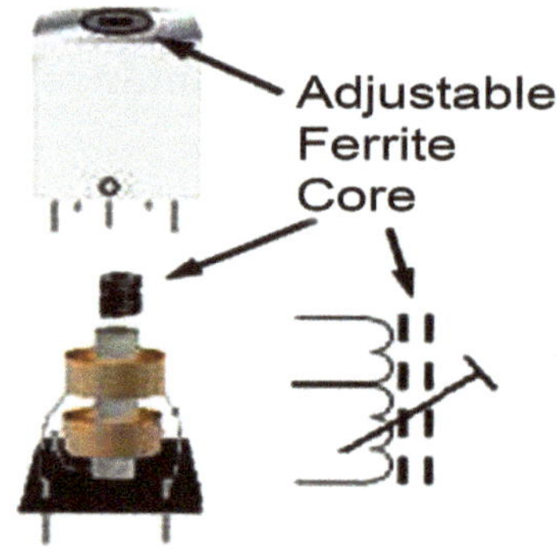

Figure 9.22. Adjustable Ferrite Core.

Source: https://learnabout-electronics.org/Oscillators/osc21.php

9.4.2. The Hartley Circuit

The load impedance of the amplifier is formed by L1/L2 and C3, which together form the frequency-determining resonant tuned circuit. As a result, the amplifier only has a high gain at the resonant frequency. In this version of the Hartley circuit, the base of TR1 is directly connected to 0V (as far as the AC signal is concerned) by C1. This is a common base amplifier. This mode has an in-phase input signal at the emitter and an output voltage waveform at the collector. This makes sure the appropriate positive feedback is provided by the portion of the output signal that is fed back to the emitter via the capacitor C2 from the tuned circuit collector load.

In order to produce an average DC voltage level proportional to the amplitude of the feedback signal at TR1's emitter, C2 additionally forms a long time constant with emitter resistor R3. This provides the required closed loop gain of 1 by automatically adjusting the amplifier's gain. Since the emitter terminal serves as the amplifier's input, the emitter resistor R3 is not decoupled. At the oscillator frequency, the base's reactance will be extremely low due to its connection to ground through C1.

9.4.3. The Tuned LC (Tank) Circuit

Because it has circulating currents significantly higher than the current supplying it, the LC circuit that regulates oscillation frequency is frequently referred to as the "tank circuit" (e.g., pulses of collector current that the amplifier produces. Its functioning is meant to be similar to that of a cistern or water tank that can provide a constant flow of water from an external supply that flows sporadically. High levels of circulating current are periodically replenished by smaller amounts of current from the amplifier in the oscillator's tank circuit.

The majority of the current in LC oscillators flows around the resonant tank circuit rather than through the oscillator's amplifier section, which results in a sine wave with very little distortion from the amplifier.

The tank circuit's ability to supply the ideal quantity of positive feedback to maintain the oscillator's operation is another feature. This is accomplished by splitting the circuit's inductive branch into two sections, each with a different value. As a result, the inductor functions similarly to an autotransformer, with the ratio of the two windings supplying the right amount of signal to be fed back into the amplifier's input.

Since in Figure 9.23, the top of L1 is connected to +Vcc; in terms of AC signals, it is connected to ground through C5's extremely low impedance. Waveform X is therefore in phase across L1, and waveform Y is in phase throughout the entire circuit. The collector and emitter signals are in phase because a common base amplifier is being used, and as a result, positive feedback is being produced by the tank circuit. Similar tank circuits with different connections are used in other Hartley designs that use common emitter amplifiers, for example, to ensure that the feedback signal is always in phase with the input signal and provides the required positive feedback.

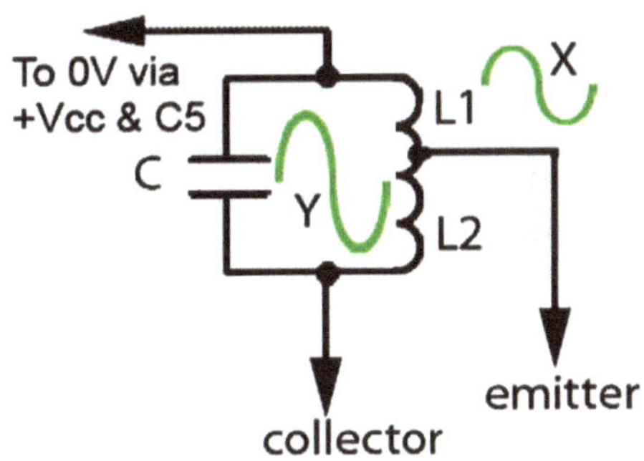

Figure 9.23. The tank circuit.

Source: https://learnabout-electronics.org/Oscillators/osc21.php

9.4.4. Automatic 'Sliding' Class C Bias

It is common in LC sine wave oscillators to use automatic class C bias. In class C the bias voltage, that is the base voltage of the transistor is more negative than the emitter voltage, making V_{be} negative so that the average (center) voltage of the input wave is located on the negative portion of the V_{be} axis of the characteristic curve shown in Figure 9.24. Therefore, only a portion of each sine wave is amplified to produce pulses of collector current. The amplifier is unable to generate an output wave that is free of distortion in class C since only the tips of the waveform are amplified. That being said, in an LC oscillator, this is irrelevant. The only thing the amplifier needs to do is pulse current into the LC resonant circuit at the frequency at which it resonantly operates. The LC circuit's resonating action is what actually generates the circuit's sine wave output. The amount of current flowing in the LC-tuned circuit will determine the amplitude of the signal that is produced. Since the amplifier supplies this current, it follows that the output sine wave's amplitude can be maintained at a constant level by automatically regulating the amount of collector current (I_c) flowing each time a pulse is generated. The base/emitter voltage of a transistor determines its collector current. To maintain a stable output wave amplitude, class C automatically modifies the base/emitter voltage.

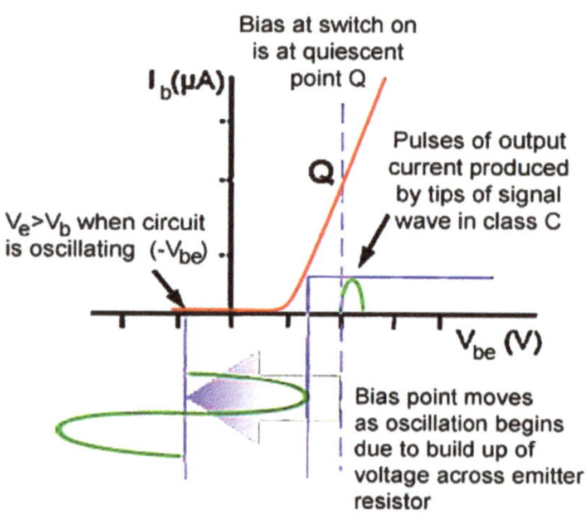

Figure 9.24. Sliding Class C Bias.

Source: https://learnabout-electronics.org/Oscillators/osc21.php

9.4.5. How the Hartley Oscillator Works

Upon initially turning on the oscillator, the amplifier operates in positive feedback class A. Collector current pulses are received by the LC tank circuit, which then starts to resonate at the intended frequency. Because of the tank circuit's high current amplification, the output amplitude is initially quite large. However, a DC voltage that is largely dependent on the time constant of C2 and R3 and is much longer than the oscillator wave's periodic period builds up across R3 once the initial pulses are present and are fed back to the emitter via C2.

As demonstrated in Fig. 9.24, the bias point of the amplifier "slides" from its class A position towards class C conditions as the emitter voltage rises, lowering the difference (V_{be}) between the increasingly positive emitter voltage and the relatively stable base voltage produced by the potential diver Rl/R2. As a result, less of the waveform can be amplified by TR1, until the closed loop gain circuit is reduced to 1 and only the tips of the waveform are generating collector current pulses through the tank circuit. In essence, there is a balance between the negative and positive feedback produced by C2 and R3 and the positive feedback from the tank circuit.

Any deviation from this balance creates a correcting effect. If the amplitude of the output wave reduces, the feedback via C2 also reduces causing the emitter voltage to decrease, making negative value of V_{be} smaller, and so creating a correcting increase in collector current and a greater output wave produced across the tank circuit. As collector current increases, then so will TR1 emitter voltage. This will cause a larger voltage across R3 making the emitter more positive, effectively increasing the amount of negative base/emitter voltage of Tr1. This reduces collector current again, leading to a smaller output waveform being produced by the tank circuit and balancing the closed loop gain of the circuit at 1.

9.4.6. Alternative Hartley Designs

The circuit depicted in Figure 9.25 employs a positive feedback system that travels from the top of the tuned circuit to the base via C2, an AC coupling capacitor and DC blocking capacitor.

The tapping point of the tank circuit is connected to the supply line, which in a common emitter amplifier is the same point as the transistor emitter due to the decoupling capacitors across the supply (not shown as they will be in the power supply) and C3 across the emitter resistor. The top and bottom ends of the tapped inductor L1/L2 are in anti-phase because of this design.

Positive feedback is produced via C2 when the base of a common emitter amplifier is likewise out of phase with the collector waveform. Automatic class C bias is once more employed in this circuit; however, unlike in a typical class A amplifier, the emitter decoupling capacitor C3's value will be crucial. R3, the time constant of R3/C3, which controls the amount of class C bias applied, will only be partially decoupled.

In Figure 9.26, the tapping on the inductor is connected to ground rather than +Vcc, and two DC blocking capacitors are used to eliminate any DC from the tuned circuit. The collector load is now provided by a RF choke which simply provides a high impedance at the oscillator frequency. Automatic class C bias is provided for the common emitter amplifier in a similar manner to Figure 9.25. In this variation however, rather than using a tuned amplifier that amplifies only at the desired frequency, the amplifier here will operate over a wide range of frequencies.

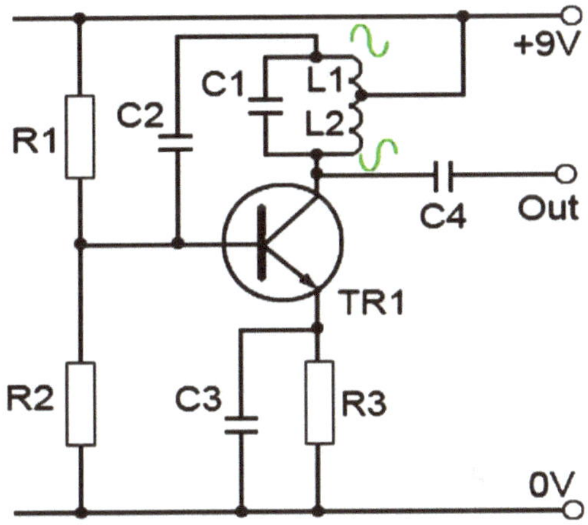

Figure 9.25. Common Emitter Hartley Oscillator.

Source: https://learnabout-electronics.org/Oscillators/osc21.php

To guarantee that positive feedback only happens at the tuned circuit's resonant frequency, the tuned tank circuit must be placed in the feedback path.

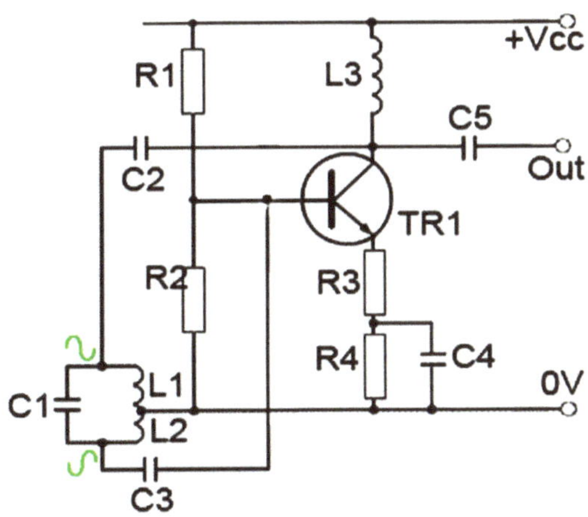

Figure 9.26. Hartley Oscillator with Tuned Feedback.

Source: https://learnabout-electronics.org/Oscillators/images/hartley-tuned-feedback.gif

9.5. CRYSTAL OSCILLATOR

An electronic oscillator circuit known as a crystal oscillator is utilized to achieve mechanical resonance in a vibrating piezoelectric crystal. It will generates an electrical signal with a specific frequency. This frequency is frequently used to monitor time; for instance, digital integrated circuits use wristwatches to supply a steady clock signal. It is also used to maintain frequencies for radio transmitters and receivers. The primary application of quartz crystal is in radio-frequency (RF) oscillators. Quartz crystals, the most popular kind of piezoelectric resonator in oscillator circuits, have therefore become synonymous with crystal oscillators. It is essential for crystal oscillators to be built with a load capacitance.

9.5.1. What is a Quartz Crystal?

One of the most significant characteristics of a quartz crystal is the piezoelectric effect. A voltage that is proportionate to the mechanical pressure across the crystal appears when mechanical pressure is applied across its faces. The crystal is distorted because of that voltage. A crystal will vibrate at its natural frequency when an alternate voltage is applied, and the distorted amount will be proportionate to the applied voltage.

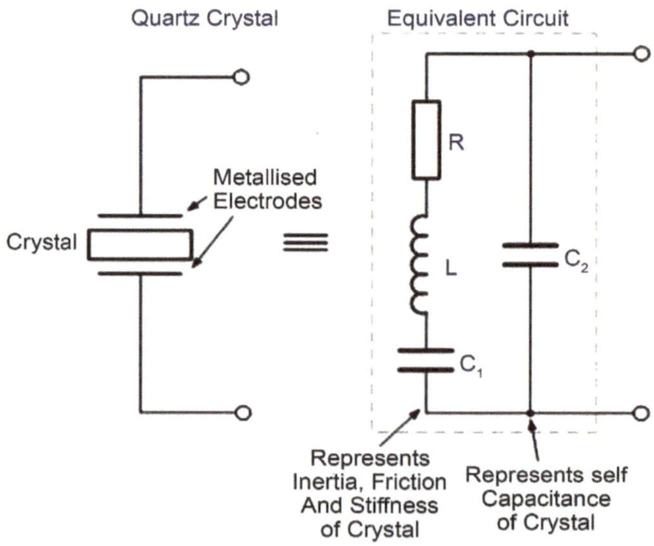

Figure 9.27. Quartz Crystal Circuit.

Source: https://www.elprocus.com/different-types-of-oscillator-circuits-its-applications/

9.5.2. Crystal Oscillator Circuit Diagram

One type of crystal oscillator that operates at a frequency of 16MHz is the 20psc New 16MHz Quartz Crystal Oscillator shown in Figure 9.27.

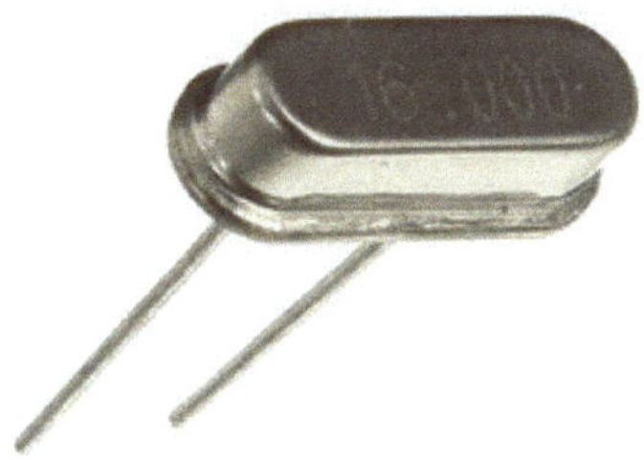

Figure 9.28. Crystal Oscillator.

Source: https://www.elprocus.com/different-types-of-oscillator-circuits-its-applications/

Circuits for crystal oscillators are typically constructed using FETs, or bipolar transistors (Figure 9.28). This is because operational amplifiers lack the bandwidth necessary to function in various low-frequency oscillator circuits below 100 KHz. At higher frequencies that are matched to crystals that are above 1MHz, there will be an issue.

Colpitts crystal oscillator is created as a solution to this issue. With higher frequencies, it will function. A quartz crystal has been used in this oscillator in place of the LC tank circuit, which was previously responsible for the feedback oscillations.

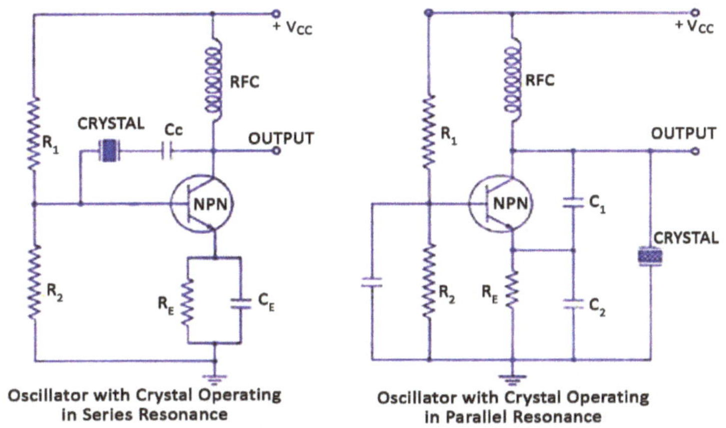

Figure 9.29. Crystal Oscillator Circuit Diagram.

Source: https://www.elprocus.com/different-types-of-oscillator-circuits-its-applications/

9.5.3. Crystal Oscillator Working

The inverse piezoelectric effect is the general operating principle of a crystal oscillator circuit. Certain materials undergo mechanical deformation when an electric field is applied. As a result, it produces an electrical signal at a specific frequency by using the mechanical resonance of the vibrating crystal, which is constructed of a piezoelectric substance.

Quartz crystal oscillators typically have the following characteristics: they are small in size, have a good quality factor (Q), are highly stable, and are economically related. Thus, when compared to other resonators such as tuning forks and LC circuits, quartz crystal oscillator circuits are superior. Typically, 8MHz crystal oscillators are used in microprocessors and microcontrollers.

The crystal's behavior is also described by its equivalent electrical circuit. Refer to the earlier displayed electrical circuit diagram for clarity. The circuit's basic components are as follows: resistance R stands for the friction within the crystal's internal structure; inductance L represents the mass of the crystal; capacitance C2 represents compliance; and C1 is used to represent the capacitance formed as a result of the mechanical molding of the crystal. The quartz crystal oscillator circuit diagram has two resonances, namely series and parallel resonance, i.e., a pair of resonance frequencies.

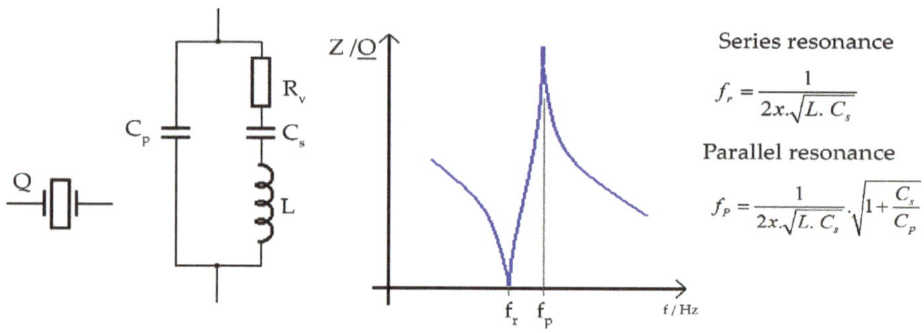

Figure 9.30. Crystal oscillator working.

Source: https://www.elprocus.com/different-types-of-oscillator-circuits-its-applications/

The series resonance occurs when the reactance produced by capacitance C1is equal and opposite to the reactance produced by inductance L.

9.5.4. Uses of Crystal Oscillator

Crystal oscillators are utilized in the design of microprocessors and microcontrollers to supply the clock signals (Figure 9.30). Take 8051 microcontrollers as an example. This controller requires an external crystal oscillator circuit that operates at 12MHz. Although the microcontroller (based on model) can operate at 40MHz (maximum), it must supply 12MHz in most cases because an 8051 requires 12 clock cycles for a machine cycle. This allows the microcontroller to provide an effective cycle rate of 1MHz (using the 12MHz clock) to 3.33MHz (using the maximum 40MHz clock). The clock pulses generated by this specific crystal oscillator, which has a cycle rate ranging from 1MHz to 3.33MHz, are necessary for the synchronization of all internal operations (Lesurf, Jim, 2006).

9.5.5. Application of Crystal Oscillator

There are various applications for crystal oscillator in various fields and some of the crystal oscillator applications are given below.

9.5.5.1. Colpitts Crystal Oscillator Application

At very high frequencies, a sinusoidal output signal is produced using a Colpitts oscillator. Because the SAW device we use in the Colpitts circuit senses directly from its surface, this oscillator can be used as a variety of sensors, including temperature sensors.

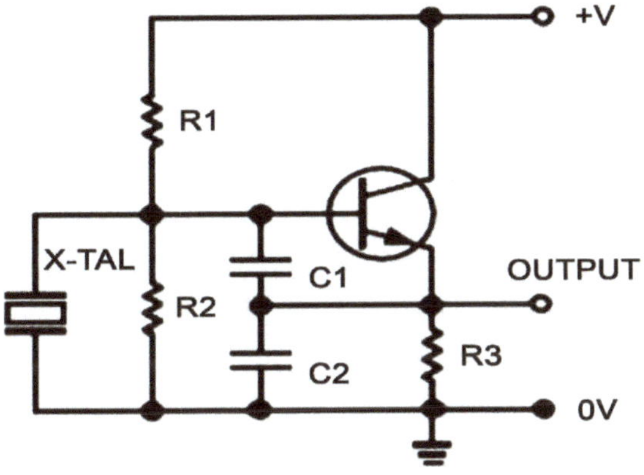

Figure 9.31. Colpitts crystal oscillator.

Source: https://www.elprocus.com/different-types-of-oscillator-circuits-its-applications/

Colpitts oscillators are primarily used in applications requiring a wide range of frequencies and in continuous, undamped oscillation conditions (Figure 9.31). By adjusting certain components of the Colpitts circuit, higher frequencies and improved temperature stability can be achieved. The Colpitts oscillator has been widely used in the development of radio and mobile communications.

9.5.5.2. Applications of Armstrong Crystal Oscillator

Up until the 1940s, this circuit was quite common. Regenerative radio receivers make extensive use of these. Through a second winding in the input, the radio frequency signal from the antenna is magnetically coupled into the tank circuit, and feedback is decreased to regain control in the feedback loop. Ultimately, it generates an amplifier and narrow-band radio frequency filter. Feedback loops are used in place of the LC resonant circuit in this crystal oscillator (Figure 9.32).

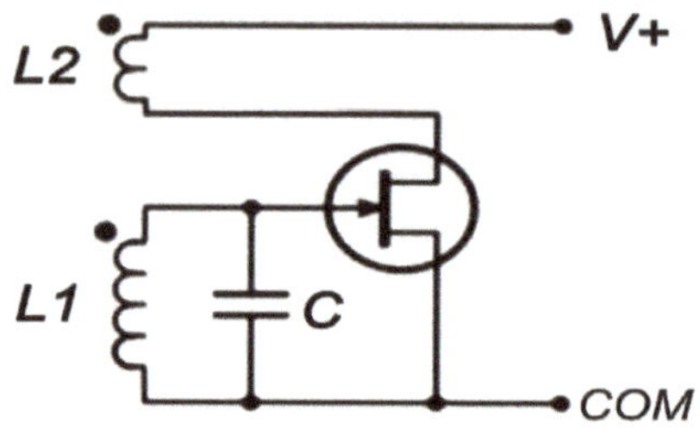

Figure 9.32. Armstrong crystal oscillator.

Source: https://www.elprocus.com/different-types-of-oscillator-circuits-its-applications/

- *In Military and Aerospace:* For the efficient communication system, Crystal Oscillators are used in military and aerospace. The communication system is to establish and for the navigation purpose and electronic warfare in the guidance systems
- *In Research and Measurement:* The crystal oscillators are used in research and measurement for the celestial navigation and the space tracking purpose, in medical devices and in the measuring instruments.

9.5.5.3. Industrial Applications of Crystal Oscillator

Crystal oscillators have many industrial applications. They are widely used in computers, instrumentation, digital systems, in phase-locked loop systems, modems, marine, telecommunications, in sensors and in disk drives. Crystal Oscillator is also used in engine controlling, clock and to trip computer, stereo, and in GPS systems. This is an Automotive application.

In addition, Crystal oscillators are used in many consumer goods. For example, cable television systems, video cameras, personal computers, toys and video games, cellular phones, radio systems. This is the Consumer Application of Crystal Oscillator.

CASE STUDY

THE ROLE OF CRYSTAL OSCILLATORS IN GPS TECHNOLOGY

Introduction

Global Positioning System (GPS) technology relies heavily on the precision and stability of oscillators. This case study examines the critical role crystal oscillators play in ensuring the accuracy and reliability of GPS systems.

Background

GPS technology is ubiquitous in modern life, used in navigation systems, smartphones, military applications, and more. The system comprises a network of satellites, ground control stations, and GPS receivers. Accurate timekeeping is fundamental to the operation of GPS, as position are determined by calculating the time it takes for signals to travel from satellites to the receiver.

The Problem

GPS technology demands highly precise time measurements. Even a small error in timing can result in significant inaccuracies in position calculation. Traditional oscillators often fail to meet the stringent requirements for stability and accuracy needed for reliable GPS performance.

Solution: Crystal Oscillators

Crystal oscillators, known for their superior frequency stability and precision, are employed to address this challenge. These oscillators use the mechanical resonance of a vibrating crystal (typically quartz) to generate an electrical signal with a very precise frequency.

1. Frequency Stability:
   - Crystal oscillators maintain a constant frequency over time and under varying environmental conditions.
   - This stability is crucial for synchronizing the satellite signals with the receiver's clock.
2. Accuracy:
   - The precision of crystal oscillators ensures that the timing errors are minimized.
   - This accuracy translates directly to the accuracy of the position calculations.

Implementation

In a typical GPS receiver, a crystal oscillator generates the clock signal that drives the timing circuits. The following components illustrate how the crystal oscillator is integrated:

1. Clock Generation:
   - o The crystal oscillator provides a stable clock signal to the receiver's processor.
   - o This clock signal is used to measure the time intervals between signal receptions from multiple satellites.
2. Synchronization:
   - o The crystal oscillator's stability ensures that the timing remains synchronized with the signals from the GPS satellites.
   - o Any drift in the oscillator's frequency can cause errors in the calculated position.
3. Error Correction:
   - o Advanced GPS receivers use algorithms to correct minor discrepancies in timing, but the baseline accuracy provided by the crystal oscillator is essential.
   - o The combination of stable crystal oscillators and sophisticated error correction algorithms enhances overall GPS accuracy.

Results

The implementation of crystal oscillators in GPS receivers has led to significant improvements in positioning accuracy. This enhanced precision is critical for various applications:

1. Navigation:
   - o Provides accurate real-time location data for automotive and personal navigation systems.
   - o Essential for autonomous vehicles that rely on precise location information.
2. Military Applications:
   - o Ensures accurate targeting and positioning for military operations.
   - o Enhances the reliability of navigation systems in challenging environments.
3. Scientific Research:
   - o Used in geolocation-based research, such as earthquake monitoring and environmental studies.
   - o Provides precise data necessary for scientific analysis.

CLASS ACTIVITY

How does varying the values of capacitors and inductors in an LC oscillator circuit affect the frequency and stability of the oscillations, and what implications do these changes have for real-world applications such as communication systems and signal generation?

SUMMARY

- An oscillator is a circuit that produces alternating current (AC) of a desired frequency. At its core, an oscillator can convert direct current (DC) from a power source into a repetitive AC signal. It sustains oscillations by providing positive feedback that compensates for losses in the circuit. When power is supplied, noise or signal triggers initial oscillations that are amplified and fed back continuously resulting in sustained output waves.
- The Oscaprobe is a network that is a short circuit at the specified frequency of oscillation and an open circuit at all other frequencies, the voltage amplitude and phase is adjusted until no current flows through the probe, under which conditions the correct oscillating amplitude is determined.
- The active device used in an oscillator is very important, factors for consideration are: output power, oscillating frequency, supply voltage and cost. The low frequency phase noise of the oscillator does depend on the flicker noise of the amplifier, so that the noise figure of the amplifier is important for low noise oscillators.
- An Armstrong oscillator, Colpitts, Clapp, Hartley, and Crystal-controlled oscillators are several types of resonant LC feedback oscillators (LC electronic oscillator). An Armstrong oscillator (also known as Meissner oscillator) is actually an LC feedback oscillator which uses capacitors and inductors in its feedback network. The
- The Armstrong oscillator is used to produce a sinusoidal wave output of constant amplitude and of fairly constant frequency within the given RF range. It is generally used as a local oscillator in receivers, can be used as a source in signal generators and as a radio-frequency oscillator in the medium- and high-frequency range.
- The tuned collector oscillator is one kind of transistor LC oscillator where the tank circuit comprises of a capacitor and a transformer, that is connected to the collector terminal of the transistor. The tuned collector oscillator circuit is the simplest & the basic kind of LC oscillator.

- The Hartley Oscillator is a particularly useful circuit for producing good quality sine wave signals in the RF range, (30kHz to 30MHz) although at the higher limits of this range and above, The Colpitts oscillator is usually preferred.
- The LC circuit that controls the frequency of oscillation is often called the "TANK CIRCUIT" because it contains circulating currents much greater than the current supplying it (e.g., pulses of collector current supplied by the amplifier). Its operation is supposed to be rather like a water tank or cistern that can supply a continuous flow of water from an intermittently flowing external supply.
- A crystal oscillator is an electronic oscillator circuit that is used for the mechanical resonance of a vibrating crystal of piezoelectric material. It will create an electrical signal with a given frequency. This frequency is commonly used to keep track of time for example wristwatches are used in digital integrated circuits to provide a stable clock signal and used to stabilize frequencies for radio transmitters and receivers.
- A quartz crystal exhibits a very important property known as the piezoelectric effect. When mechanical pressure is applied across the faces of the crystal, a voltage which is proportional to mechanical pressure appears across the crystal.

REVIEW QUESTIONS

1. What is the fundamental principle behind the operation of an oscillator in electronic circuits?
2. Describe the role of positive feedback in sustaining oscillations within an oscillator circuit.
3. Explain the working principle of an Armstrong oscillator. What are its main components and how do they contribute to its functionality?
4. How does a tuned collector oscillator differ from an Armstrong oscillator in terms of circuit design and application?
5. What is the main function of a Hartley oscillator, and how does it achieve frequency stability?
6. What are the advantages of using a crystal oscillator in electronic devices? How does it achieve high frequency stability?
7. Discuss the importance of the piezoelectric effect in the operation of crystal oscillators.
8. In what types of applications are crystal oscillators most commonly used, and why are they preferred over other types of oscillators in these applications?

REFERENCES

1. Carr, J. (2002). *RF Components and Circuits*. Newnes. pp. 125–126. ISBN 0080498078.
2. Chattopadhyay, D. (2006). *Electronics (Fundamentals and Applications)*. New Age International. pp. 224–225. ISBN 978-81-224-1780-7.

3. Fleming, J. A. (1919). *The Thermionic Valve and Its Developments in Radiotelegraphy and Telephony*. London: The Wireless Press. pp. 148–155.
4. Hempstead, C., & Worthington, W. E. (2005). *Encyclopedia of 20th-Century Technology* (Vol. 2). Taylor & Francis. p. 648. ISBN 978-1579584641.
5. Huijsing, J., van de Plassche, R. J., & Sansen, W. (2013). *Analog Circuit Design*. Springer Scientific and Business Media. p. 77. ISBN 978-1475724622.
6. Lesurf, J. (2006). Feedback oscillators. In *The Scots Guide to Electronics*. School of Physics and Astronomy, University of St. Andrews, Scotland.
7. Razavi, B. (2001). *Design of Analog CMOS Integrated Circuits*. The McGraw-Hill Companies. pp. 482–484. ISBN 7302108862.
8. Snelgrove, M. (2011). Oscillator. In *McGraw-Hill Encyclopedia of Science and Technology* (10th ed.). Science Access online service. McGraw-Hill.

CHAPTER 10

Digital Electronic Principles

LEARNING OBJECTIVES

After studying this chapter, you will be able to:

- Define analog vs. digital;
- Explain number systems;
- Discuss about digital electronics;
- Describe Boolean algebra and logic gates;
- Explain combinational and sequential circuits.

KEY TERMS FROM THIS CHAPTER

Digital electronic	Analog signal	Number system
Binary number system	Binary digit	Decimal number
Digital circuits	Unicode	Redundancy
Analog sensor	Logic symbol	Logic circuit
Electrical circuit-		

INTRODUCTION

Digital electronic principles form the foundation of modern computing and electronic devices, encompassing the study and application of digital signals to represent and process information. Unlike analog systems that use continuous signals, digital systems use discrete binary signals, represented by the digits 0 and 1. This binary representation, rooted in Boolean algebra, allows for the development of robust, efficient, and versatile electronic circuits.

At the core of digital electronics is the concept of logic gates, which perform basic logical functions like AND, OR, and NOT. These gates are the building blocks of more complex digital circuits, such as multiplexers, demultiplexers, encoders, decoders, and arithmetic circuits like adders and subtractors. Combinational logic circuits, which combine multiple logic gates to achieve specific functions, operate without memory; their output depends solely on the current inputs.

Sequential logic circuits, on the other hand, incorporate memory elements, enabling them to store and process information over time. These circuits include flip-flops, counters, and shift registers, which are essential for creating state machines and other memory-dependent devices. The interplay between combinational and sequential logic allows for the design of sophisticated systems, such as microprocessors and digital signal processors.

Digital electronic principles also encompass data representation and storage, including binary numbers, hexadecimal notation, and various coding schemes like ASCII and Unicode. Furthermore, the principles of timing and synchronization are crucial, involving clock signals that coordinate the operation of digital circuits, ensuring data integrity and correct sequencing of events.

Advancements in digital electronics have led to the miniaturization of components through integrated circuits (ICs) and very-large-scale integration (VLSI) technology. This miniaturization has paved the way for powerful, compact, and energy-efficient devices, from smartphones and laptops to embedded systems and IoT devices.

Digital electronic principles are fundamental to understanding and designing the vast array of modern electronic devices and systems. They provide the tools and frameworks necessary for converting raw data into meaningful information, enabling the digital revolution that defines contemporary technology.

10.1. ANALOG VS. DIGITAL

Signals that have continuous values are called analog signals, i.e., they exhibit continuity in both magnitude and time. Many systems still use analog signals, but as inexpensive digital signals have become available, their use has decreased. Time and magnitude discreteness characterize digital signals. These signals are denoted by binary numbers, either "0" or "1."

Digital signals are representations of continuous time signals that have been sampled, quantized, and encoded. Furthermore, certain methods encrypt the signal to increase the system's tolerance to the channel.

Analog systems use bandwidth efficiently, are less resistant to noise, and are simple to work with mathematically. However, hardware transmitters and receivers that are specifically made to match the specific transmission are needed for analog signals. Transmitters and receivers must be completely changed to alter an analog signal. In contrast, digital signals are more resilient to noise; however, excessive noise can cause them to become completely corrupted. Noise in digital signals can lead a "1" to be read as a "0" or vice versa, altering the original data that was sent and the data that was received. The main advantage of digital signals is their handling by straightforward, standardized transmitters and receivers. The signal can then be processed in software, which can be changed relatively inexpensively.

Discrete time signals are distinct from digital signals because they are a discrete only in terms of time, not in terms of magnitude. While discrete can take on any numerical value, digital quantity can only take on values of 0 or 1, 9.

Digital signals can be created from analog signals. Analog signals can also be converted from digital signals. "Sampling" is the process of transforming analog data into digital data. "Reconstruction" is the process of creating an analog signal from a digital one.

10.2. NUMBER SYSTEMS

Number systems involve the relationship between quantities and symbols. The quantity of digits or basic symbols in a number system is known as the base or radix. Humans commonly use the decimal (base 10) number systems for counting and measurements. In contrast, computers utilize binary (base 2), hexadecimal (base 16), or octal (base 8) number systems to efficiently represent binary numbers.

10.2.1. Decimal (Base 10) Number System

There are ten symbols: 0, 1, 2, 3, 4, 5, 6, 7, 8, and 9, in the decimal number system. This system 0 uses positional notation. That is, the least-significant digit (right-most digit) is of the order of 10^0 (units or ones), the second right-most digit is of the order of 10 (tens), the third right-most digit 2 is of the order of 10 (hundreds), and so on. For example,

$$8735 = 8x10^3+7x10^2+3x10^1+5x10^0$$

If ambiguity arises, an optional suffix D will be used in the chapter to denote a decimal number.

10.2.2. Binary (Base 2) Number System

There are just two basic symbols used in the binary number system. In binary numbers, each digit is referred to as a bit. It can be either 0 or 1. A binary number's leftmost bit is referred to as the MSB (most significant bit), and its rightmost bit is known as the LSB (least significant bit). The LSB has a weight or value of 2^0, the second from the right has a value of 2^1, the next 2^2 etc. In case of fractions, the weights starting from the right of the fraction point are $2^{-1}, 2^{-2}, 2^{-3}$ and so on. Eight bits are called a byte.

The binary equivalent for some decimal numbers is:

Decimal	0	1	2	3	4	5	6	7	8	9
Binary	0	1	10	11	100	101	110	111	1000	1001

10.2.3. Octal (Base 8) Number System

An octal number consists of eight symbols, the first eight decimal digits 0, 1, 2, 3, 4, 5, 6, 7. Each digit in the octal system represents a different power of 8. The weight of positions in octal number is as follows ... $8^4 8^3 8^2 8^1 8^0 . 8^{-1} 8^{-2} 8^{-3}$...

10.2.4. Hexadecimal (Base 16) Number System

The hexadecimal number system has 16 symbols from 0-9 and A-F. The weight of each digit of hexadecimal number is ... $16^3 16^2 16^1 16^0 . 16^{-1} 16^{-2} 16^{-3}$....

Decimal	Hexadecimal	Binary
0	0	0000
1	1	0001
2	2	0010
3	3	0011
4	4	0100
5	5	0101
6	6	0110
7	7	0111
8	8	1000
9	9	1001
10	A	1010
11	B	1011
12	C	1100
13	D	1101
14	E	1110
15	F	1111

10.2.5. Number System Conversions

There are often occasions when there is a requirement to convert between various systems. The next section will delve into the intricacies of inter-conversions.

10.2.5.1. Binary to Decimal Conversion

Each binary digit is multiplied by its corresponding weight in order to convert it to decimal, and the result for each product is then added.

$(1101)_2 = (X)_{10}$

$1101 = 1x2^3 + 1x2^2 + 0x2^1 + 1x2^0$

$= 8+0+2+1$

$(1001)_2 = (11)_{10}$

$(0.0011)_2 = (X)_{10}$

$= 0x2^{-1} + 1x2^{-2} + 0x2^{-3} + 1x2^{-4}$

$= 0 x0.5 + 0x0.25 + 1x0.125 + 1x0.0625$

$= (0.1875)_{10}$

$(1001.101)_2 = (X)_{10}$

$= 1x2^3 + 0x2^2 + 0x2^1 + 1x2^0 + 1x2^{-1} + 0x2^{-2} + 1x2^{-3}$

$= 8 + 0 + 0 + 1 + 0.5 + 0 + 0.125$

$= (9.625)_{10}$

10.2.5.2. Decimal to Binary Conversion

To convert a decimal number to binary, start by dividing the decimal number by 2 repeatedly and then write down the remainders in reverse order to create the binary number.

$(12)_{10} = (X)_2$

Start by dividing it by 2 and keep noting the remainder (R) at each stage.

12/2=6 Remainder=0

6/2=3 Remainder=0

3/2=1 Remainder=1

1/2=0 Remainder=1

Now reading the remainders in reverse order gives the binary equivalent number.

$(12)_{10} = (1100)_2$

$(21)_{10} = (X)_2$

$(21)_{10} = (10101)_2$

A fraction is converted to binary by multiplying it by two, and the corresponding binary equivalent is obtained by writing the carry in the integer position after each multiplication in the forward order.

$(0.4375)_{10} = X_2$

Multiply the decimal fraction by 2 and keep noting the integral part (I).

$2 \times 0.4375 = 0.8750$	Integer= 0
$2 \times 0.8750 = 1.750$	Integer=1
$2 \times 0.750 = 1.5$	Integer=1
$2 \times 0.5 = 1.0$	Integer=1

Now reading the integers in forward order we get the binary equivalent

$(0.4375)_{10} = (0.0111)_2$

10.2.5.3. Octal to Decimal Conversion

Multiplying the weight of each octal digit and then adding the resultant products translates a number between the octal and decimal systems.

$(47)_8 = (X)_{10}$

$47 = 4 \times 8^1 + 7 \times 8^0$

$= 32 + 7$

$= 39$

$(47)_8 = (39)_{10}$

$(22.34)_8 = (X)_{10}$

$22.34 = 2 \times 8^1 + 2 \times 8^0 + 3 \times 8^{-1} + 4 \times 8^{-2}$

$=16 + 2 + 3 \times 1/8 + 4 \times 1/64$

$= 18.4375$

$(22.34)_8 = (18.4375)_{10}$

10.2.5.4. Decimal to Octal Conversion

A decimal number can be converted to octal by gradually dividing it by 8 and noting the remainder each time. In order to form the octal number, write the remainders in reverse order at the end. If there is a fractional portion of the number, that fraction is multiplied by eight, and the integer portion is taken. In the forward order, the carries are finally taken.

$(19.11)_{10} = (X)_8$

19/8=2	Remainder = 3
2/8=0	Remainder = 2
$0.11 \times 8 = 0.88$	Integer = 0
$0.88 \times 8 = 7.04$	Integer = 7
$0.04 \times 8 = 0.32$	Integer = 0
$0.32 \times 8 = 2.56$	Integer = 2
$0.56 \times 8 = 4.48$	Integer = 4

To obtain the octal equivalent number, write the remainders in reverse order prior to the octal point and the integers in forward order following the octal point.

$$(19.11)_{10} = (23.07024)_8$$

10.2.5.5. Octal to Binary Conversion

Each octal number is converted into its equivalent binary digit of length three as the base $8 = 2^3$.

$(21.357)_8 = (X)_2$

2	1	.	3	5	7
010	001	.	011	101	111

$(21.357)_8 = (010001.011101111)_2$

10.2.5.6. Binary to Octal

The given binary number is divided into groups of three bits beginning at the octal point, and each group is then translated into its corresponding octal value.

$(1100111.1011)_2 = (X)_8$

001	100	111	.101	100
1	4	7	.5	4

$(1100111.1011)_2 = (147.54)_8$

10.2.5.7. Binary to Hexadecimal

Since base 16 equals 2^4, the binary number is divided into bits from the binary point, and the corresponding hexadecimal equivalent is then written.

$(100111011.10011)_2 = (X)_{16}$

0001	0011	1011	.1001	1000
1	3	B	.9	8

$(100111011.10011)_2 = (13B.98)_{16}$

10.2.5.8. Hexadecimal to Binary

Each hexadecimal number is converted into its equivalent binary digit of length four.

$(6D.4A)_{16} = (X)_2$

6	D	.4	A
0101	1101	.0100	1010

$(6D.4A)_{16} = (01101101.01001010)_2$

10.2.5.9. Decimal to Hexadecimal

The process of converting a decimal number to hexadecimal involves multiplying it by 16 and recording the remainders in the opposite order after each division. The integer portion is taken in the forward order after the fractional part of the decimal number is multiplied by 16.

$(3287.645))_{10} = (X)_{16}$

3287/16=205	Remainder=7
205/16=12	Remainder=13(D)
12/16=0	Remainder=12(C)
16 x 0.645 = 10.320	Integer = 10 (A)
16 x 0.320 = 5.120	Integer = 5
16 x 0.120 = 1.92	Integer = 1
16 x 0.92 = 14.72	Integer = 14(E)
16 x 0.72 = 11.52	Integer = 11(B)

$(3287.645)_{10} = (CD7.A51EB)_{16}$

10.2.5.10. Hexadecimal to Decimal

Each digit of the hexadecimal number is multiplied by its weight and then added.

$(61.24)_{16} = (X)_{10}$

$=6 \times 16^1 + 1 \times 16^0 + 2 \times 16^{-1} + 4 \times 16^{-2}$

$=6 \times 16 + 1 \times 1 + 2/16 + 4/162$

$= (97.1497)_{10}$

$(61.24)_{16} = (97.1497)_{10}$

10.2.6. 1's and 2's Complement

The binary number system does not include the notion of a negative sign. Despite the development of other techniques, 1's complement and 2's complement remain the most widely used. Even though 2's complement is more widely used than 1's complement, 1's complement is still in use because it has simpler hardware and concepts.

10.2.6.1. 1's Complement

The most significant bit (MSB) in 1's compliment, the sign bit, is set to 1 for negative integers and 0 for positive integers. A binary number can be made to have its 1's complement by deducting every bit from 1. In other words, a binary number's 1's complement can be found by flipping each 0 to a 1 and each 1 to a 0. The magnitude of the positive integer is given by the absolute integral value of the next n-1 bits after the sign bit. The magnitude of the 1's complement of the (n-1) bit binary pattern equals the absolute value of a negative integer.

Binary number	1's Complement	Decimal Number
0101	1010	+5
1001	0110	-6

The MSB in the first instance of the aforementioned examples is 0. As a result, +5 is the decimal equivalent. The second example has an MSB of 1, denoting a negative number. –6 is the equivalent number in decimals.

10.2.6.2. 2's Complement

Similar to 1's complement, the sign bit in 2's complement representation is the most significant bit (MSB), with a value of 0 denoting positive integers and a value of 1 denoting negative integer. When we add "1" to the given number's 1's complement, we get the number's 2's complement, i.e., 1's complement + 1 equals 2's complement. The magnitude of the positive integer is given by the absolute integral value of the next n-1 bits after the sign bit. The magnitude of the 2's complement of the (n-1) bit binary pattern equals the absolute value of a negative integer.

Binary Number	1's complement	2's complement	Decimal Number
1010	0101	0110	-6
0101	1010	1011	+5

Since the MSB in the first example is 1, it denotes a negative number. There are 010 bits left. Thus, the complement of its 2 is its negative absolute value, i.e., 110. Hence, the figure is -6. In the following example, 0 in the MSB denotes +5, a positive number.

10.2.7. Fixed Point Number Representation

Fractional and integer numbers are both considered real numbers. We use the binary number system in computers. Similar to how a decimal point divides a decimal number's integral and fractional parts, a binary point divides a binary number's integral and fractional parts. All digits to the left of a binary point carry the weights $2^0, 2^1, 2^2$ and so on and those to the right have weights $2^{-1}, 2^{-2}, 2^{-3}$ and so on.

Consider the bit pattern of 9.

$9_{10} = 1001_2$

Next, consider $4.5_{10} = 100.1_2$

Generally speaking, dividing a number by two is the same as shifting its bit pattern one bit to the right if the binary point position is fixed. Similarly, multiplying a number by two results from shifting it one bit to the left. The comprehension of fixed-point number representation relies on the shifting process mentioned earlier. By defining a fixed binary point number type and then following this implicit convention, real numbers are represented. A fixed-point type number can only be defined with two parameters. These are the binary point position within the number and the width of the number representation.

We use a notation called fixed, in which b represents the binary point counted from the least significant bit (counting from 0), and w indicates the total number of bits used (a number's width). As an illustration, fixed <5,4> represents a 5-bit fixed point number, with four fractional bits at the rightmost position. Thus, this is the bit pattern:

1.001_2 represents 1.0625_{10}

Fixed <5,2> for the same bit pattern is 100.01_2 representing 4.25 and as already seen 1000.1_2 which is fixed <5,1> is 8.5 and when treated as an integer is 17.

10.2.8. Floating Point Number Representation

Real numbers are numbers containing fractional part such as 3.0, -111.5, ½, 3E-5, etc. The last example is computer shorthand for scientific notation implying 3×10^{-5}.

Computers are essentially integer machines that represent real numbers with complex codes like IEEE Floating Point Standard. The term "floating point" describes the lack of a standard for the fixed number of digits preceding and following the decimal point, i.e., decimal point is floatable. Fixed-point representations, on the other hand, have a predetermined number of digits before and after the decimal point. Larger number ranges are typically associated with floating-point representations, which are slower and less precise than fixed-point representations.

An integral and fraction part of a real number in binary have each bit set to either 0 or 1. Three fields make up a number in floating point representation: sign(s), exponent(e), and significand (m). A real number's numerical value is calculated as.

$$(-1)^s * m * 2^e \qquad \text{where } m < |2|$$

Binary32 is a type of base 2 format for single precision, and Binary64 is a type for double precision. The number of bits used to encode different components determines the difference (Table 10.1).

Table 10.1. Precision Representation

Precision	Base	Sign	Exponent	Significand
Single precision	2	1	8	23+1
Double precision	2	1	11	52+1

Source: https://cbseacademic.nic.in/web_material/Curriculum/Vocational/2018/Digital_Electronics_820.pdf

10.2.8.1. Single Precision Format

There are 23 significand bits (1 implies bit), 8 exponent bits, and 1 sign bit in the 32-bit single precision format. In a single precision format, the exponent's range is –126 to +127. *For example, conversion of the rational number 21÷4 to single precision float format is done as follows,

$21_{(10)} \div 4_{(10)} = 5.25_{(10)} = 101.01_{(2)}$

$= 1.0101_{(2)} \times 2^2$ (normalized form with leading 1 bit)

Bits following the binary point indicate the float number's mantissa. A normalized number can represent an even more accurate significand (23 + 1 = 24 bits) than a corresponding de-normalized number because of the implied MSB. We refer to this as subnormal representation. Normalized form is used to represent floating point numbers.

10.2.8.2. Double Precision Format

There are 52 significand bits (1 implies bit), 10 exponent bits, and 1 sign bit in the 64-bit double precision format. All other definitions for double precision format are the same as for single precision format, with the exception of component sizes.

10.2.9. Binary Arithmetic

10.2.9.1. Binary Addition

Addition of two or more binary numbers follows the following rules:

0 + 0 = 0

0 + 1 = 1

1 + 0 = 1

1 + 1 = 0 and a carry ahead of 1

Carry ahead is in a manner like decimal arithmetic.

111	1010	011.01
+110	+1101	+101.11
1101	10111	1001.00

10.2.9.2. Binary Subtraction

0Binary subtraction follows the following rules:

0 − 0 = 0

1 − 0 = 1

1 − 1 = 0

0 − 1 = 1 with a borrow of 1 i.e., equivalent to 10 − 1 = 1

111	110.01
-010	-100.10
101	001.11

10.2.9.3. 1's Complement Subtraction

In order to perform 1's complement subtraction, the following steps must be followed:

1. Write the minuend as is.
2. Write the 1's complement of subtrahend.
3. Add the two numbers.
4. Add the carry that arises from the addition (called "end around carry") with the sum obtained in step 3 to get the result.
5. If there is no end around carry, compute the 1's complement of the sum obtained in step 3 and put a negative sign before the result.

We know that

9	1001
−5	-0101
4	0100

Now repeating the same using 1's complement, we have

```
 1001
+1010 (1's compliment of 5)
10011
```

Here the end carry is 1 which is added to 0011 to get the result as

0011+1 = 0100

The absence of END AROUND CARRY in the 1's complement subtraction, is indicative of a negative result requiring 1's complement to get the final result.

6	0110
-9	+0110 (1's complement)
-3	0100

Now since there is no END AROUND CARRY, the answer would be 1's Complement of 0100, i.e., 1011 which is -3.

10.2.9.4. 2's Complement Subtraction

In order to perform 2's complement subtraction, we must follow the under mentioned steps:

1. Write the minuend as such.
2. Write down the 2's complement of the subtahend.

3. Add the two numbers.
4. Discard carry, if any. The sum is the result (positive).
5. If there is no carry, find out the 2's complement of the sum and put negative sign before the result as the result is negative.

Consider,

9	(1001)	1001
-3	(0011)	+1101 (2's Complement of 3)
6		10110

Now, since we have a carry, we discard it (neglect it) and the result is 0110, i.e., 6.

Next consider,

3	(0011)	0011
-9	(1001)	+0111
-6		1010

In the absence of a carry, we determine the 2's compliment of the result obtained and place a negative sign. So here we have 1010 for which the 2's compliment would be 0110 which is 6. Hence the answer after prefixing the negative sign is –6.

10.2.10. Digital Codes

Each character is given a 7 or 8 bit binary code that is used by computers and other digital circuits for information processing in order to identify it as a numeric, alphabetic, or special symbol. As an illustration, the binary number 1000001 denotes 41 (decimal) in BCD code, 65 (decimal) in ASCII code, and A in straight binary code.

10.2.10.1. BCD (Binary-Coded Decimal) Code

The number system known as Binary Coded Decimal, or BCD, uses binary digits or numbers to represent decimal numbers. There are ten digits in a decimal number (0–9). In a BCD coded representation, each decimal digit is represented by its four-bit binary equivalent. Therefore, we can use 0000-1001 as the binary equivalents of 0-9 in BCD. In BCD code, 1010, 1011, 1100, 1101, 1110, and 1111 are INVALID CODE.

In BCD code, a decimal number with one digit is represented by its four-bit equivalent. A decimal number with two digits (e.g., 12) will be expressed in BCD as eight bits, with four bits assigned to the first decimal digit and four more to the second. Therefore, in BCD, (12) will be represented as 0001 10 0010. Four fragments from L. S. The next four are the binary equivalent of 1, and B is the binary equivalent of 2.

Binary and BCD are equal for decimal numbers ranging from 0 to 9, but they diverge when the number has more than one digit. Example: In binary code, (37) is represented as (100101), but in BCD 10 code, it is 0011 0111. Therefore, compared to binary code, BCD code needs more bits. The binary and BCD codes for the decimal numbers 0 to 15 are displayed in Table 10.2.

Table 10.2. The Binary and BCD Codes for the Decimal Numbers 0 to 15

Decimal number	Binary number	Binary Coded Decimal(BCD)
0	0000	0000
1	0001	0001
2	0010	0010
3	0011	0011
4	0100	0100
5	0101	0101
6	0110	0110
7	0111	0111
8	1000	1000
9	1001	1001
10	1010	0001 0000
11	1011	0001 0001
12	1100	0001 0010
13	1101	0001 0011
14	1110	0001 0100
15	1111	0001 0101

Source: https://cbseacademic.nic.in/web_material/Curriculum/Vocational/2018/Digital_Electronics_820.pdf

10.2.10.2. ASCII (American Standard Code Information Interchange) Code

Alphanumeric codes, also known as character codes, are binary codes that represent alphanumeric data, including letters, numbers, symbols, and punctuation. These codes are commonly used for interfacing input-output devices like keyboards, monitors, and mice due to their simplicity for computer processing. The most widely used alphanumeric codes include ASCII, EBCDIC, and Unicode.

Seven-bit code based on the English alphabet, ASCII was first published in 1967. The following table lists a few ASCII characters.

DEC	OCT	HEX	BIN	Symbol	Description
0	000	00	00000000	NUL	Null char
1	001	01	00000001	SOH	Start of Heading
2	002	02	00000010	STX	Start of Text
3	003	03	00000011	ETX	End of Text
4	004	04	00000100	EOT	End of Transmission
5	005	05	00000101	ENQ	Enquiry
6	006	06	00000110	ACK	Acknowledgment
7	007	07	00000111	BEL	Bell
8	010	08	00001000	BS	Back Space
9	011	09	00001001	HT	Horizontal Tab
10	012	0A	00001010	LF	Line Feed
11	013	0B	00001011	VT	Vertical Tab
12	014	0C	00001100	FF	Form Feed
13	015	0D	00001101	CR	Carriage Return
14	016	0E	00001110	SO	Shift Out / X-On
15	017	0F	00001111	SI	Shift In / X-O

Source. https://cbseacademic.nic.in/web_material/Curriculum/Vocational/2018/Digital_Electronics_820.pdf

ASCII is a 7-bit alphanumeric code that can represent 127 characters. Control characters range from 00h to 1Fh, while graphics symbols range from 20h to 7Fh. Extended ASCII is an 8-bit code that can represent 256 symbols, including special graphics and math symbols. The Extended ASCII series starts at 80h and goes up to FFh.

10.2.10.3. Gray Code

Gray code is a non-weighted code in which there is only a one-bit difference between two adjacent code numbers. Certain input/output devices and ADCs both use this code. The binary and gray code differences for the corresponding decimal numbers are shown in the gray code table below.

Decimal Number	Binary code	Gray code
0	0000	0000
1	0001	0001
2	0010	0011
3	0011	0010
4	0100	0110
5	0101	0111
6	0110	0101
7	0111	0100
8	1000	1100
9	1001	1101
10	1010	1111
11	1011	1110
12	1100	1010
13	1101	1011
14	1110	1001
15	1111	1000

Source. https://cbseacademic.nic.in/web_material/Curriculum/Vocational/2018/Digital_Electronics_820.pdf

Gray coding involves changing only one bit at each step, starting from the right side (L.S.B) towards the M.S.B changes. The first three bits are initially held constant (000) and the fourth bit changes from 0 to 1. Then the next bit (3rd bit) is changed from 0 to 1, which is the decimal equivalent for 2. Now, keeping the first three bits constant (001), the 4th bit is changed from 1 to 0. Thus, the gray code for decimal number 3 is 0010. Next, we change the second bit as all possible combinations for bit 3 and 4 are over. It may be observed that we can't change the third bit again (which is also a one-bit change from its previous), as changing it would give the equivalent gray code 0000 which has occurred earlier and must not be repeated. So, the equivalent code for 4 will be 0110. Now, again holding the first and second bit constant, we find the possible combinations of the third and fourth bit by only changing 1 bit in each step. All other equivalent representations are found in a similar manner.

10.3. DIGITAL ELECTRONICS

Digital electronics is the field of electronics that focuses on the study of digital signals and the devices that generate or utilize them. This branch of electronics is concerned with the manipulation and representation of data in digital format, using components like transistors, diodes, and microcontrollers to process and transmit these signals. Digital electronics finds applications in various areas such as computer systems, communication systems, and control systems. Key concepts in digital electronics include Boolean algebra, logic gates, digital filters, and flip-flops.

10.3.1. Digital Electronic Circuits

In the modern age, digital electronics play a crucial role. They operate using a digital format to process data and codes. Digital electronics serve as the foundation for contemporary computers and digital communications, spanning devices such as phones and computers. This field encompasses intricate digital logic circuits that combine to form a single circuit, commonly referred to as a microprocessor. These circuits can carry out numerous operations simultaneously.

There are many types of digital circuits, including:

- *Combinational Circuits*: Combinational circuits are digital circuits that output a value based on the current input values. They do not have any internal memory and do not retain any information from one input to the next. Examples of combinational circuits include decoders, multiplexers, and adders.
- *Sequential Circuits:* Sequential circuits are digital circuits that output a value based on both the current input values and the previous output values. They have internal memory and can store information from one input to the next. Examples of sequential circuits include registers, counters, and flip-flops.
- *State Machines:* State machines are digital circuits that output a value based on the current state and the current input values. They have internal memory and can store information from one input to the next. The state of the machine is determined by the current input values and the previous state. State machines are used to implement complex logic and control systems.
- *Synchronous Circuits:* Synchronous circuits are digital circuits that operate in discrete time intervals, using clock signals to synchronize the operation of the circuit. All components in a synchronous circuit are triggered by the

same clock signal, which ensures that all operations are performed in a coordinated and predictable manner.

10.3.2. Asynchronous Circuits

Digital circuits known as asynchronous circuits function without a clock signal, relying on internal signals to regulate the data flow. While asynchronous circuits are typically slower than synchronous circuits, they offer greater versatility and are often easier to design.

Digital circuits can be implemented using various types of digital logic, including:

1. **Boolean logic:** This logic is based on the principles of Boolean algebra, using AND, OR, and NOT gates to perform logical operations.
2. **K-map logic:** This logic uses Karnaugh maps to simplify Boolean expressions and minimize the number of gates needed to implement a circuit.
3. **Arithmetic logic:** This logic performs arithmetic operations such as addition and subtraction, using digital circuits known as adders and subtractors.
4. **Memory circuits:** These circuits store and retrieve data, using components such as flip-flops and registers.
5. **Microprocessor circuits:** These circuits are the central processing units (CPUs) of computers, responsible for executing instructions and controlling the operation of the entire system.

10.3.2.1. Applications

1. **Computing:** Digital electronics are used in computers to store, process, and transmit data. They are used in the central processing units (CPUs) of computers, as well as in other components such as memory, storage, and input/output devices.
2. **Communication:** Digital electronics are used in a variety of communication systems, including cell phones, satellite systems, and the internet. They are used to transmit and receive data, as well as to process and decode signals.
3. **Entertainment:** Digital electronics are used in a variety of entertainment devices, including TVs, video game consoles, and music players. They are used to process and display images, as well as to store and playback audio and video content.
4. **Transportation:** Digital electronics are used in transportation systems, including vehicles and traffic control systems. They are used to control the operation of engines and other systems, as well as to navigate and communicate with other vehicles.
5. **Industrial control:** Digital electronics are used in industrial control systems to automate and control manufacturing processes. They are used to monitor and control the operation of machines, as well as to process and transmit data.
6. **Medical equipment:** Digital electronics are used in medical equipment, such as x-ray machines and patient monitoring systems. They are used to process and display medical images, as well as to monitor and analyze patient data.

7. **Military:** Digital electronics are used in military systems, such as radar systems and missile guidance systems. They are used to detect and track objects, as well as to communicate and control military operations.
8. **Home appliances:** Digital electronics are used in a variety of household appliances, such as refrigerators, washing machines, and thermostats. They are used to control the operation of the appliance, as well as to monitor and display data such as temperature and energy usage.
9. **Environmental monitoring:** Digital electronics are used in environmental monitoring systems to measure and analyze data on air quality, water quality, and other environmental factors. They are used to collect and transmit data, as well as to analyze and display the results.
10. **Security:** Digital electronics are used in security systems, such as alarm systems and surveillance cameras. They are used to detect and monitor activity, as well as to communicate and respond to security threats.

10.3.3. Properties

One benefit of digital circuits over analog circuits is the ability to transmit signals represented digitally without experiencing noise-induced degradation. If the noise detected during transmission is insufficient to obscure the identification of the 1s and 0s, a continuous audio signal, for instance, can be accurately reconstructed.

In a digital system, a signal can be represented with more binary digits to get a more accurate representation. The system is easily scalable because each digit is handled by the same type of hardware, even though processing the signals requires more digital circuits. Further resolution in an analog system necessitates significant advancements in the linearity and noise properties of every signal chain link.

Computer-controlled digital systems allow for the addition of new functions through software updates without the need for hardware changes. These updates can be completed outside of the factory, making it possible to correct design errors even after the product is in the hands of the customer.

Digital systems have the potential to store information more easily than analog ones. Digital systems' immunity to noise allows data to be stored and retrieved without deterioration. Age-and wear-related noise deteriorates the stored information in an analog system. In a digital system, information can be perfectly recovered if the overall noise is below a certain threshold. Redundancy allows for the recovery of the original data even in the presence of more significant noise, as long as there aren't too many errors.

When digital circuits are used to perform the same tasks as analog circuits, they can consume more energy and produce more heat, which makes the circuits more complex and necessitates the addition of heat sinks. This may restrict the use of digital systems in battery-operated or portable devices. For instance, a low-power analog front-end is frequently used by battery-operated mobile phones to tune and amplify radio signals from the base station. But, a base station can use software radios that are highly configurable but power-hungry, as long as it has grid power. Reprogramming such base stations to process the signals used in new cellular standards is a simple process.

The conversion of continuous analog signals to discrete digital signals is a necessary step in many practical digital systems. This leads to errors in quantization. If the system retains enough digital data to accurately represent the signal to the required level of fidelity, quantization error can be minimized. An essential rule of thumb for determining how much digital data is required to faithfully represent a given analog signal is the Nyquist-Shannon sampling theorem.

In certain systems, a single lost or misinterpreted digital piece of data can lead to a minor error, but in other systems, it can completely alter the meaning of large blocks of related data. For instance, in audio data stored directly as linear pulse-code modulation, a single-bit error results in, at most, one audible click. However, a single bit error may result in a much more significant disruption when audio compression is used to reduce storage space and transmission time.

Users may find it challenging to determine whether a system is about to fail or if it can withstand a greater amount of noise before failing due to the cliff effect. By building a digital system with robustness in mind, digital fragility can be decreased. One possible solution to handle errors is to add a parity bit or another technique to the signal path. These protocols assist the system in identifying errors so that it can either fix them or ask for a retransmission of the data.

10.3.3.1. Digital Electronics – Advantages and Disadvantages

Advantages

Digital electronics have several advantages over analog electronics:

1. Digital signals are more accurate and reliable than analog signals, as they are less susceptible to noise and interference.
2. Digital signals can be easily stored, processed, and transmitted, as they are easy to represent with 1s and 0s.
3. Digital electronics are more precise and can perform more complex operations than analog electronics.
4. Digital electronics are easier to design and manufacture than analog electronics, as they can be implemented using standard components such as transistors and gates.
5. Digital electronics are more energy efficient than analog electronics, as they can perform the same operations using less power.

Disadvantages

However, digital electronics also have some disadvantages:

1. Digital electronics require a source of electricity to operate, which may not be available in some situations.
2. Digital signals can suffer from a phenomenon known as quantization error, where the signal is not perfectly represented by a finite number of bits.
3. Digital electronics may require additional hardware to convert

analog signals into digital signals, and vice versa.

4. Digital electronics may be more expensive to produce than analog electronics, as they may require more complex manufacturing processes.
5. Digital electronics may have a higher latency than analog electronics, as they may require additional time to process and transmit signals.

10.3.4. Difference between Digital and Analog Electronics

If we first divide the world of electronics into two distinct categories—the "analog" world and the "digital" world—it becomes much easier to understand. Any natural phenomenon that changes its own properties over time is referred to as existing in the analog world. Consider the temperature outside as an example. We observe that the temperature varies gradually throughout the day, and we can always use a basic thermometer to find out exactly how hot or cold it is.

Other natural phenomena, like barometric pressure, wind speed, solar radiation, etc., can also be observed, measured, and recorded to exhibit similar fluctuating characteristics. One common feature that would emerge from recording and graphing all of the aforementioned occurrences over a 24-hour period is that they all undergo physical properties that change with time.

In each case, if you connect all the dots on a graph, the results will always form some type of "analog signal" (wave) like these examples (Watson, 1986):

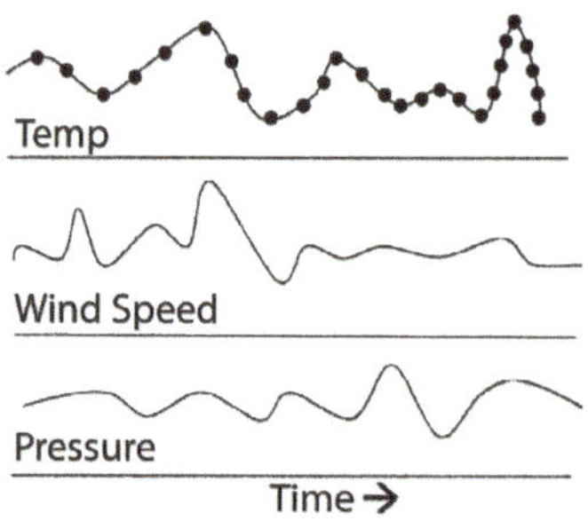

Source: https://www.nutsvolts.com/uploads/wygwam/NV_0416_Secura_Intro.jpg.

Many of these "signals"—such as thermal, magnetic, optical, acoustical, biological, chemical, or electrical—can be handled by analog electronics engineers by creating the proper analog sensors and control circuitry. Information can be processed and transmitted in two different ways: analog electronics and digital electronics.

Table 10.3. Difference between Analog and Digital Electronics

	Analog Electronics	Digital Electronics
Signal Type	Continuous	Discrete
Accuracy	High	High (with some loss)
Data Capacity	Low	High
Data Transmission	Limited	Unlimited
Noise Immunity	Low	High
Example Applications	Radio and audio equipment, control systems	Computers, smartphones, data storage and transmission

Source. https://www.electronicsforu.com/technology-trends/learn-electronics/analog-electronics-basics#:~:text=Analog%20electronics%20can%20be%20used,to%20represent%20and%20process%20information.

10.3.4.1. Difference between Analog and Digital Electronics

Analog electronics utilize continuous signals for information representation and processing. These systems are commonly employed in applications that necessitate a continuous range of values, like in radio and audio equipment, and control systems. Analog electronics are utilized for amplifying signals, filtering noise, and executing a diverse range of other functions. Resistors, capacitors, inductors, and transistors are some of the standard components found in analog electronics.

Discrete signals, on the other hand, are used by digital electronics to represent and process data. Although digital systems are frequently chosen for high degree of accuracy in data storage and transmission, they are not ideal for all tasks, such as processing continuous signals. Transistors, gates, and flip-flops are a few of the components of digital systems that are used to manipulate binary data.

Digital electronics are generally more suited to tasks involving large amounts of data and the ability to tolerate some accuracy loss, whereas analog electronics are better suited to tasks involving continuous signals and high accuracy requirements.

10.4. BOOLEAN ALGEBRA AND LOGIC GATES

A fundamental component of every digital circuit is a logic gate. These logic gates decide whether to combine different inputs in accordance with a logical operation and then output the result. Logic gates use Boolean algebra to perform logical operations. Assume A and B are our two inputs. Assuming that R is the output, the different kinds and symbols of logic gates are listed below.

Gaining knowledge of digital logic gates is essential to understanding all digital electronic circuits. The AND gate, OR gate, and NOT gate are the three basic kinds of electronic "logic gates." Just keep in mind that the fundamental components of all digital logic circuits are logic gates.

10.4.1. De-Morgan's Theorem

To solve Boolean algebraic expressions, we apply De Morgan's theorems. These are an extremely potent tool for digital design. The complements of the products of all terms are equal to the sums of the complements of each term, as this theorem explains. Similarly, the products of the complements of each term equal the complements of the sums of all the terms.

Two of the theorems were suggested by De Morgan that are extremely useful for Boolean Algebra. These two theorems have been discussed below.

10.4.1.1. Theorem 1

$\overline{A.B}=\bar{A}+\bar{B}$

NAND = Bubbled OR

- The LHS (left-hand side) of this theorem represents the NAND gate that has inputs A and B. On the other hand, the RHS (right-hand side) of this theorem represents the OR gate that has inverted inputs.
- The OR gate here is known as a Bubbled OR (Watson, 1986).

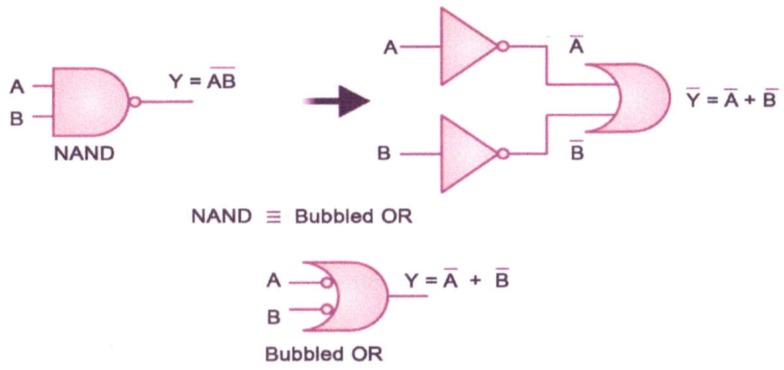

Here is a table that shows the verification of the first theorem of De Morgan:

A	B	$\overline{A+B}$	$\overline{A}$	$\overline{B}$	$\overline{A}+\overline{B}$
0	0	1	1	1	1
0	1	1	1	0	1
1	0	1	0	1	1
1	1	0	0	0	0

10.4.1.2. Theorem 2

$\overline{A+B}=\overline{A}.\overline{B}$

NOR = Bubbled AND

- The left-hand side of this theorem represents the NOR gate that has inputs A and B. On the other hand, the right-hand side represents the AND gate that has inverted inputs.
- The AND gate here is known as Bubbled AND (Watson, 1986).

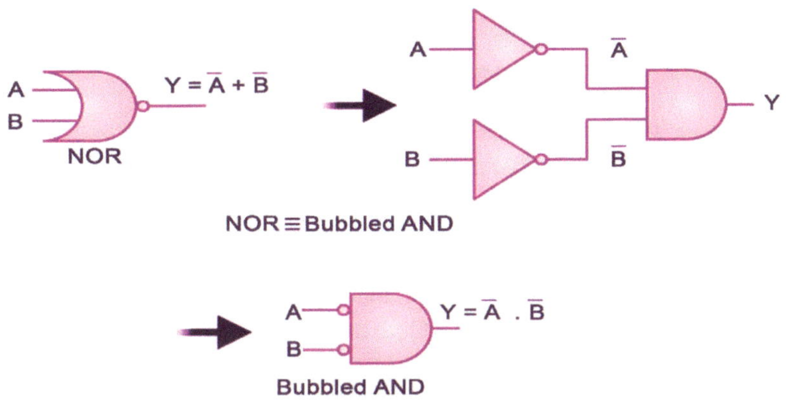

Here is a table that shows the verification of the second theorem of De Morgan:

A	B	$\overline{A+B}$	$\overline{A}$	$\overline{B}$	$\overline{A}.\overline{B}$
0	0	1	1	1	1
0	1	0	1	0	0
1	0	0	0	1	0
1	1	0	0	0	0

10.4.2. The AND Gate

Let's analyze the AND gate in Figure 10.1 first. Take a look at the logic symbol for an AND gate in Figure 10.1a (Watson, 1986).

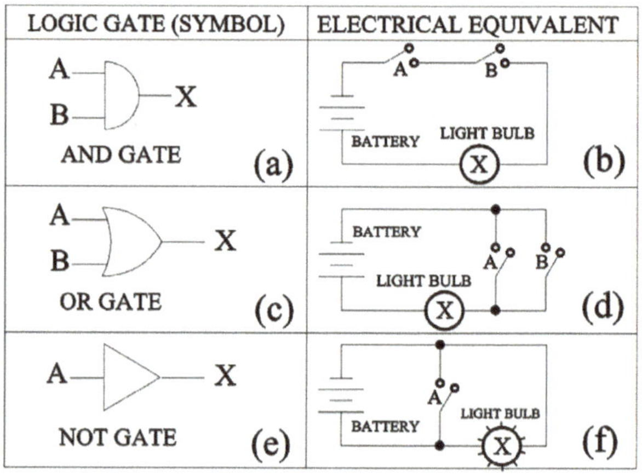

Figure 10.1. The logic symbol for an AND gate.

Source: https://www.nutsvolts.com/uploads/wygwam/NV_0416_Secura_Figure06.jpg

It has one output (X) and two inputs (A and B). The binary output (X) of the AND gate will equal 1 if there are binary 1 value on inputs A and B. This is how the AND gate functions (logically). Output (X) equals a 0 if input A has a 1 and input B has a 0. If input A is a 0 and input B is a 1, the result (X) will also be a 0. Finally, output (X) will likewise be a 0 if inputs A and B both have 0s.

Now that we have a working electrical circuit, let's replicate this AND gate logic. Examine the circuit diagram in Figure 10.1b. We have two switches, a lightbulb, and a nine-volt battery. Observe that since switch B is open, the light won't turn "ON" if we close switch A (i.e., the battery is not connected to the electric light).

The light stays off if switch A is open and switch B is closed. Keep in mind that the light will only turn on when both switches A and B are closed.

Now, if you look at Figure 10.2b, you should be able to see that our electrical light circuit in Figure 10.2a functions in the same manner as our old friend the AND logic gate. Examining the two logic tables in Figure 10.2 will show you this crucial point.

Notice in Figure 10.2b that the ones (1) in the logic truth table match (same position) the "closed" switches in the circuit truth table of Figure 10.2a (Watson, 1986).

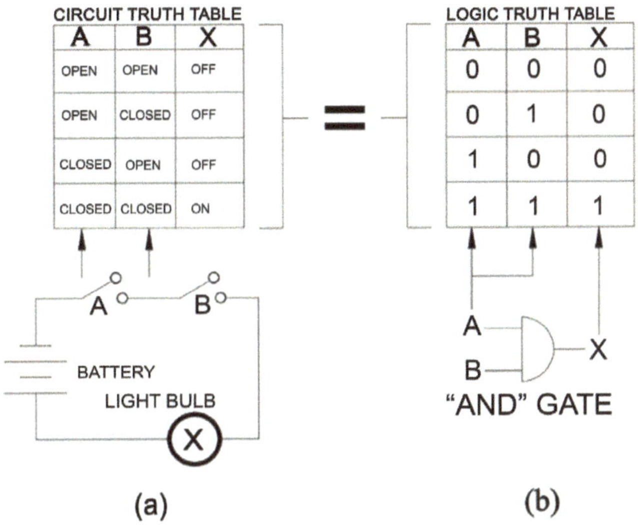

Figure 10.2. AND gate logic symbol.

Source. https://www.nutsvolts.com/uploads/wygwam/NV_0416_Secura_Figure07.jpg

Understanding the relationship between the AND gate logic symbol in Figure 10.2b and the electrical circuit in Figure 10.2a is crucial. The binary number system is highlighted in Figure 10.2b, where the output (X) equals 1 only when both A and B inputs are 1. Similarly, in Figure 10.2a, the light only turns on when both A and B switches are closed.

Now that we have the ones and zeros in Figure 10.2 converted to digital electronic signals (1 = five volts, 0 = zero volts), what happens? The answer to this question can be found in Figure 10.3. Watson, J. (1986).

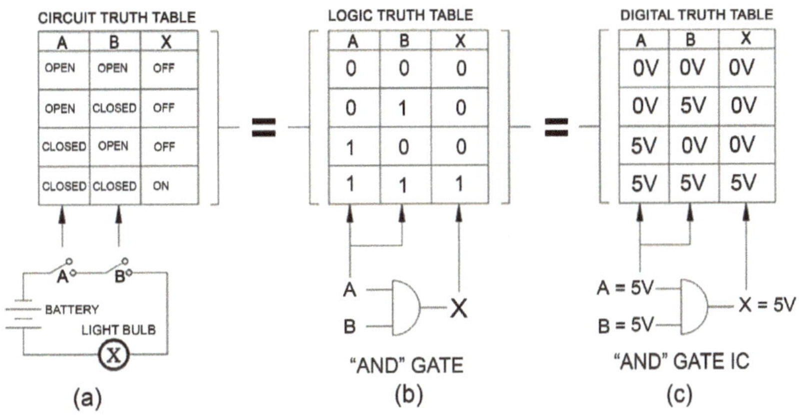

Figure 10.3. Digital electronic signals.

Source: https://www.nutsvolts.com/uploads/wygwam/NV_0416_Secura_Figure08.jpg

As you can see in Figure 10.3c, the truth tables remain the same even though we changed the binary ones and zeros to five volts and zero volts. This indicates that Figure 10.3a, the electrical circuit with two switches, A and B, functions similarly to Figure 10.3b and Figure 10.3c's AND gate logic truth table.

To understand digital electronics, you will undoubtedly wonder if you really need all these truth tables, AND gates, and 1 and 0 logic. The answer is absolutely yes! Understanding the input and output values of a digital logic circuit is the foundation of the digital world.

For example, look at the 7400 integrated circuit (IC) in Figure 10.4. You can purchase this IC on the Internet and verify that it works exactly like the truth tables we've constructed (Watson, 1986).

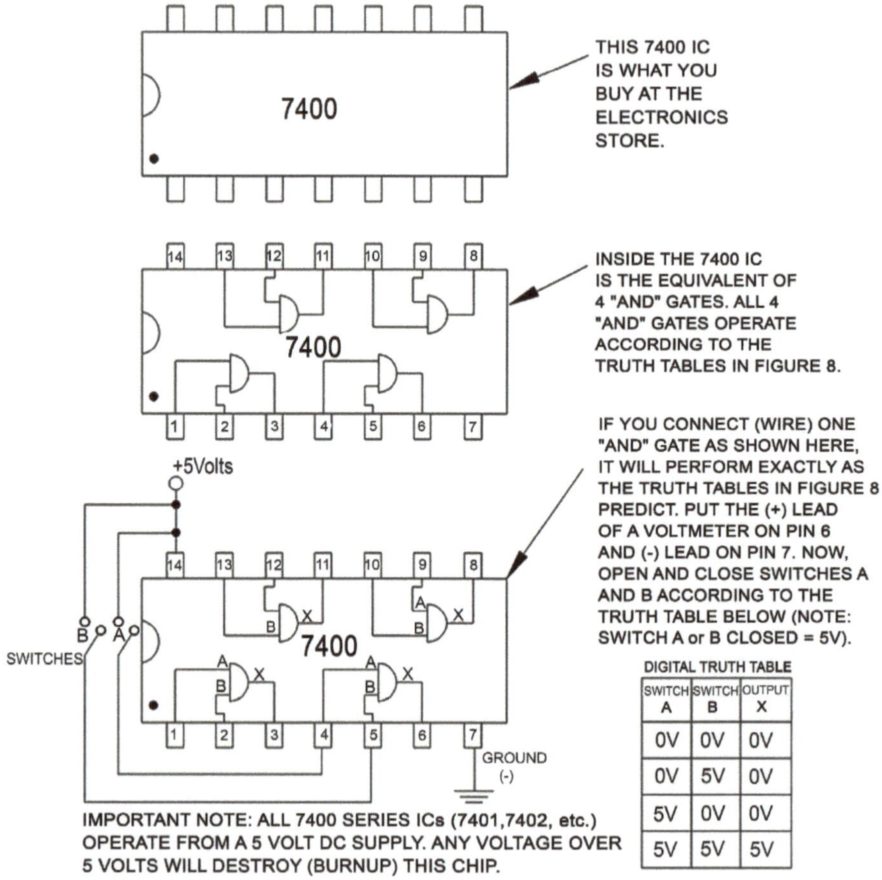

Figure 10.4. Integrated circuit (IC).

Source: https://www.nutsvolts.com/uploads/wygwam/NV_0416_Secura_Figure09.jpg

Within the 7400 IC, there is an AND gate discussed in previous figures. By following the connections shown in Figure 10.4, you can demonstrate the accuracy of the truth table in Figure 10.3c. This exercise also illustrates the use of digital logic gates and truth tables in constructing digital electronic circuits.

Now that you know the operation of an AND gate, let's apply this electronic version of the gate to a practical situation. Assume for the moment that there are two doors on an elevator that must be closed in order for the elevator to move up or down. An AND gate is connected to electrical switches (A and B) on each door. An electric motor that raises and lowers the elevator is connected to the AND gate's output (X).

It is our responsibility to ensure that the elevator can only move when both doors are fully closed. Let's arrange the functions we want the elevator to perform in a truth table. Take note of the table creation process in Figure 10.5. We calculated the response of the output (X) (motor ON or motor OFF) to each door's position (OPENED or CLOSED) (Watson, 1986).

ELEVATOR TRUTH TABLE

ELEVATOR DOOR SWITCH A	ELEVATOR DOOR SWITCH B	AND GATE OUTPUT X
OPEN (LOW)	OPEN (LOW)	STOP MOTOR
OPEN (LOW)	CLOSED (HIGH)	STOP MOTOR
CLOSED (HIGH)	OPEN (LOW)	STOP MOTOR
CLOSED (HIGH)	CLOSED (HIGH)	START MOTOR

NOTE THAT THE ONLY TIME THAT THE ELEVATOR MOTOR WILL TURN ON IS WHEN DOOR SWITCH A AND DOOR SWITCH B ARE 'BOTH' CLOSED.

=

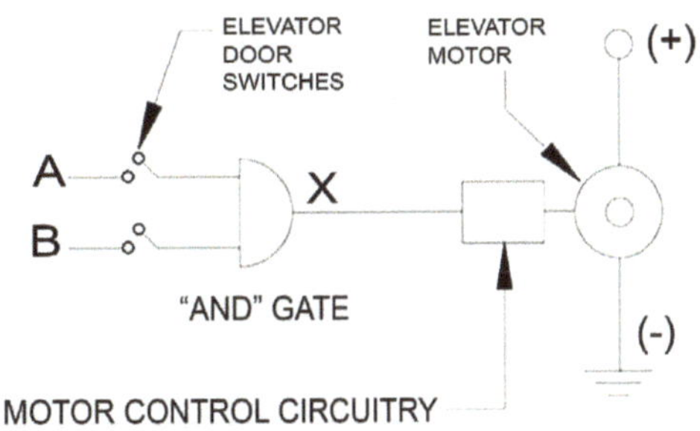

Figure 10.5. The output (X) (motor ON or motor OFF) should respond according to the position of each door (OPENED or CLOSED).

Source: https://www.nutsvolts.com/uploads/wygwam/NV_0416_Secura_Figure10.jpg

Also notice that the elevator truth table is the same as the AND gate truth table and our logic (binary) truth table.

ELEVATOR TRUTH TABLE

ELEVATOR SWITCH A	ELEVATOR SWITCH B	ELEVATOR MOTOR X
OPEN	OPEN	OFF
OPEN	CLOSED	OFF
CLOSED	OPEN	OFF
CLOSED	CLOSED	ON

=

AND GATETRUTH TABLE

INPUT A	INPUT B	OUTPUT X
0V	0V	0V
0V	5V	0V
5V	0V	0V
5V	5V	5V

=

BINARY TRUTH TABLE

A	B	X
0	0	0
0	1	0
1	0	0
1	1	1

Figure 10.6. The AND gate truth table.

Source: https://www.nutsvolts.com/uploads/wygwam/NV_0416_Secura_Figure11.jpg

By now, it should be clear that digital electronic circuits with real-world applications can be designed and constructed using the AND gate and its truth table.

Now, let's look at our next gate: the "OR" gate.

10.4.3. The OR Gate

The AND gate is one of the three fundamental logic gates (AND, OR, and NOT), as was previously mentioned. Let's examine the differences between the AND and OR gates.

It is evident from examining the electrical circuit in Figure 10.7 that two switches (A, B) are linked in parallel. Observe that light (X) will turn on and the electrical circuit will be completed if switch A is CLOSED (Watson, 1986).

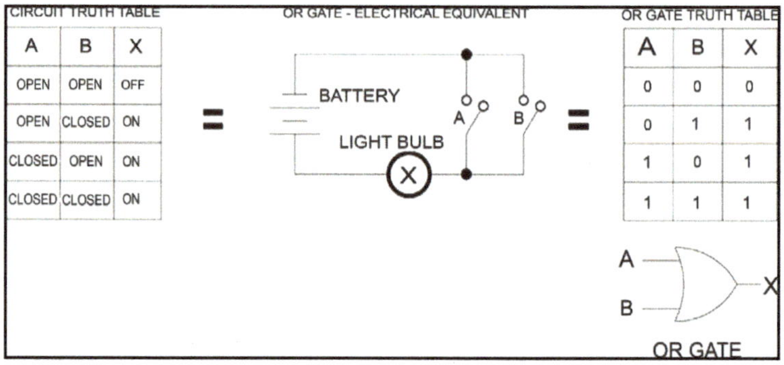

Figure 10.7. The electrical circuit and light (X) will turn ON.

Source: https://www.nutsvolts.com/uploads/wygwam/NV_0416_Secura_Figure12.jpg

That means that the light will turn on as well if switch A is OPEN and switch B is CLOSED. The light is ON when both switches are closed. It should be noted that the light will turn on as long as one switch (A OR B) is closed.

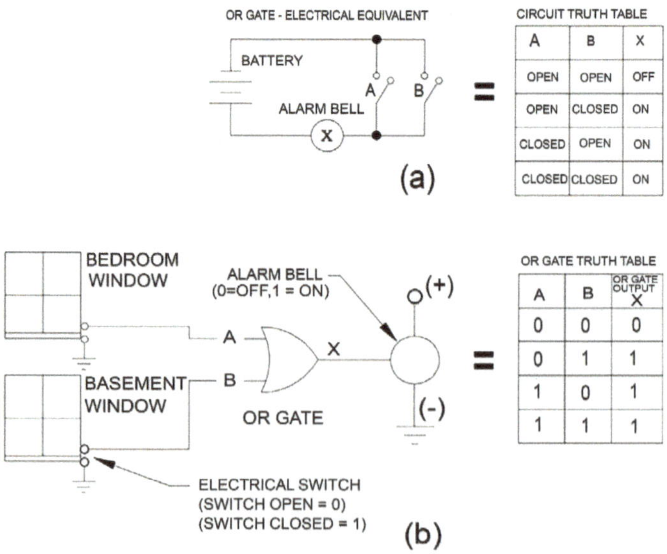

Figure 10.8. A simple alarm system based on the OR gate.

Source: https://www.nutsvolts.com/uploads/wygwam/NV_0416_Secura_Figure13.jpg

When both switches are OPEN, that's the only moment when the light goes out. Examine Figure 10.7 until you can see how the Electrical Circuit Truth Table and the OR Gate Truth Table align.

Let's look at a practical application for the OR gate. An OR gate circuit can be used to protect a house from a burglar. Figure 10.8 shows a simple alarm system based on the OR gate (Watson, 1986).

Figure 10.8b makes it clear that the alarm will sound if either the basement window switch (B) or the bedroom switch (A) is CLOSED (window is open). Take note that switch A OR B can activate the alarm in the OR gate truth table (Figure 10.8b). As you can see, just as the AND gate logic truth table was used to start and stop an elevator motor, the OR gate logic truth table can be used to design a burglar alarm system. Look now at Figure 10.9 so you can see the difference between an AND gate and an OR gate (Watson, 1986).

OR GATE TRUTH TABLE

A	B	X
0	0	0
0	1	1
1	0	1
1	1	1

AND GATE TRUTH TABLE

A	B	X
0	0	0
0	1	0
1	0	0
1	1	1

"OR" GATE

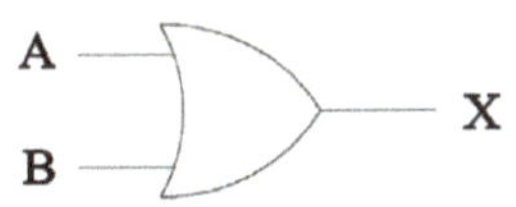

"AND" GATE

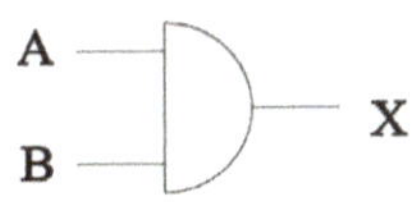

Figure 10.9. OR gate logic truth table.

Source: https://www.nutsvolts.com/uploads/wygwam/NV_0416_Secura_Figure14.jpg

Take note of the primary distinction between the AND and OR gates. Whereas the AND gate only outputs a 1 when both inputs A AND B are 1, the OR gate will output a one (1) when either input A OR B is a 1.

10.4.4. The NOT Gate (INVERTER)

Let's turn our attention to the last of the three basic gates: the NOT gate (Watson, 1986).

NOT GATE TRUTH TABLE

A (INPUT)	X (OUTPUT)
1	0
0	1

IF A = 1 THEN X = 0
IF A = 0 THEN X = 1

=

NOT GATE (INVERTER)

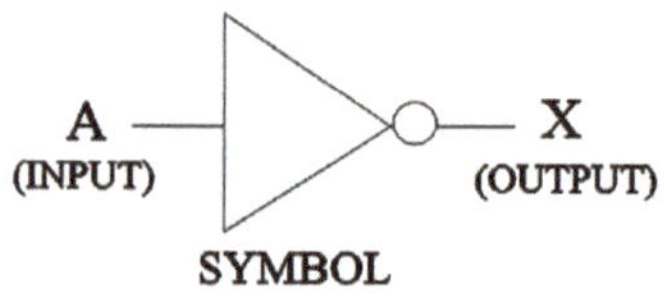

CIRCUIT TRUTH TABLE

A (SWITCH)	X (LIGHT)
CLOSED	OFF
OPEN	ON

=

NOT GATE ELECTRICAL EQUIVALENT

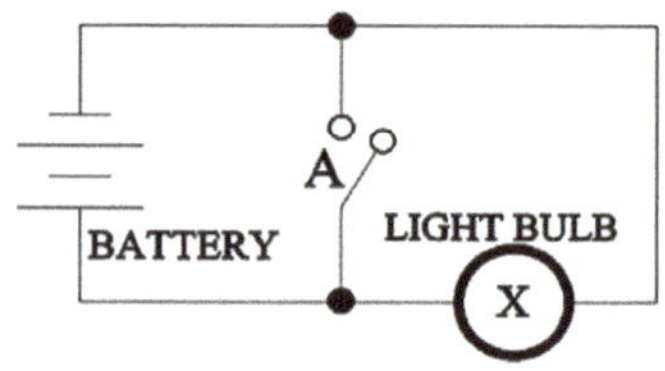

Figure 10.10. The NOT gate.

Source: https://www.nutsvolts.com/uploads/wygwam/NV_0416_Secura_Figure15.jpg

An "inverter" is the term used to refer to the NOT gate. It has a single output (X) and a single input (A). A NOT gate's (A) output will be 0 if its input is a 1. In contrast, a NOT gate's (A) output will be a 1 if its input is a 0. It is now clear why the NOT gate is referred to as an "inverter."

Examine the electrical circuit in Figure 10.10 to gain a better understanding of how the NOT gate functions. In this case, if switch A is CLOSED, the light will not receive enough electrical current to turn on because the battery's current will bypass it.

The light will stay ON, though, if switch (A) is OPEN because electrical current will go straight to the light and not through the switch. Digital logic circuits make extensive use of the NOT gate.

10.4.5. The NAND, NOR, EX-OR, and EX-NOR Gates

You brought up the significance of the NOT gate in digital electronic circuits. You will now see how to create both NAND and NOR gates using the NOT gate. We'll also discuss the EX-OR and EX-NOR gates, two additional unique gates that are utilized in the digital electronic world.

10.4.5.1. The NAND Gate

Let's start with the NAND gate. The NAND (Not AND) gate is basically an AND gate with a NOT gate (inverter) connected to its output (Watson, 1986).

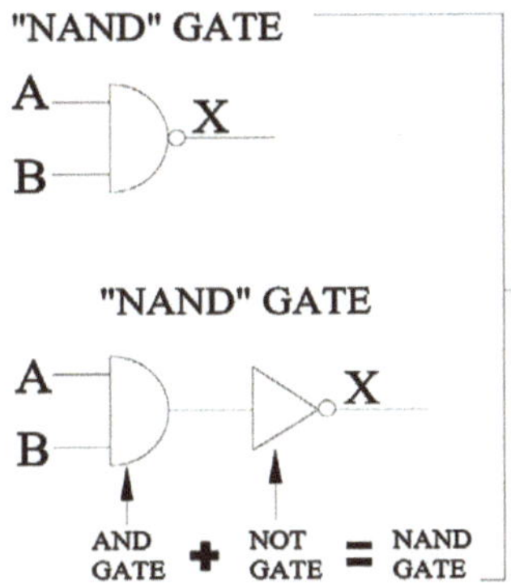

NAND GATE TRUTH TABLE

A	B	X
0	0	1
0	1	1
1	0	1
1	1	0

Figure 10.11. The NAND gate.

Source: https://www.nutsvolts.com/uploads/wygwam/NV_0416_Secura_Figure16.jpg

In Figure 10.11, take note of the circle at the NAND gate's output (X). In digital electronics, this circle is the standard symbol for inversion (NOT = INVERTER). With its triangle and circle symbols, it is logically equivalent to the NOT gate.

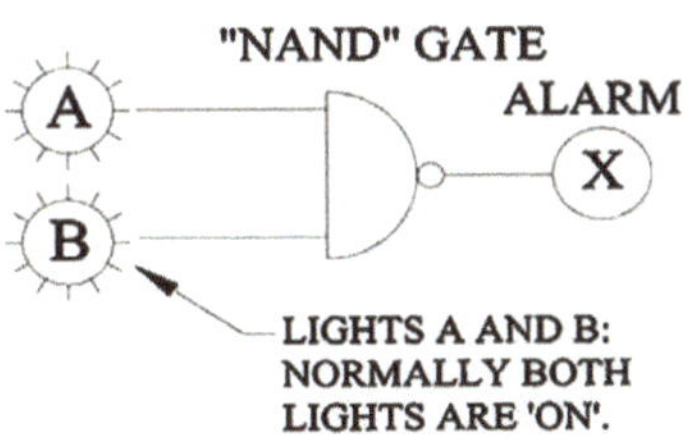

NAND GATE CIRCUIT TRUTH TABLE

A (LIGHT)	B (LIGHT)	X (ALARM)
OFF	OFF	ON
OFF	ON	ON
ON	OFF	ON
ON	ON	OFF

NOTICE ABOVE THAT AS LONG AS BOTH LIGHTS A AND B ARE TURNED 'ON' THE ALARM DOES 'NOT' SOUND. ONLY WHEN EITHER LIGHT (A,B) OR BOTH LIGHTS ARE 'OFF' WILL THE ALARM SOUND.

NAND GATE LOGIC TRUTH TABLE

A	B	X
0	0	1
0	1	1
1	0	1
1	1	0

Figure 10.12. A truth table of NAND gate.

Source: https://www.nutsvolts.com/uploads/wygwam/NV_0416_Secura_Figure17.jpg

Henceforth, inverting the signal or logic (1 to 0, 0 to 1, zero volts to five volts, five volts to zero volts, HIGH to LOW, LOW to HIGH) is indicated whenever a circle appears on the input or output of a gate.

The truth table, as shown in Figure 10.11, demonstrates that a NAND gate's output will only ever be 0 (or LOW) when both of its inputs, A and B, are 1 (HIGH). Observe that any 0 (LOW) on input A or B will cause the output (X) to be a 1. This is also shown in the truth table. Put differently, a NAND gate will yield a 1 (HIGH) at the output (X) for any 0 (LOW) on the input. Let's apply the NAND gate to a real-world scenario, just as we did with the AND and OR gates that were previously covered. Assume we construct an electrical circuit for a NAND gate that watches two outdoor security lights and sounds an alarm if one of them goes out (OFF). Once again, the first thing we do is set up a truth table for our light circuit (Watson, 1986).

Recall from the truth table that the alarm will activate if "either" light A or light B turns off. It should be noted that the alarm will only stay off if "both" lights are on.

It is impossible to overestimate the significance of truth tables in the design of electronic circuits. If a digital electronic circuit uses logic gates (AND, NOT, OR, NAND, etc.), you need to specify the function of the circuit. Setting up a truth table is, of course, the best way to see what the input and output logic of your circuit will be. Now let's proceed to the NOR gate, which is our next gate.

10.4.5.2. The NOR Gate

Once more, the NOR gate's precise operation is explained by the truth table in Figure 10.13. A NOR gate's output (X) will never equal 1 unless "both" inputs A and B are 0. Another way to say it is when both inputs to a NOR gate are 0 (LOW), the output of a NOR gate will be 1 (HIGH) Watson, J. (1986).

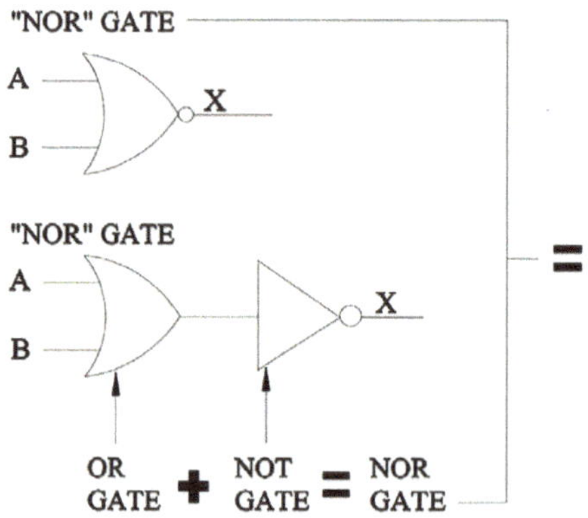

NOR GATE LOGIC TRUTH TABLE

A (INPUT)	B (INPUT)	X (OUTPUT)
0	0	1
0	1	0
1	0	0
1	1	0

Figure 10.13. A NOR gate are 0 (LOW), the output of a NOR gate will be 1 (HIGH).

Source: https://www.nutsvolts.com/uploads/wygwam/NV_0416_Secura_Figure18.jpg

Naturally, there is an alternative perspective on the truth table, just like with our other gates. Be aware that the NOR gate's output (X) is 0 if "either" of its inputs, A or B, is 1. We can also state that any 1 on a NOR gate's input (A or B) will result in 0 (LOW) at the gate's output (X).

Let's move on to the next gate: the EX-OR.

10.4.5.3. The EX-OR Gate

When switch A is in the 0 position and switch B is in the 1 position, as shown in Figure 10.14c, the light will turn ON. The reverse is also true (Watson, 1986).

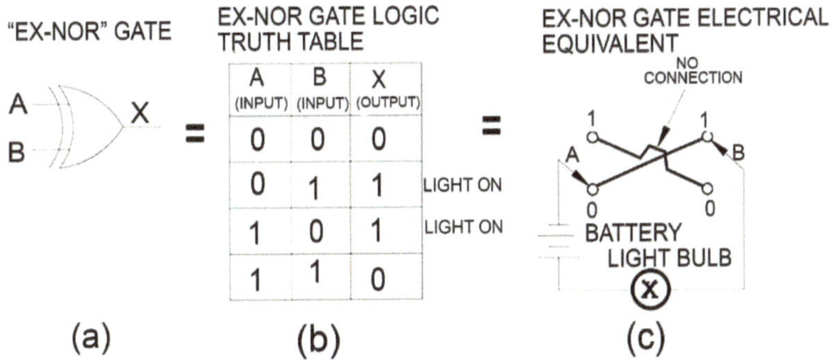

A (INPUT)	B (INPUT)	X (OUTPUT)	
0	0	0	
0	1	1	LIGHT ON
1	0	1	LIGHT ON
1	1	0	

Figure 10.14. The EX-OR Gate reverse is also true.

Source. https://www.nutsvolts.com/uploads/wygwam/NV_0416_Secura_Figure19.jpg

The light will turn on only when switch B is in the 0 position and switch A is in the 1 position. If both switches are in the same position (1, 1 or 0, 0), the light will not come on. Another way of phrasing it is: Either 1 or 0, but not both. Now, let's talk about EX-NOR gates.

10.4.5.4. The EX-NOR Gate

As you can see in Figure 10.15c, the light will only turn ON when switch A and switch B are both in either the 1 position or the 0 position (Watson, 1986).

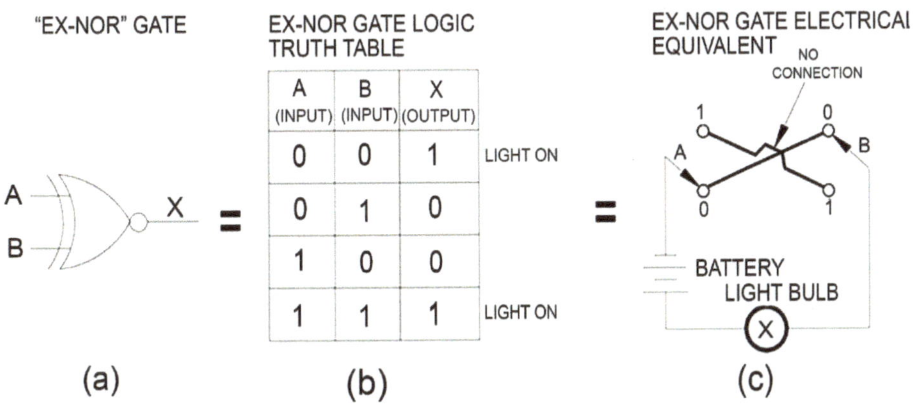

A (INPUT)	B (INPUT)	X (OUTPUT)	
0	0	1	LIGHT ON
0	1	0	
1	0	0	
1	1	1	LIGHT ON

Figure 10.15. EX-NOR gates.

Source. https://www.nutsvolts.com/uploads/wygwam/NV_0416_Secura_Figure20.jpg

Stated differently, when "both" switches A and B are in the same position, you have an output (light ON); however, when A and B are either 1, 0 or 0, 1, you do not.

In digital circuits, the EX-OR and EX-NOR gates are frequently utilized. One use for the EX-NOR, which is widely employed in communications, is confirming the equality of two binary numbers.

Recall that we can convert binary numbers into decimal numbers and compare two decimal numbers by comparing their binary equivalents using EX-NOR gates.

For example, if we needed to compare two binary numbers like 10 and 01 to verify that they are equal, we could use the EX-NOR gate circuit in Figure 10.16 (Watson, 1986).

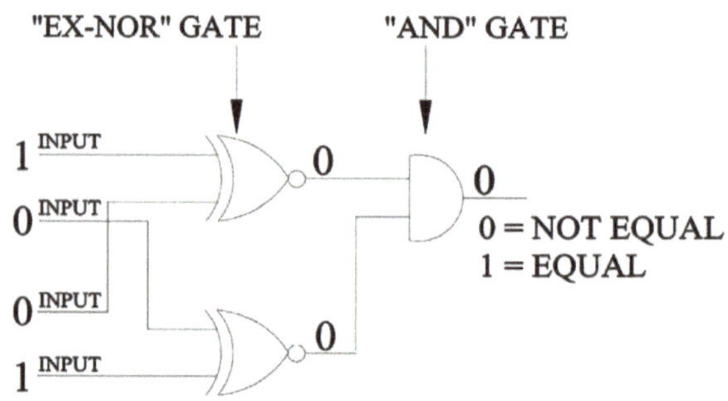

Figure 10.16. EX-NOR gate circuit.

Source: https://www.nutsvolts.com/uploads/wygwam/NV_0416_Secura_Figure21.jpg

When binary 1, 0 and 0, 1 are fed into the inputs of both EX-NOR gates, the AND gate's output is 0. Hence, binary numbers 1, 0 and 0, 1 are not equal. Attempt to input 1, 0 and 1, 0 into both EX-NOR gates now. The AND gate's output will equal 1, indicating the equality of the two binary numbers.

10.5. COMBINATIONAL AND SEQUENTIAL CIRCUITS

Sequential and combinational circuits are the most commonly used circuits in digital electronics. These two broad categories of circuits—one independent of time and the other dependent on it—are defined in digital electronics.

In digital electronics, the most common types of digital circuits are combinational and sequential circuits. According to the definition, a sequential circuit's outputs depend on its previous state, whereas in a combinational circuit, each output depends only on its inputs. Therefore, their methods for producing different outputs are what really set them apart. The fact that sequential circuits contain the fundamental memory elements for digital circuits while combinational circuits do not is another important distinction between the two types of circuits.

10.5.1. Combinational Circuit

A circuit that bases its output at any given time only on the current combination of inputs is known as a combinational circuit. NAND, NOR, and NOT logic gates constitute a combinational logic circuit. The fundamental components of a combinational circuit's circuitry are these logic gates. A combinational circuit can perform a variety of tasks, such as sharing data, converting code, and performing arithmetic operations on two operands

In combinational circuits, the input at any given time determines the output that is produced. Data transmission, code conversion, arithmetic and logical functions are the three main uses for combinational logic circuits. Combination logic circuits include, among others, adders, subtractors, and comparators.

1. In this output depends only upon present input.
2. Speed is fast.
3. It is designed easily.
4. There is no feedback between input and output.
5. This is time independent.
6. Elementary building blocks: Logic gates

7. Used for arithmetic as well as Boolean operations.
8. Combinational circuits don't have the capability to store any state.
9. As combinational circuits don't have clocks, they don't require triggering.
10. These circuits do not have any memory element.
11. It is easy to use and handle.

10.5.2. Sequential Circuit

Sequential circuits are a type of digital logic circuit where the output is dependent on both the previous and current inputs. Therefore, a sequence of circuits that are connected in a feedback path of combinational circuits and memory elements is known as a sequential circuit.

Devices with the capacity to store binary data are known as memory elements. The state of a sequential circuit is indicated by the binary data that is stored in the memory elements at any given time. Therefore, the time sequence of inputs and previous outputs determines the output of a sequential circuit.

Digital circuits are made up primarily of sequential circuits. A finite state machine is a popular illustration of a sequential circuit. Two categories—synchronous and asynchronous sequential circuits—can be used to further categorize sequential circuits.

1. In this output depends upon present as well as past input.
2. Speed is slow.
3. It is designed tough as compared to combinational circuits.
4. There exists a feedback path between input and output.
5. This is time dependent.
6. Elementary building blocks: Flip-flops
7. Mainly used for storing data.
8. Sequential circuits have the capability to store any state or to retain earlier state.
9. As sequential circuits are clock dependent they need triggering.
10. These circuits have memory elements.
11. It is not easy to use and handle.

10.5.3. Difference between Combinational and Sequential Circuit

Combinational circuits have outputs that depend only on the current inputs, which is the main distinction between them and sequential circuits. On the other hand, sequential circuits consist of a sequence of inputs and outputs, where the output depends on both the inputs present at that time and the outputs that came before it. Both types of circuits are fundamental parts of digital electronics. Sequential and combinational circuits operate differently from one another in specific ways. Following our discussion of the functions and duties of each circuit separately, we will move on to the differences between combinational and sequential circuits. We'll begin by examining the fundamental distinctions between sequential and combinational circuits before taking a closer look at each one for a quick overview. In digital electronics, the most widely used electric circuits are combinational and sequential circuits. Combinational circuits are time-independent, while sequential circuits are time-dependent.

The following table highlights how a combinational circuit is different from a sequential circuit:

Key	Combinational Circuit	Sequential Circuit
Definition	A Combinational Circuit is a type of circuit in which the output is independent of time and only relies on the input present at that particular instant.	A Sequential circuit is a type of circuit where output not only relies on the current input but also depends on the previous output.
Feedback	Since output does not depend on the time instant, no feedback is required for its next output generation.	The output relies on its previous feedback so output of previous input is being transferred as feedback used with input for next output generation.
Performance	As the input of current instant is only required in case of Combinational circuit, it is faster and better in performance as compared to that of Sequential circuit.	Sequential circuits are comparatively slower and has low performance as compared to that of Combinational circuit.
Complexity	No implementation of feedback makes the combinational circuit less complex as compared to sequential circuit.	The implementation of feedback makes sequential circuit more complex as compared to combinational circuit.
Elementary Blocks	The elementary building blocks of a combinational circuit are its logic gates.	The building blocks of a sequential circuit are the logic gates along with flip flops.
Operation	Combinational circuits are mainly used for arithmetic as well as Boolean operations.	Sequential circuits are mainly used for storing data.

10.5.4. Flip-Flops and Registers

In electronics, bistable multivibrators, known as flip-flops and latches, are circuits with two stable states that can store state information. Signals applied to one or more control inputs can cause the circuit to change state, and it will then output that state (often along with its logical complement). In sequential logic, they are fundamental storage component. Fundamental components of digital electronics systems used in computers, communications, and numerous other types of systems are flip-flops and latches.

A single bit of data, or binary digit, is stored using flip-flops and latches as data storage components; one of their two states denotes a "one," and the other a "zero." Such a circuit is known as sequential logic in electronics, and it can be utilized to store states. When utilized in a finite-state machine, the current state (and thus, prior inputs) as well as the machine's current input determine both the output and the subsequent state. In addition, they can be used to synchronize variable-timed input signals with a reference timing signal and count pulses. In the past, the term flip-flop was used to describe both level-triggered and edge-triggered circuits that stored a single bit of data using gates. However, modern authors now use the term flip-flop only for edge-triggered storage elements, reserving the term latch for level-triggered circuits. This distinction helps avoid confusion, as the terms "edge-triggered" and "level-triggered" can be used to clarify the type of circuit being discussed. In a level-triggered latch, the output changes when enabled, while an edge-triggered flip-flop only changes its output on a clock edge.

CASE STUDY

ELECTRONIC DESIGN: AMPIC – PUBLIC DEMONSTRATION DESIGN

In a bid to prove our design process expertise (always difficult when often heavily under NDA) DSL have taken the unprecedented step of publishing all stages of one of our internal product development designs. The AMPIC.

Background

The concept that originally spawned the **"Functional Building Block"** approach to electronic design, the AMPIC, was conceptualized in reaction to feedback from DSL's client base; craving the benefit from emerging ARM technology but without the necessary in-house expertise to develop a solution from an off-the-shelf development board. Those same customers, desperate to get their product to market ahead of their competitors, were less keen on 'from scratch' design time-scales, risk and of course, cost.

The **"Functional Building Block"** approach was born. Simplistically translating to DSL 'pre-designing' a fully functional development board, designed from the ground up to be fully featured yet segmented easily into its functional parts, from schematics through to circuitry. Any client specific design derived from this starting platform then effectively starts near the end, with the risk, cost and design time-scales long behind us.

Challenge

The challenge of course was to create a development platform that whilst feature rich, wasn't unnecessarily complex and could be used in parallel to support a client's software development (further expediting overall completion time) whilst their specific derivative is being built.

The second challenge was to incorporate relevant functionality for the target industries, which not only covered multiple industries but also multiple applications within, including Remote Sensors, Standalone Control Systems and Industrial I/O conversion, though of course equally not restricted to them.

Solution

DSL's internal engineering, sales and marketing teams collaborated throughout the project, the result – the **AMPIC 'Functional Building Block.'**

Process

The AMPIC development was undertaken rigidly inline with DSL's standard design process. Click each hyperlinked section for the applicable document.

Step 1 – Requirements Definition – Requirements definition compiled via multiple internal departmental input.

Step 2 – Design Proposal (Outline)

Step 3 – Design Proposal: Quotation (Costs/Time-scales)

Step 4 – Dynamic Project Plan

Step 5 – Design Specification

Step 6 – Schematic Design

Step 7 – PCB Design – (3D PCB Layout. Note: requires Adobe PDF reader only)

Step 8 – Prototypes – (links to product datasheet, though feel free to come and view the prototypes!)

CLASS ACTIVITY

Students will explore NAND, NOR, XOR, and XNOR. Boolean algebra and methods for deriving combinational logic circuits are introduced. Students will build a variety of basic circuits to explore the function of different logic gates.

SUMMARY

- Digital electronic principles form the foundation of modern computing and electronic devices, encompassing the study and application of digital signals to represent and process information.
- Analog signals are signals with continuous values, i.e., they are continuous in both time and magnitude.
- Digital signals on the other hand are more tolerant to noise but can be completely corrupted in the presence of excess noise.
- Number systems relate quantities and symbols. The number of digits or basic symbols in a particular number system is represented by the base or radix of the number system.
- The concept of negative sign is not there in binary number system. Although various methods have been developed, the most popular of them all are 1's complement and 2's complement.
- Real numbers refer to both integers and fractional numbers. In computers, we use the binary number system.
- BCD or Binary Coded Decimal is that number system that employs binary numbers or digits to represent a decimal number.
- Asynchronous circuits are digital circuits that operate without a clock signal, using signals from other parts of the circuit to control the flow of data.
- A logic gate is a building block for any digital circuit. These logic gates need to make the decision of combining various inputs according to some logical operation and produce an output.
- In any digital electronic circuit that employs logic gates (AND, NOT, OR, NAND, etc.), you must define what you want the circuit to do.

REVIEW QUESTIONS

1. What do you understand by number systems in digital electronics?
2. Express the floating-point number representation.
3. List the difference between digital and analog electronics.
4. Describe the digital electronics Boolean algebra and logic gates.
5. Explain about combinational and sequential circuits.

REFERENCES

1. Nair, B. S. (2002). *Digital Electronics and Logic Design*. PHI Learning Pvt. Ltd. p. 289.
2. Farhat, H. A. (2004). *Digital Design and Computer Organization* (Vol. 1). CRC Press. p. 274.

3. Maini, A. K. (2007). *Digital Electronics Principles, Devices and Applications*. Chichester, England: John Wiley & Sons Ltd.
4. Mano, M. M., & Kime, C. R. (2004). *Logic and Computer Design Fundamentals* (3rd ed.). Upper Saddle River, NJ, USA: Pearson Education International. p. 283.
5. Pedroni, V. A. (2008). *Digital Electronics and Design with VHDL*. Morgan Kaufmann. p. 329.
6. Watson, J. (1986). Introduction to digital electronics. In *Mastering Electronics. Macmillan Master Series*. Palgrave, London. https://doi.org/10.1007/978-1-349-08533-0_18.

Index

A

Alternating current (AC) 36, 63
Amplification 71, 76
Amplifier 189, 199, 200, 204, 207, 209, 220, 222, 224, 226
Amplifier configuration 300
Amplitude modulation 77
Analog electronic circuits 266, 268
Analog sensor 331
Anode 37, 40, 62, 68
Apparatus functions 161
Armstrong circuit 308
Armstrong oscillator 295, 307, 308, 310, 327, 328
Audio amplifiers 76, 77
Audio frequencies 297
Automotive industries 227

B

Base-collector voltage 71
Base-emitter voltage 71
Battery s capacity 233
Biasing chain 300
Bidirectional devices 160
Binary digit 331, 335, 337, 338, 369
Binary number system 331
Binary representation 332
Binary signals 228
Bipolar devices 157, 163
Bipolar junction transistor (BJT) 72, 78, 109
Bipolar transistor 72, 73, 94
Body diode 113
Breakdown voltage 36, 39, 48, 55, 59
Broadcast transmitters 114

C

capacitors 114, 154
Capacitors 270, 273, 275, 283
carbon nanotube field-effect transistor (CNTFET) 117
Carrier wave 295
Cell phones 73, 75
Channel length modulation 113
channel resistance 119
Circuit designer 195
Circuit operation 229
Closed-loop gain 189, 197, 225
Collector-base junction 71
Collector-base junction (CBJ) 78
Common-base configuration 71
Common-collector configuration 71
Common-emitter configuration 71
Common-mode rejection ratio (CMRR) 269, 271
Communication equipment 296
Communication networks 228, 263
Communication systems 296, 327

Comparator 189
Complementary metal-oxide-semiconductor (CMOS) 72
Complex communication systems 36
Computational purposes 194
Computing 228, 230, 240
Conduction band 3, 4, 5, 12, 13, 14, 17, 18, 23, 24, 33
Conductivity 2, 4, 5, 7, 8, 17, 19, 22, 25, 33
Configuration 266, 270, 271, 275
Consumer electronics 2
Contemporary technology 228, 332
Continuing development 228
conventional drain current 119
Countless devices 296
Crystal oscillator 295, 322
Current flow 3, 14, 17, 22, 23, 25
Current Series Feedback 199, 200

D

Darlington pair 227
Darlington Pair 232
Ddigital electronics 296
Decimal number 331, 334, 336, 337, 338, 340, 344, 347, 373
Decoupling capacitor 232
Dependability 117, 266
Depletion mode 113, 141
Depth of Discharge (DOD) 233
Desensitivity 189
Differential amplifier 195
Differential inputs 266, 268, 291
differential voltage 194, 195, 225
Digital circuits 227, 263, 266
Digital electronic systems 266
Digital revolution 332
Digital systems 228
Diode 35, 36, 37, 38, 39, 40, 46, 47, 48, 49, 52, 55, 57, 58, 59, 61, 62, 63, 64, 65, 67, 68, 69, 70
Diode For alternating current (DIAC) 158
Direct current (DC) 36
Directed current flow 38
Discrete time signals 333
Discrete transistors 229
Dominant-pole frequency 272
Doping 1, 2, 5, 32

E

Electrical conductivity 2, 4, 12, 20, 21, 22, 23, 24, 29, 32, 33
Electrical energy 119
Electrical engineering 269
Electrical signals 229, 230
Electric field 3, 5, 13, 15, 23, 24, 36, 48, 56, 57, 58
Electricity 2, 3, 20, 33
Electrodes 114
Electronic band structure 4
Electronic circuit 191, 225
Electronic circuit design 36
Electronic circuitry 227
Electronic circuit simulator 229
Electronic component 36, 37, 62, 67, 68
Electronic devices 2, 3, 5, 16, 20, 23, 29, 33
Electronic oscillators 307
Electronics technology 117
Electronic switches 76
Electrons 3, 15, 27, 33, 34
Electrons flow 75
Emitter-base junction 71
Energy absorption 3
Energy level 42, 50, 51, 52
Engineering principle 268
Enhancement mode 116, 120, 122, 127, 131, 132, 139
Equilibrium state 44
External devices 231

F

Feedback 189, 190, 191, 192, 193, 194, 195, 196, 197, 198, 199, 200, 201, 203, 205, 206, 208, 209, 210, 211, 212, 213, 214, 215, 216, 217, 220, 221, 222, 223, 224, 225, 226
Feedback network 298, 307, 327
Fermi level 42, 43, 49, 50
Field-effect transistor (FET) 72
Filtering 266, 268

Flat-band voltage 113
Forward blocking mode 160
Free charge carriers 5
Frequency Selective Network 297
Frequency stability 295

G

gallium arsenide 158, 186
Gallium arsenide (GaAs) 6, 7, 33
Gate capacitance 113
Gate leakage 113
Gate terminal 119, 120, 125, 128, 147, 149
Gate voltage 160
Global positioning systems (GPS) 296
Grounded emitter 274

H

Hardware transmitters 333
Harmonic balance analysis 304
Harmonic oscillator 295
Harmonics 299, 311
Healthcare 228
Hetero Bipolar Junction 75
Heterojunction 75
High Electron Mobility Transistor 118
High-frequency applications 36, 66, 67
High-frequency operation 76
High power control 162
High radio waves 297
Holding current 160, 163, 168, 170, 173

I

Impedance scaling 208
Impurities 2, 4, 7, 16, 18, 21, 24
Indifference 73
Inductors 228, 229
Industrial controls 73
Industrial control systems 267
Information signals 296
Instrumentation 266, 268, 284, 292, 293
Insulated-gate bipolar transistor (IGBT) 157, 158
Insulated-gate FET (IGFET) 113
Integrated circuits 2, 3, 8, 29
Integration 115, 119, 153
Interconnecting circuit 158
Intrinsic semiconductor 1, 5, 15, 16, 23, 24, 33
Isolation amplifiers 268

J

Junction FET (JFET) 113
Junction transistor 74

K

Kirchhoff's Current Law (KCL) 75

L

Latching current 160, 167, 168
Lattice position 4
Light emission 6
Light-emitting diodes (LEDs) 5
Linear analysis 298, 305, 306
Load current information 201
Load resistor 189
Logic symbol 331
Loop transmission 203, 205, 206

M

Magnetism 227
Magnitude 265
Manipulation 2
Mathematical equation 75
Mathematical functions 266
Measurement systems 266
Mechanical deformation 158
Mechanical resonance 295
Medical devices 2
memory chip 122
metal-oxide-semiconductor FET (MOSFET) 114
Metal oxide semiconductor (MOS) 157
Microcontroller 231, 232
Microprocessor 122
Microwave Office 301
Miniaturization 2, 29, 30, 227
Modulation 295

N

Nanotechnology 228
Network combination 201
Null detector 189
Number system 331

O

Octal number 334, 337
Offset circuit 297
Offset voltage 269, 271, 279, 283, 289
Open-loop amplifier 268
Operational amplifiers 266, 270, 271, 272, 280, 281, 282, 287, 290, 291, 292
Organic field-effect transistors (OFET) 113
Oscaprobe 298, 327
Oscillator circuit 297, 299, 307, 308, 312, 320, 321, 322, 327, 328
Oscillator design cost 306
Oscillators 37
Oxide layer 114, 118, 120, 125, 126, 149, 152, 154

P

Parallel input-signal connections 201
Parameters 269, 270, 271, 274, 283, 292
Passband 197
Periodic waveform 296
Phase-locked loops (PLLs) 296
Phase noise 298, 299, 303, 327
Piezoelectric material 295
Potentiometer 189
Power electronics 115, 122, 146, 147, 152
Power engineers 159
Power supply 196, 212, 222
Power supply rejection ratio (PSRR) 271
Printed circuit boards (PCBs) 227
Product development cycle 233

R

Radio-frequency (RF) 72
Rail voltage 189
Receivers 333
Rectification 39, 63, 66
Redundancy 331, 350
Relaxation oscillator 295
Renewable energy systems 2
Resistivity 3, 4, 11, 21, 25, 33
Resistors 114, 134, 154
Resonator 301, 302, 303, 320
Reverse bias collector junction 75
Reverse blocking mode 160, 161
Reverse connection error 38
Reverse voltage 38, 52, 67

S

Saturation current 47
Schematic diagrams 229, 230
Schottky barrier diodes 41, 68
Schottky diodes 35, 36, 66, 67
Scientific instruments 267
Self-stabilizing system 189, 193
Semiconductor channel 115, 117
Semiconductor crystal 2, 12, 13, 15, 16
Semiconductors 2, 3, 4, 5, 6, 11, 12, 20, 27, 29, 30, 32, 33, 34
Semiconductor technology 2, 29, 30, 31
Sensitive components 36
Serial communication 227
Signal conditioning 266, 268
Signal cycle 114
Signal demodulation 36
Signal frequency 272
Signaling circuit 230
Signal processing functions 268
Silicon Bond Model 4
Silicon controlled rectifier (SCR) 158
Simulation Control 306
Simulation software 227
Sine wave 295
Sinusoidal waves 296
Smartphones 2, 29, 30
Source impedance 196
Stabilization 39
Standalone components 268
State of Charge (SOC 233
Straightforward devices 228

Subthreshold region 113
Switch mechanism 230
Synonymous 159

T

Telecommunications 2
Televisions 73, 105
Terminal device 36
Thermal energy 5, 12, 13, 14, 15, 18, 23
Thermal noise 273, 282
thermionic emission 158
Threshold voltage 37, 68
Thyristor 157, 161
Timekeeping 295
Traction choppers 161
Transfer resistor 71, 79
Transistor parameter variations 300
Transition 160
Triode for alternating current (TRIAC) 158

U

Unijunction transistor (UJT) 157, 158
Unipolar transistor 72

V

Valence band 3, 4, 12, 13, 14, 17, 23, 24, 25, 33
Valence electrons 5, 12, 13, 14, 16, 17, 18, 19, 22, 23
Very-large-scale integration (VLSI) technology 332
Voltage buffers 189
Voltage-controlled oscillators (VCOs) 296
Voltage follower 189
Voltage regulation 36, 64
Voltage regulators 37, 267
Voltage Series Feedback 199
Voltage Shunt Feedback 199, 200
Voltage source 195, 196, 203, 205, 212, 213

W

Waveform generators 266
Wave rectifier circuits 39
Wireless communications systems 114

FSC
www.fsc.org
MIX
Papier | Fördert gute Waldnutzung
FSC® C083411

Kolibri 360 GmbH
Ferdinand-Jühlke-Straße 7
99095 Erfurt, Deutschland
produktsicherheit@kolibri360.de